Fungal Biology

Series Editors

Vijai Kumar Gupta, Biorefining and Advanced Materials Research Center
Scotland's Rural College (SRUC), SRUC Barony Campus, Parkgate
Dumfries, Scotland, UK

Maria G. Tuohy, School of Natural Sciences
National University of Ireland Galway
Galway, Ireland

Fungal biology has an integral role to play in the development of the biotechnology and biomedical sectors. It has become a subject of increasing importance as new fungi and their associated biomolecules are identified. The interaction between fungi and their environment is central to many natural processes that occur in the biosphere. The hosts and habitats of these eukaryotic microorganisms are very diverse; fungi are present in every ecosystem on Earth. The fungal kingdom is equally diverse, consisting of seven different known phyla. Yet detailed knowledge is limited to relatively few species. The relationship between fungi and humans has been characterized by the juxtaposed viewpoints of fungi as infectious agents of much dread and their exploitation as highly versatile systems for a range of economically important biotechnological applications. Understanding the biology of different fungi in diverse ecosystems as well as their interactions with living and non-living is essential to underpin effective and innovative technological developments.

This series will provide a detailed compendium of methods and information used to investigate different aspects of mycology, including fungal biology and biochemistry, genetics, phylogenetics, genomics, proteomics, molecular enzymology, and biotechnological applications in a manner that reflects the many recent developments of relevance to researchers and scientists investigating the Kingdom Fungi. Rapid screening techniques based on screening specific regions in the DNA of fungi have been used in species comparison and identification, and are now being extended across fungal phyla. The majorities of fungi are multicellular eukaryotic systems and therefore may be excellent model systems by which to answer fundamental biological questions.

A greater understanding of the cell biology of these versatile eukaryotes will underpin efforts to engineer certain fungal species to provide novel cell factories for production of proteins for pharmaceutical applications. Renewed interest in all aspects of the biology and biotechnology of fungi may also enable the development of "one pot" microbial cell factories to meet consumer energy needs in the 21st century. To realize this potential and to truly understand the diversity and biology of these eukaryotes, continued development of scientific tools and techniques is essential. As a professional reference, this series will be very helpful to all people who work with fungi and should be useful both to academic institutions and research teams, as well as to teachers, and graduate and postgraduate students with its information on the continuous developments in fungal biology with the publication of each volume.

Dan Bahadur Pal • Swarn Lata Bansal
Editors

Fungal Waste Biomass Management for Energy, Environment and Value-Added Products

Springer

Editors
Dan Bahadur Pal
Department of Chemical Engineering
Harcourt Butler Technical University
Kanpur, Uttar Pradesh, India

Swarn Lata Bansal
DAV College
Kanpur, Uttar Pradesh, India

ISSN 2198-7777 ISSN 2198-7785 (electronic)
Fungal Biology
ISBN 978-3-031-82598-9 ISBN 978-3-031-82599-6 (eBook)
https://doi.org/10.1007/978-3-031-82599-6

This Springer imprint is published by the registered company Springer Nature Switzerland AG
The registered company address is: Gewerbestrasse 11, 6330 Cham, Switzerland

Preface

Bioconversion of the wastes is a natural way for the recovery of resources, and the natural recycling process can be facilitated by biotechnology. Biotechnological treatments of food processing wastes that are found in large quantities can produce useful end products, including microbial biomass protein while the wastes are also purified during the process. Fungi as microorganism and as fungal biomass have been used for many applications, such as enzyme production for biomedical, biorefinery and several industrial applications. In ensuring the vast application of fungal biomass in the piezoelectric field, increasing the quality and ease of operation for the fungal cultivation and extraction of chitosan is significant.

Environmental pollution is a crucial problem for the entire world, and it is growing continuously. Continuous growth of pollution is resulting in harmful changes like global warming and pollution of air, water and soil. These changes are directly associated with various activities like uncontrolled agricultural practices, deforestation, urbanization, accumulation of huge amount of agricultural and food waste, improper dumping of naturally occurring waste and forest residues, etc. Therefore, proper utilization of these wastes may be the better solution to this problem. These waste materials are fungal-based biomass materials and are good sources of carbohydrates, oil and fats, cellulosic content and other useful chemical components that can be converted into value-added products like the production of clean energy, bio-energy, bio-adsorbents and useful chemicals.

Chapter 1 discusses "Sustainable Hybrid Composites, Constructed from Cellulose Nanofibrils and Wood Fungal Mycelium". Chapter 2 discusses "Fungal Waste Biomass as a Feedstock for Polylactic Acid (PLA) Production". Chapter 3 discusses "Utilizing Fungal Mycelium in Bio-nanocomposites as a Replacement for Chemically Synthesized Light Weight Polymers". Chapter 4 discusses "Life Cycle Assessment and Techno-economic Analysis on Setting Up of Pilot Scale Plants". Chapter 5 discusses "Importance of Utilizing Fungus Rather Than Bacteria for Biomass Valorization". Chapter 6 discusses "Extraction and Purification of Essential Oils from Food Waste Biomass". Chapter 7 discusses "Exploring the Potential of Fungal Biomass in the Evolution of Polylactic Acid (PLA) Biocomposites". Chapter 8 discusses "Biochar from Biobased Materials: Applications, Clean Technologies, and Novel Production Methods". Chapter 9 discusses "Advanced Materials from Fungal Mycelium: Fabrication and Tuning of Physical Properties". Chapter 10 discusses "Rice Polishing as a Substrate for Fungal Biomass and Protein Production by *Trichoderma harzianum* and *Candida utilis*". Chapter 11 discusses with "Enzymes Playing Their Specific Roles in Biomass Waste Management". Chapter 12 discusses "Production of Hemicellulose from Biomass and Agricultural/Food Waste Contains Lignocellulosic Materials". Chapter 13 discusses "Extraction and Purification of Essential Oils from Food Waste-Derived Waste Biomass".

Kanpur, Uttar Pradesh, India

Dan Bahadur Pal
Swarn Lata Bansal

Acknowledgements

Editors are thankful to the host institution for providing all necessary facilities. Editors also thankfully acknowledge our parents and family members for their support during these activities.

About the Book

Bioconversion of the wastes is a natural way for the recovery of resources, and the natural recycling process can be facilitated by biotechnology. Biotechnological treatments of food processing wastes that are found in large quantities can produce useful end products, including microbial biomass protein while the wastes are also purified during the process. Fungi as microorganism and as fungal biomass have been used for many applications, such as enzyme production for biomedical, biorefinery and several industrial applications. In ensuring the vast application of fungal biomass in the piezoelectric field, increasing the quality and ease of operation for the fungal cultivation and extraction of chitosan is significant.

Environmental pollution is a crucial problem for the entire world, and it is growing continuously. Continuous growth of pollution is resulting in harmful changes like global warming and pollution of air, water and soil. These changes are directly associated with various activities like uncontrolled agricultural practices, deforestation, urbanization, accumulation of huge amount of agricultural and food waste, improper dumping of naturally occurring waste and forest residues, etc. Therefore, proper utilization of these wastes may be the better solution to this problem. These waste materials are fungal-based biomass materials and are good sources of carbohydrates, oil and fats, cellulosic content and other useful chemical components that can be converted into value-added products like the production of clean energy, bio-energy, bio-adsorbents and useful chemicals. This book will provide waste information, processes and ideas for the conversion of waste into useful and consumable enzymes through biological approaches. Recently, several scientists, researchers and technologists of different countries have been trying to say that the food and agricultural waste biomass has the potential to produce value-added products. Technological information for the production of nutraceuticals and organic acids from the agro-waste will also be covered in this book.

Contents

About the Editors

Dan Bahadur Pal is currently working as an assistant professor in the Department of Chemical Engineering, Harcourt Butler Technical University, Kanpur, UP, India. He has received his MTech in 2011 and PhD in 2017 in the field of chemical engineering from the Indian Institute of Technology (BHU), Varanasi, Uttar Pradesh, India. Before that, he had completed his BTech in Chemical Engineering from UPTU, Lucknow, in 2006. He has completed a doctorate degree in the field of nanotechnology and catalysis. These nanofibres have very promising potential to provide benefits to nanotechnologies, energy, environment, catalysts, sensors, etc. He has published 92 articles in the international referred journals, 6 books and 30 book chapters. From all the publications, he has got 3027 citations with 24 h-index and 54i-10-indexes. His research interests are nano-technology, catalysis, energy, environment and waste management with a special focus on developing processes and materials by using waste as raw materials. He also prefers to work on bio-waste processing and value addition.

Swarn Lata Bansal is working as a research associate at Kanpur DAV College, Kanpur, UP, India. She completed her doctorate degree in the field of biotechnology and environment. She has published 12 articles in international referred journals as well as conference papers and book chapters.

Sustainable Hybrid Composites, Constructed from Cellulose Nanofibrils and Wood Fungal Mycelium

Dilip Kumar, Manish Singh Rajput, Apoorva Anshu Jha, and Mahendra Kumar

Abstract Sustainable materials sourced from renewable resources and developed using eco-friendly technologies possess the ability to impact the environmentally harmful generation and utilization patterns established during the industrial age. This literature investigates novel hybrid composite panels composed of wood, fungal mycelium, and cellulose nanofibrils. Conventional wood composites, which rely heavily on synthetic adhesives for mechanical strength, are reimagined here through the use of mycelium to enhance their properties. Additionally, the integration of experimental methods with machine learning is explored to optimize mechanical performance. This study focuses on the creation and analysis of fully bio-based hybrid composites composed of wood, fungal mycelium, and cellulose nanofibrils. These materials aim to merge the sustainability and biodegradability of mycelium with the high physical properties of cellulose nanofibrils. The hybrid composites are assessed for their physical, mechanical, and thermal characteristics, highlighting their potential use in environmentally conscious material applications. Nanocellulose improves the structural strength of biocomposite films, while mycelium lessens the usual water interaction with cellulose, leading to faster-drying fibrous slurries and films with superior water resistance compared to nanocellulose alone. These mycelium-nanocellulose biocomposites have a range of applications, such as purification, packing, and health products, and they are compatible with conventional papermaking techniques.

D. Kumar (✉) · M. S. Rajput
Dr. Ambedkar Institute of Technology for Divyangjan, Awadhpuri, Kanpur, Uttar Pradesh, India
e-mail: dilip.rs.che14@itbhu.ac.in

A. A. Jha
Harcourt Butler Technical University, Nawabganj, Kanpur, Uttar Pradesh, India

M. Kumar
Indian Institute of Technology, Kanpur, Uttar Pradesh, India

D. B. Pal, S. L. Bansal (eds.), *Fungal Waste Biomass Management for Energy, Environment and Value-Added Products*, Fungal Biology,
https://doi.org/10.1007/978-3-031-82599-6_1

Keywords Biocomposite · Mycelium · Nanocellulose · Filtration · Hygiene and nanofibrils

1 Introduction

Lignocellulosic-developed composites have garnered growing attention in recent years owing to their renewable, biodegradable, and environmentally friendly nature in comparison to artificial materials. However, the extensive use of formaldehyde-containing adhesives in binding these natural particles poses a challenge to the creation of fully natural composites (Mantanis et al., 2018). Formaldehyde emissions, known to be carcinogenic and harmful to human health, have led to concerns (Ferdosian et al., 2017; Hemmilä et al., 2017). As a result, minimizing formaldehyde emissions in wood composite products is highly desirable in most industries (ANSI, 2016; Diop et al., 2017; EN, 2015; Zhao et al., 2024).

Cellulose nanofibrils (CNFs) have recently gained attention as an efficient binder in both conventional and advanced composite systems. Their large surface area allows CNFs to connect wood particles and fibers via hydrogen bonds and mechanical interlocking, improving the strength and stability of composites (Amini et al., 2017; Arevalo & Peijs, 2016; Leng et al., 2017; Lu et al., 2024; Tajvidi et al., 2016). Due to the varied physiological roles of plant cells, thick-walled cells, such as fibers, are commonly utilized in production to minimize fiber fragments in cellulose nanofiber (CNF) products. These fibers can help remove other structures like parenchyma cells, stone cells, and reticular wall cells. Typically, a higher proportion of raw fiber materials used for CNF preparation results in longer, more uniform single fibers, fewer unwanted cells, and improved strength and fibrillation of the CNFs. As a result, high-quality nanofibers are predominantly derived from pure cotton fibers (Atalla & Isogai, 2010). Table 1 presents the physical properties of the cellulose raw materials primarily used for CNF production. The production of self-binding

Table 1 Content of fiber and fiber diameter of common raw materials (Tan et al., 2020)

Raw material	Content of fiber (%)	Content of lignin (%)	Fiber diameter (μm)	Fiber length (μm)	References
Wood-hardwood	41–44	20–25	14–20	1	Wang (1999)
Herb	35–55	15–25	10–30	1–2	Pei (2014)
Bast fiber	65–80	4–20	10–25	–	Winandy (2017)
Cotton fiber	95–97	–	–	25–65	Eero (1993)
Lint	90–91	<3	12–38	10–20	Mishra et al. (2017)

composites typically involves processes such as chemical or enzymatic pretreatments (Kharazipour et al., 1998; Robles et al., 2016), steam explosion (Luo et al., 2014), and steam injection pressing (Widyorini et al., 2005). Enzymatic pretreatment techniques use phenol-oxidizing enzymes (e.g., laccases, peroxidases) obtained from white rot fungi or by other means to treat wood or agricultural fiber residues (Widsten & Kandelbauer, 2008).

In addition, biopolymer composites derived from mycelium have recently been commercialized (Tudryn et al., 2017). Mycelium from filamentous fungi breaks down and bonds with lignocellulosic materials, forming interconnected networks that enhance mechanical strength, fire resistance, and sound absorption in composite panels. While the majority of studies have concentrated on lightweight, foam-like, raw mycelium-based composites (MBCs) (Jiang et al., 2017; Jone et al., 2017; Pelletier et al., 2013; Tudryn et al., 2017; Yang et al., 2017), their potential continues to expand.

The densified composites exhibited five different density levels while also showcasing acoustic absorption capabilities. If this innovative technology can be adapted for use in traditional wood-derived composites, such as fiberboard and particleboard, it could open up new markets and applications for this promising natural material. This study presents a panel system that integrates wood particles treated with fungal mycelium, where additional adhesion is achieved through the use of cellulose nanofibrils (CNF). The aim was to investigate the physical characteristics of a fully bio-based hybrid composite, similar to particleboard, by blending the binding properties of both CNF and mycelium. The study analyzed the material's morphology, as well as the physical and mechanical properties of composites with different compositions and densities. Cellulose fiber is a tightly cohesive, multilayered structure made up of beta-1,4 glycosidic bonds within a macromolecular polysaccharide rigid chain. These chains interlink through intermolecular hydrogen bonds, creating nanofibers by merging with larger hemicellulose microfibrils (Broxterman & Schols, 2018). Figure 1 depicts the fibrillation process from lignocellulose to cellulose nanofibrils, along with the crystalline structure found inside the cellulose nanofibrils.

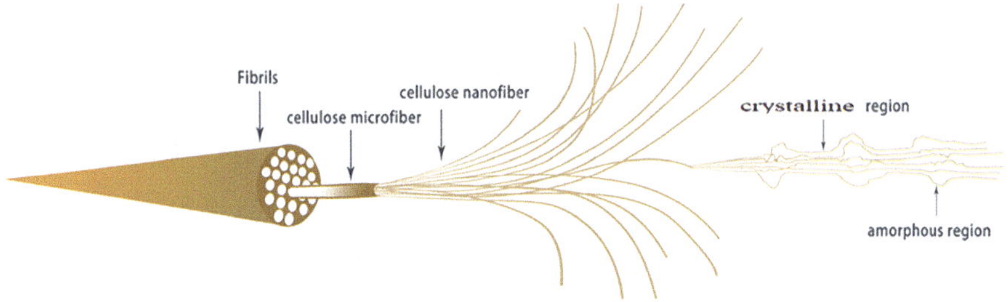

Fig. 1 Cellulose internal structure (Tan et al., 2020)

2 Biocomposites

Composites are mainly constructed using a matrix and reinforcement as the fundamental elements, although additional fillers or additives may be included. The majority of these composites rely on artificial fibers and polymers. However, over the recent decades, attention has shifted toward the development of biocomposites derived from renewable sources. Traditionally, biocomposites are made by using natural fibers like jute, kenaf, or ramie, with synthetic resins such as polyethylene, polypropylene, and epoxy acting as the matrix. These biocomposites tend to be partially biodegradable, containing either the reinforcement or the matrix as a biodegradable component. Although these partially degradable materials offer good performance, the presence of a nonbiodegradable matrix component limits their overall biodegradability (Saha et al., 2024).

Contrary, fully biodegradable composites, where both the matrix and reinforcement are sourced from renewable materials, have been developed. However, these fully degradable biocomposites often lack the desired mechanical properties and are less stable in high humidity or wet atmosphere, which inhibit their use in certain applications. Moreover, natural fibers and resins employed in biocomposites come with disadvantages and can be challenging to process. To address the drawbacks associated with using a single bio-derived reinforcement or matrix, hybrid composites—incorporating multiple reinforcements or matrix materials—have been developed. These hybrid composites are produced either by thoroughly blending two or more fibers or other reinforcements or by layering the reinforcements and shaping them into composites with one or more than one resin.

Historically, the term "blended composites" has referred to materials that incorporate metallic and ceramic reinforcements, matrices, and fillers. While blended biocomposites are attaining considerable interest, the complexity of multiple reinforcements and/or matrices makes them more challenging to process and to anticipate their characteristics. Therefore, it is essential to attain a deeper knowledge of the properties and potential of combining multiple renewable-resource-based reinforcements and matrices in hybrid biocomposites. This chapter explores the materials used in hybrid composites made predominantly from renewable sources, as well as their potential applications. Not only do these hybrid composites offer enhanced characteristics, but they could also give rise to significant reductions due to the use of affordable raw materials. These materials hold promise for use in a wide range of industries, including aeronautics, sports equipment, wind rotor blades, protective headgear, and civil engineering projects like pedestrian walkways.

3 Preparation of Hybrid Composites

Hybrid composite materials are essentially a blend of both hybrid and composite materials. In composite materials, two or more distinct substances are combined to enhance each other's properties. The effectiveness of hybrid composites is significantly influenced by the characteristics of their constituent materials and their additive properties (Bunsell & Harris, 1974; Mishra & Yagci, 2009; Summerscales & Short, 1978).

Herzog et al. (2005) investigated the durability of hybrid products that combine fiber-reinforced polymer (FRP) composites with wood. They used a composite pressure resin infusion system to apply E-glass/vinyl ester FRP directly onto a wood surface, forming a hybrid composite. Their research looked into shear stress, the extent of wood failure during shear, and delamination at the FRP–wood interface under accelerated aging conditions. The results showed that the hybrid composites exhibited shear strengths comparable to or exceeding those of traditional reinforced FRP-glulam methods. The researchers suggested that the FRP should be sufficiently thick to prevent failure during testing and recommended a post-curing procedure (Herzog et al., 2005; Mishra & Yagci, 2009).

In a different study, Abdul Khalil et al. (2009) examined the mechanical and physical characteristics of vinyl ester strengthened with oil palm empty fruit bunch (OPEFB) fibers. They analyzed different layer configurations with glass fiber (GF) hybrid composites and found that these hybrids exhibited better physical properties, reduced water absorption, and improved density compared to traditional composites.

3.1 *Jute-Based Hybrid Composites*

Jute fibers, prized for their affordability and strong tensile properties, are commonly used in textiles and composites. However, they are relatively brittle, with only 2–3% elongation, and tend to absorb water, which can impact how well they bond with the matrix. To enhance the properties of jute fiber composites, both chemical and physical modifications have been employed. Composites were made by layering jute and glass fibers in different configurations, utilizing epoxy as the binder and incorporating up to 15% micro fly ash as an additive. The incorporation of fibers of jute enhanced the tensile strength of epoxy to 74 MPa from 42 MPa, and when glass fibers were included, it enhanced the tensile strength to 188 MPa (Raghavendra et al., 2014). Including fly ash further improved the strength of the jute-reinforced composites. Similar enhancements were observed in the flexural modulus, with tensile strength increasing to 54 MPa from 21 MPa with jute and 115 MPa with both jute and glass fibers. The properties varied with the fiber proportions, but the data were limited, making it challenging to draw definitive conclusions or identify precise failure modes (Karaduman et al., 2015).

Rather than combining two natural fibers, an alternative method involved using stitched jute fiber as the organic component and mats of glass fiber as the inorganic component in epoxy hybrid composites (Gujjala et al., 2014). While natural fiber composites are used in many applications, they often have bad electrical characteristics. To enhance the dielectric characteristics of these composites, surface-modified jute-bamboo fibers (treated with sodium hydroxide) were combined with unsaturated polyester and polypropylene to create hybrid composites (Jayamania et al., 2014).

3.2 Ramie Fiber Blended Composites

Ramie, derived from plants cultivated specifically for their fibers, is recognized for its elevated strength, crystallinity, and delicacy. Although traditionally used in textiles, extensive research has explored the use of ramie fibers in composite materials. Blended composites were created incorporating ramie and glass fibers as strengthening agent in a polyester matrix (Romanzini et al., 2012; Romanzini et al., 2013). Analysis of these composites revealed that raising the quantity of ramie led to greater moisture absorption. Additionally, resistance and shear strength between layers improved with increased levels of ramie and glass fibers. However, an increased proportion of glass fibers resulted in a noticeable decrease in adhesion due to greater fiber–fiber contact, which reduced bonding between the fibers and the matrix. High amount of glass fiber resulted in aggregation, worsening compatibility with the matrix, while ramie fibers demonstrated a strong bond with the resin (Romanzini et al., 2013).

3.3 Sisal Fiber Blended Composites

Sisal fibers, known for their coarse texture and brittleness, are typically used in cords and ropes rather than fabrics. Nevertheless, they have been extensively investigated for composite applications. Composites were developed using sisal and glass fibers merged with polyester resin that was unsaturated, and their chemical resistance was evaluated (John & Naidu, 2007). Various proportions of glass and sisal fibers were used, with the resulting composites showing excellent tolerance to most chemicals, with the exception of carbon tetrachloride. The weight increase in the composites was attributed to gelling. It was proposed that sisal/glass fiber composites might be appropriate for uses including tanks of water (John & Naidu, 2007). The addition of compatibilizers and glass fibers improved the composite strength but reduced degradability. The pretreatment of fibers and their stacking arrangement were shown to considerably influence the properties of polyester matrix composites (Amico et al., 2010). Notably, hybrid composites with natural fibers showed superior properties per unit area compared to those with inorganic fibers like glass, because of the lower density of natural fibers.

3.4 Curaua Fiber Blended Composites

Curaua fibers are extracted from the plants of Curaua through physical and chemical methods. Hybrid composites were formed by stacking eight layers of curaua and glass fibers with various configurations, using polyester resin (Angrizani et al., 2014). These blended composites exhibited intermediate properties compared to pure composites. The improvement in properties was largely influenced by physical loading and the arrangement of the fibers. Positioning glass fibers on top produced the highest flexural strength and modulus. However these properties were not significantly affected by different laminate lay-up methods (Angrizani et al., 2014). The hybridization with glass fibers also enhanced tensile properties (Silva et al., 2012).

3.5 Agave Fiber Blended Composites

Agave plants are cultivated for their fibers, jewelry use, and sometimes as weedicides. Agave fibers are primarily extracted through mechanical processing and soaking. Many studies have explored the use of agave fibers as reinforcements in composites, with significant property improvements achieved through self-hybridization with different particle sizes. Agave reinforcement with pine dust led to a 30% improvement in impact strength and a 15% enhancement in tensile modulus (Perez-Fonseca et al., 2014). However, agave fibers showed poor compatibility with synthetic matrices, requiring compatibilizers to improve composite properties. With 30% content of fiber and 80/20 agave pine to fiber ratio, the composites exhibited 41% improvements in impact, 22% improvements in flexural properties, and 13% improvements in tensile properties, compared to those with only pine. Comparing hybrid composites to pine sawdust is difficult due to variations in aspect ratio, which significantly affects composite properties (Perez-Fonseca et al., 2014).

3.6 Kapok Fiber Blended Hybrid Composites

Kapok fibers, which come from the seeds of kapok trees, look similar to cotton. Although not widely produced, these fibers have several uses. Studies combining kapok fibers with sisal and glass fibers have resulted in the creation of hybrid polyester composites (Reddy et al., 2008). It was observed that composites containing glass fibers had better properties compared to those with a mix of kapok and sisal fibers, because of the superior inherent capacity of the glass fibers (Reddy et al., 2008).

3.7 Coir Fiber Blended Hybrid Composites

Coir fibers, a common agricultural by-product, are coarse and contain up to 40% lignin, which poses challenges for their use in conventional textiles. A study by Kumar et al. (Kumar et al., 2009) explored hybrid composites made from short coir fibers and glass fibers, with polypropylene as the matrix. While the addition of cork powder with a bonding agent did not notably improve tensile properties, incorporating 10% of coir fibers and 2% of blending agent led to a notable 46% enhancement in tensile strength. Additionally, multiple layers of coir pith blended with nylon fabric and epoxy demonstrated improved mechanical properties. Various chemical treatments were applied to coir pith prior to its integration into the composites (Narendar et al., 2016).

3.8 Palm Fiber Blended Composites

Palm fibers, which are mainly obtained from the fruit bunches, trunk, and leaves of palm trees, are utilized in composite applications. These fibers, known as oil palm empty fruit bunch (OPEFB) fibers, were combined with glass fibers in a study to create composites using polyester resin and 43% styrene (Khalil et al., 2007). By layering glass fiber mats between palm fiber mats, the content of fiber in the composites varied between 14% and 40%, with the ratio of palm to glass fibers ranging from 9:1 to 3:7 (Khalil et al., 2007).

3.9 Husk Blended Composites

Husk blended composites, with particle sizes less than 20 mm of powdered wheat husk and wood flour, were utilized as reinforcements in composites of polypropylene. The study evaluated the mechanical properties and water absorption of these composites (Udayanga & Miriyagalla, 2021). The reinforcements were first treated with sodium hydroxide solutions and benzoyl chloride, then melt compounded with the matrix using a twin-screw extruder and molded into test specimens.

3.10 Sugarcane Bagasse Fiber Composites

Bagasse, a major by-product of sugarcane, is predominantly utilized for the production of bioethanol or fuel. However, its large supply and cost effectiveness make it suitable for composite applications. Bagasse fibers, processed with furfuryl alcohol to generate lignin quinones, were mixed with fibers of jute to produce blended epoxy thermoset resins (Saw & Datta, 2009). Alkali-treated jute fibers were used alongside bagasse fibers in epoxy novolac resin. Incorporating up to 40% bagasse fibers enhanced the tensile strength to 23 MPa from 11 MPa and the tensile modulus to 492 MPa from 302 MPa. Further increases in fiber content led to reduced mechanical properties due to poor dispersion (Saw & Datta, 2009).

3.11 Banana Fiber Hybrid Composites

Banana plants, grown for their fiber-rich stems and leaves, generate substantial by-products. Banana fibers, extracted from the stems, are used in textiles and composites. Hybrid composites utilizing glass fibers (6 mm in length) and short banana linked with polystyrene were developed (Haneefa et al., 2008). Additionally, banana fibers and glass fibers were utilized to produce composites containing polypropylene, and their decomposition and thermal properties were studied (Nayak, 2009). It was noticed that banana fibers exhibited significantly increased decomposition (weight loss) compared to glass fiber composites, with the degree of decomposition increasing with higher irradiation exposure. Studies also showed that layering banana fibers with glass fibers affected water absorption, with intimate mixing resulting in higher sorption, while placing glass fibers as the outer layer reduced water absorption due to discontinuity in water flow (Pothan et al., 2007). The banana–jute–banana layered composites showed better compatibility but experienced delamination of outer layers before the core layer (Venkateshwaran & Elaya Perumal, 2012). Additionally, hybrids incorporating sisal and banana fibers with silica microparticles and epoxy resin were explored (Silva et al., 2012).

3.12 Pineapple Fiber Hybrid Composites

Pineapple leaf fibers (PALF), obtained from the leaves after fruit harvesting, have been utilized in both textile and composite applications, either in pure form or blended with other fibers. PALF and glass fibers (GFs) were utilized as reinforcements in the matrices of unsaturated polyester, with PALF/GF ratios varying from 90/10 to 50/50, resulting in a total fiber content of 40% (Devi et al., 2010). The fibers were thoroughly blended and employed as reinforcements for the composites. Additionally, composites were developed with different layering configurations of PALF and glass mats (Devi et al., 2010). The incorporation of PALF and glass fibers impacted the dynamic mechanical properties of the polyester. The use of banana and jute fibers without chemical modifications resulted in noticeable voids and fiber pullout, indicating inadequate compatibility; surface modifications or compatibilizers could potentially enhance the properties (Boopalan et al., 2013).

Another study explored composites made from leaf fibers of sisal-banana and pineapple layered between mats of glass fiber, using polyester resin as the matrix (Rathika et al., 2014). Additionally, municipal waste, primarily vegetable waste, was blended with banana fibers and used in composites with the resin of urea formaldehyde. Sodium chlorite was utilized for the waste treatment, with the waste further ground into powder before integrating it into the composite (Aseer et al., 2015; Kistaiah et al., 2014).

3.13 Switchgrass-Derived Composites

Switchgrass, a sustainable and high-producing grass, is known to be an excellent source of biomass for cellulosic ethanol. Research has focused on extracting natural cellulosic fibers from switchgrass, which exhibit characteristics similar to linen and cotton (Reddy & Yang, 2007). Thermal analysis revealed that hybrid composites made from switchgrass fibers showed enhanced degradation resistance and heat deflection temperature.

3.14 Bamboo Fiber Reinforced Composites

Bamboo, despite requiring exclusive cultivation, is a sustainable and rapidly growing resource. Both natural and regenerated cellulose fibers from bamboo have been utilized in composites. Reports indicate differences in properties between natural cellulosic bamboo fibers and recycled fibers. Blended composites for various applications have been developed utilizing bamboo fibers as mixture and nanoclay (Cloisite B) dispersed in polylactic acid as the matrix (Kumar et al., 2013). The incorporation of bamboo fibers alongside glass fibers in both unmodified and modified into maleic anhydride polypropylene composites demonstrated a range of performance properties (Samal et al., 2009).

3.15 Soy Oil as Resin for Blended Composites

Epoxy-treated soybean oil was explored as a substitute for traditional thermoset resins and synthetic thermoplastics in blended composites. Various cellulose fabrics and fibers, including woven glass fibers, were used as reinforcements. Blended composites were formed by joining lyocell, viscose rayon, and jute fibers with fibers of glass, with jute and lyocell fibers treated with 4% alkali to improve their properties and performance (Ramamoorthy et al., 2012). Composites with a mix of 66% jute and 34% glass fibers showed better properties compared to other combinations.

In a separate study, soybean oil treated with epoxy was used to half replace unsaturated polyester in hybrid composites reinforced with raw hemp fibers and filled with nanoclay (Cloisite 30B) (Haq et al., 2008). Additionally, a biodegradable blended biocomposite was created by blending cassava starch, pulp of cassava or bagasse, and sugar palm fibers (Edhirej et al., 2017).

3.16 Natural–Natural Composites

Asim et al. (2017) investigated the impact of fiber loading on the mechanical characteristics of chopped pineapple leaf fiber (PALF) and kenaf fiber (KF) composites with phenolic (PF) matrix. The composites, including PALF/PF, KF/PF, and PALF/KF/PF, were produced using a hot press molding technique with 50% fiber by weight. Poor fiber wetting can lead to stress concentration at interfaces (Chen et al., 2006). Fiber wettability impacts the rigidness, tensile strength, and bending resistance of composites, and chemical treatments like alkali, acetyl, silane, maleated anhydride, and enzyme treatments can enhance wettability and interfacial strength (Faruk et al., 2014).

4 Understanding Mycelium-Developed Elements

Material science and engineering examine the relationships between elemental structures, properties, and processing conditions (Karana et al., 2018). The cultivation of mycelium significantly affects the properties of mycelium-developed substances. Factors such as moisture and temperature play a crucial role in the final product's performance (Majumder et al., 2021). Research has focused on various substrates for growing mycelium, including agricultural and landscape waste, by-products obtained from food processing, and industrial waste from paper and fabric production (Karana et al., 2018; Robertson, 2020; Sydor et al., 2022a, b). Certain fungal species and substrates may hinder mycelium growth, with variability in growth rate and density (Holt et al., 2012; Jones et al., 2020).

5 Developing Material/Product Designs

Material experience designs are utilized to develop products aligned with material visions (Parisi et al., 2016). For instance, Karana et al. (Karana et al., 2018) developed five different ideas using mycelium, including a planter, roots, coffin packaging, an outer layer, furniture, and footwear. The second skin concept, selected as an innovative packaging solution for wine bottles, entails cultivating a mycelium-based material around the bottle to conform to its shape, offering protection until removed. Mycelium-based products span various applications, from packaging and textiles to furniture and building materials, offering long-term, eco-friendly solutions to environmental challenges.

6 Pure Mycelium Materials

Pure mycelium materials (PMMs) are composed completely of mycelial biomass, distinguishing them from mycelium-based composites, which are typically made by inoculating lignocellulosic waste with saprotrophic fungi within molds (Alexopoulos et al., 1996; Alves et al., 2020; Elsacker et al., 2020; Hawksworth, 2001). PMMs offer versatile characteristics and have the ability to replace conventional materials such as petrochemical-based polymers and animal-derived leathers (Raman et al., 2022). The fundamental characteristics of PMMs are known by the biological parameters of the fungi and their growing environment (Attias et al., 2021; Manan et al., 2021; Vandelook et al., 2021). By growing fungi independently of their substrate, PMMs can be produced with adjustable features, including foam, paper, leather, and polymer-like characteristics (Vandelook et al., 2021). Typically, PMMs are obtained from liquid mycelium cultures (Raman et al., 2022). The final properties, such as translucency, color, and rigidness, can vary depending on the supplements added during growth, such as ethanol or glycerol.

Recent advancements in fermentation processes are enhancing the functionality of PMMs when combined with correct post-growth treatments (Angelova et al., 2021). The expanding range of properties in mycelium-developed materials has led to an increase in consumer products, including packaging, kitchen utensils, furniture, bio-cement, blocks, masonry units, and synthetic leather, created from low-cost agricultural by-products as growth media and possible co-cultivation with multiple fungal species (Vasatko et al., 2022; Meng et al., 2021). PMMs can be produced via various fermentation processes, with liquid fermentation traditionally preferred for metabolite and enzyme extraction. However, solid-state fermentation (SSF) has shown promise, particularly in cultivating extensive aerial hyphae, as demonstrated with *Ganoderma* sp. Commercial mushroom cultivators typically maintain lower CO_2 concentrations and cooler temperatures to inhibit the differentiation of mycelium into fruiting bodies (Alemu et al., 2022).

7 Wood Fungal Mycelium

Wood-decaying fungi, such as *Ganoderma lucidum*, are cultivated on suitable media to explore their effects on wood and other substrates. Mycelium, the network of hyphae, can degrade wood by consuming its material, leading to a reduction in weight and strength. Additionally, mycelium forms mycorrhizal networks that connect plants, facilitating the exchange of water, nitrogen, carbon, and other minerals between them.

7.1 Mycelium-Based Nanomaterials

The use of mycelium in nanomaterial production has gained attention, particularly with coarse materials like sawdust and rice husks. Recent studies have highlighted the potential of mycelium-blended composites incorporating nanomaterials (Jones et al., 2019; Sun et al., 2019; Trabelsi et al., 2021). Nanotechnology has transformed material science, with microorganisms now being explored for their ability to develop biopolymers. These biopolymers, often derived from Ascomycetes (molds), differ functionally from regular structural mycelium-based composites. Bacterial cellulose-based materials have been utilized in applications like drug delivery, wound healing, and bio-sensing (Farooq et al., 2020). More recently, fungi are increasingly used to develop bio-based materials (Cerimi et al., 2019; Trabelsi et al., 2021). For example, *Pleurotus ostreatus* cultivated on polyacrylonitrile (PAN) nanofiber mats produces mycelium/PAN nanocomposites with applications in biotechnology, medicine, water purification, architecture, and vertical farming.

A bio-ceramic blend designed by Ahn et al. (2020) enhances water treatment by utilizing fungal mycelia of *Aspergillus fumigatus* with halloysite nanotubes, which improves the mycelium's water treatment capabilities. Another study explored using polyethylene oxide (PEO) nanofibers to coat bean seeds, facilitating the transport of arbuscular mycorrhizal fungi (AMF) (Campana & Arias, 2020). Jones et al. (2019) developed nano-papers from mycelium-derived chitin-chitosan or chitin-glucan, exhibiting great mechanical characteristics compared to present mycelium substances. These nano-papers, with increased hydrophobicity due to lipid residues, were refined through treatments with HCl or H_2O_2 to enhance their mechanical, thermal, and surface properties.

Fungi, such as *Phanerochaete chrysosporium*, are also utilized in synthesizing nanoparticles for various applications, including packaging, biomedical fields, cosmetics, and construction (Chang et al., 2019). The production of fungi-mediated nanoparticles involves introducing metal pioneers into fermentation cultures, where they are integrated by mycelium and changed into nanoparticles (Manan et al., 2021). The integration of nano-based mycelium-based composites (MBCs) and pure mycelium materials offers significant potential for enhancing the characteristics of mycelium-developed composites.

8 Mycelium-Developed Composite Manufacturing Process

The production of mycelium-based composites (MBCs) involves various key steps (Alemu et al., 2022; Lelivelt et al., 2015; Udayanga & Miriyagalla, 2021; Marzluf, 2014). The process begins with the sterilization of the substrate, which is then integrated with fungal spawn (mycelium). Suitable substrates include cellulose and lignin-rich materials such as straw, wood, and hemp. These substrates are preferred due to their high cellulose content, which provides a carbon and energy source for fungal growth and imparts tensile strength to the material (Haneef et al., 2017; Ziegler et al., 2016). The substrate must offer essential nutrients, including carbon, nitrogen, vitamins, minerals, and sufficient moisture.

Mycelium growth occurs in the dark to stop fruiting body formation (Lelivelt et al., 2015). The optimal growth temperature ranges between 25 °C to 30 °C, a range that can be maintained cost-effectively, especially in tropical climates. High moisture content (60–90%) and elevated CO_2 levels are necessary to support mycelium growth while inhibiting mushroom formation. After initial growth, the composite is shaped and left to mature before being dried, heat-pressed, or cold-pressed to achieve the desired properties (Alemu et al., 2022). Post-growth, the material can be remolded and coated to enhance its quality and durability (Lelivelt et al., 2015).

MBCs offer several advantages over artificial composites, including reduced manufacturing costs, lower energy consumption, and reduced density. They also have a less environmental effect and generate fewer CO_2 emissions in comparison to traditional wood composites or bio-plastics (Javadian et al., 2020; Karana et al., 2018). As a result, establishing industrial-scale manufacturing plants for MBCs presents an environment-friendly opportunity for investors, with low continuous production costs due to minimal energy requirements (Sadashiva & Purushothama, 2023).

9 Characteristics of Mycelium-Developed Composites

When evaluating MBCs, it is essential to consider various material properties, including physical strength, heat conductivity, heat decomposition, sound absorption, moisture absorption, fire resistance, and termite protection (Jones et al., 2020; Pelletier et al., 2013). MBCs typically exhibit heat conductivities comparable to or less than conventional thermal insulation materials. They also demonstrate 70–75% greater sound absorption compared to old ceiling tiles, plywood, and polyurethane foams (Manan et al., 2021). However, great moisture absorption and the presence of voids within the material restrict its applications. Ongoing research is needed to address these limitations. Despite this, the growing interest in MBCs and their advantageous characteristics highlight their potential as a low-cost, effective, and eco-friendly renewable technology for future green construction (Sutkowska et al., 2024).

10 Characterization Methods

Once prepared, mycelium-based composites are characterized using various analytical techniques to assess their morphological and spectroscopic properties.

10.1 Heat Resistance Assessment

The heat resistance of raw materials is evaluated using a TA Instruments TGA Q500 under nitrogen gas. The Hi-Res option of the thermogravimetric analysis (TGA) allows for dynamic and continuous adjustment of the thermal rate from 0.001 °C min^{-1} to 20 °C min^{-1}, depending on the sample's decomposition rate. This method helps differentiate overlapping degradation peaks. The resolution and sensitivity parameters are set at 4.0 °C and 1.0 °C, respectively. The results are presented as variations in sample mass (thermogravimetric curve, TG) or derivative weight loss (derivative thermogravimetric curve, DTG) curves relative to temperature.

10.2 Particle Size Range and Spatial Dimensions

Analysis of wood and wood-mycelium particulates are spread on paper and scanned using a Canon Document Feeder (DADF-AP1, Canon, Inc., Tokyo, Japan) at 600 dpi. The images are processed in Photoshop (Photoshop CC, Adobe Systems, Mountain View, CA, USA) to convert them to black background and then to binary images using ImageJ software (ImageJ 1.48 v, National Institutes of Health, USA). Particle properties, consisting of length, width, area, and perimeter, are examined depending on the best-fitting ellipse. Dimension factors such as aspect ratio, circularity, and roundness are calculated. At least 500 particles from each sample are examined.

10.3 Moisture Uptake and Thickness Expansion

Moisture uptake and thickness expansion are measured based on ASTM D1037 (2012) with changes. Specimens (3 × 3 cm, 8 replicates per group) are dipped in distilled water, and their weights and thicknesses are measured after 2 and 24 h. Moisture uptake and thickness expansion are determined from the differences in weight and thickness relative to the initial measurements.

10.4 Mechanical Testing

Modulus of rupture, modulus of elasticity, and internal bond strength are involved in the evaluation of mechanical properties in mycelium-based composites (MBCs). These properties are assessed following ASTM D1037 (2012) standards, with changes applied utilizing an Instron 5966 universal testing machine (Instron, Norwood, MA, USA) that has a 10-kN load cell capacity. For the three-point bending test, rectangular specimens measuring 12 × 3 cm are analyzed with a span of 80 mm and a cross-head speed of 3 mm/min. In the interior bond strength test, specimens of 3 × 3 cm dimensions are tested with a cross-head speed of 0.4 mm/min. To ensure accuracy and reliability, eight samples are tested for each parameter.

10.5 Morphological Study

Electron microscopic techniques, such as scanning electron microscopy (SEM) and transmission electron microscopy (TEM), are extensively utilized to explore the structure of MBCs. SEM provides detailed insights into surface morphology with nanometer resolution, making it an effective tool for examining surface topography. On the other hand, TEM offers even higher resolution down to atomic levels, ideal for observing internal structures and microstructures, particularly in core-shell nanomaterials. TEM requires samples to be thinned to approximately 100 nm. Atomic force microscopy (AFM) is another technique employed for analyzing surface morphology under ambient conditions, which avoids the vacuum environment required by SEM and TEM, making it suitable for direct sample observation.

10.6 Spectroscopic Analysis

Spectroscopic techniques are essential for characterizing various aspects of MBCs, including elemental composition, chemical bonds, optical properties, and crystallinity. Energy dispersive X-ray spectroscopy (EDX), often together with SEM or TEM, provides information about the material composition of specific areas. Electron spectroscopy for chemical detection (ESCA or X-ray photoelectron spectroscopy, XPS) offers both elemental and chemical bonding details. Fourier transform infrared (FTIR) and Raman spectroscopies are used to study chemical bonds within the composites. Ultraviolet (UV)-visible absorption and fluorescence spectroscopy techniques are employed to study optical characteristics and electronic transitions. X-ray diffraction (XRD) analysis is particularly useful for determining crystallinity, measurements of crystalline regions, and the structural conformation of crystalline domains within the composites.

10.7 High-Resolution Scanning Electron Microscopy (HR-SEM)

High-resolution scanning electron microscopy (HR-SEM), specifically the Zeiss Ultra Plus model, having a Schottky field emission electron gun and Gemini electron beam column, is utilized for detailed imaging. The microscope operates at working distances of 3.5–4.5 mm and low acceleration voltages of 1.5–2 kV and. Imaging is performed using either Everhart–Thornley secondary electron detectors or high-resolution in-the-lens secondary electron detectors. To enhance electrical conductivity, samples are coated with a few nanometers of carbon using a Quorum Q150TES system, followed by ethanol treatment and critical point drying before imaging.

10.8 Cryo-Scanning Electron Microscopy (Cryo-SEM)

Cryo-scanning electron microscopy (cryo-SEM) using the Zeiss Ultra Plus, combined with a Bal Tec VCT100 cold stage maintained below −145 °C, provides valuable insights into the internal structure of MBCs. Specimens are pictured, with working distances of 3.5–4.5 mm and low acceleration voltages of 1–1.4 kV, employing Everhart–Thornley secondary electron detectors. The preparation involves freezing the samples in liquid ethane at −183 °C, followed by transfer to liquid nitrogen. The frozen specimens are subsequently broken in a BAF060 freeze fracture system before imaging. To minimize radiation damage, low-dose imaging techniques are applied, and gentle sublimation may be performed if necessary.

10.9 Polarized Optical Microscopy

Polarized optical microscopy is conducted using a Zeiss Axioplan microscope having a 540-nm waveplate. This technique is employed to visualize and contrast various materials, such as nanocellulose, pellets, encompassing liquids, homogenized mixtures, films, and aerogels. It helps differentiate between birefringent and nonbirefringent components within the samples.

10.10 Atomic Force Microscopy (AFM)

Atomic force microscopy (AFM) pictures of nanocellulose and mycelium elements are captured in suitable conditions utilizing a MultiMode 8 (Bruker, Santa Barbara, CA) in Tapping Mode, equipped with RTESP-150 cantilevers that have a resonant frequency of 150 kHz and a spring constant of 5 N/m. Specimens are made by spin coating onto plasma-cleaned silica wafers. For anionic nanocellulose specimens, the wafers are first wrapped with poly(allylamine hydrochloride) (PAH) at a 0.1-wt% concentration before rinsing. The mean roughness of the hybrid films is measured using topographical images captured in tapping mode with a Bruker Fast Scan AFM. Nanomechanical properties are measured using the Peak Force QNM method with a Tap 300 DLC probe. The cantilever spring constant is measured using the thermal tune method, and the tip's outer radius is calibrated by performing elastic modulus mapping on a reference specimen. In Peak Force QNM measurements, the imaging set point is set to 75 nN, and the images consist of 256 × 256 data points. The DMT modulus is fitted to create nanomechanical maps, which are used to calculate the average elastic modulus of the specimens.

10.11 X-Ray Photoelectron Spectroscopy (XPS)

X-ray photoelectron spectroscopy (XPS) analyses were performed utilizing a Kratos AXIS Ultra DLD X-ray photoelectron spectrometer (Kratos Analytical, Manchester, UK). The samples were examined with a monochromatic Al X-ray source, which provided insights at a depth of approximately 10 nm and a lateral resolution of around 1 mm^2, focusing on an area of about 700 × 300 μm. Survey spectra were acquired to identify elements on the surface layer, with surface composition quantified in atomic percentages from detailed spectra of each detected element. High-resolution carbon spectra were also curve-fitted to reveal chemical shifts associated with different functional groups.

10.12 Attenuated Total Reflectance-Fourier Transform Infrared Spectroscopy (ATR-FTIR)

The chemical characteristics of films manufactured from homogenized mixtures, following fermentation and dialysis, were analyzed using a Perkin Elmer Spectrum One FTIR Spectrometer equipped with an ATR accessory and diamond ATR crystal. This method provided valuable information on the chemical signatures of the films.

10.13 Tensile Analysis of Films

The tensile characteristics of the films were assessed with an MTS ISA 3125 system, featuring a 2-kN load cell. Films were prepared by cutting them into parts (6 mm wide × 45 mm long) using a stamp. The mean mass of a single part was calculated from the mass of four parts. Parts were clamped at 40 psi to achieve a 30-mm specimen length and were extended at a rate of 5 mm/min until failure. The tensile characteristics were averaged over six film strips to ensure accuracy.

10.14 Ultraviolet-Visible (UV-Vis) Spectroscopy

UV-Vis spectroscopy was employed to evaluate cloudiness and consistency by measuring absorbance at 550 nm over time (0 and 12 h), with and without dilution. The liquids were kept in standard plastic cuvettes with a 10-mm path length, and absorbance readings were taken using a LAMBDA 265 UV/Vis spectrophotometer, with water serving as a blank reference.

11 Various Applications of Hybrid Composites

11.1 Environmental Applications

Environmental protection involves two key considerations: reducing the impact of industrial and household waste on the Earth's atmosphere and optimizing energy use. The former focuses on minimizing pollution and waste through sustainable practices, while the latter involves efficient energy production from renewable sources, energy storage to prevent losses, and energy conservation using advanced materials and techniques to meet the increasing demand for industrial and domestic activities.

11.2 Uses of Fungal Biomaterials and Emerging Business Startup Trends

The shift toward a bio-based economy emphasizes replacing traditional fossil-based resources with renewable biomass for producing necessary materials. This trend is advancing fungal biotechnology, which is driving the development of biomaterials obtained from fungi. Fungal enzymes and bioactive compounds are extensively utilized across various sectors, including biofuels, food, detergents, and medicine.

Recent innovations in fungal biomaterials include their application in protective packing, automotive materials, electrical circuits, and construction materials.

11.3 Protective Packing Uses

Green biocomposites made entirely from bio-based elements present a renewable alternative to petroleum-based artificial packing. Mycelium's ability to grow into mold shapes forms a versatile option for making various protective packing elements. By keeping in control the growth parameters and substrate selection, mycelium-based materials can be tailored for structural characteristics like strength and density. These materials are lightweight and biodegradable, offering significant environmental benefits. In 2015, Ecovative developed a packing material featuring an orange-red hue using the *Pycnoporus cinnabarinus* strain, achieving coloration without artificial pigments.

12 Conclusion

The growing interest in biocomposite materials highlights their potential to reduce carbon emissions and energy consumption associated with traditional, finite, and non-decomposable materials utilized in the building and construction industries, like cement, concrete, metals, and polymers. This review has evaluated the potential of mycelium-based composites to address these needs by examining their biofabrication processes, substance characteristics in comparison with conventional materials, and their potential applications in building and construction.

Mycelium-developed composites offer various benefits from technical, environmental, and sustainable modes, making them attractive for building purposes. Notably, the growth of mycelium on wood does not alter particle dimensions or shape but enhances surface coverage, positively affecting bonding. Despite these benefits, the low mechanical strength, excessive water uptake, and absence of a standardized development process currently restrict their use to partial load-bearing and decorative applications, including wall paneling, furniture, and decking.

The potential for producing outdoor-type composites using mycelium-treated wood, particularly with water-resistant resins, is promising due to their impressive dimensional stability. Future research should focus on overcoming existing limitations by improving the properties of mycelium through species selection, genetic modification, and optimization of feed preparations. Additionally, extensive testing and systematic cost–benefit analyses are necessary to address factors such as weathering, hydrophilic tendencies, and scalability. Addressing these challenges will be crucial for achieving long-lasting commercial success and broader application of mycelium-based composites.

13 Future Direction

The development of mycelium-based materials is an evolving field with significant potential. Future research should focus on optimizing growth conditions, improving material properties, and expanding applications to meet the increasing demand for sustainable solution. Collaboration between researchers, industries, and policymakers will be crucial in advancing this technology and integrating mycelium-based materials into mainstream applications. Further studies are needed to address the current limitations and enhance the scalability and performance of mycelium-based composites for broader adoption in various sectors.

References

Abdul Khalil, H., Kang, C., Khairul, A., Ridzuan, R., & Adawi, T. (2009). The effect of different laminations on mechanical and physical properties of hybrid composites. *Journal of Reinforced Plastics and Composites, 28*, 11231137.

Ahn, H., Rehman, J. U., Kim, T., Oh, M. S., Yoon, H. Y., Kim, C., Lee, Y., Shin, S. G., & Jeon, J. R. (2020). Fungal mycelia functionalization with halloysite nanotubes for hyphal spreading and sorption behavior regulation: A new bio-ceramic hybrid for enhanced water treatment. *Water Resouces, 186*, 116380.

Alemu, D., Tafesse, M., & Mondal, A. K. (2022). Mycelium-based composite: The future sustainable biomaterial. *International Journal of Biomaterials, 2022*, 1.

Alexopoulos, C. J., Mims, C. W., & Blackwell, M. (1996). *Introductory mycology* (4th ed.). Wiley.

Alves, R. M. E., Alves, M. L., & Campos, M. J. (2020). Morphology and thermal behaviour of new mycelium-based composites with different types of substrates. In H. A. Almeida & J. C. Vasco (Eds.), *Progress in digital and physical manufacturing*. Springer.

Amico, S. C., Angrizani, C. C., & Drummond, M. L. (2010). Influence of the stacking sequence on the mechanical properties of glass/sisal hybrid composites. *Journal of Reinforced Plastics and Composites, 29*(2), 179–189.

Amini, E., Tajvidi, M., Gardner, D. J., & Bousfield, D. W. (2017). Utilization of cellulose nanofibrils as a binder for particleboard manufacture. *Bio Resources, 12*, 4093–4110.

Angelova, G. V., Brazkova, M. S., & Krastanov, A. I. (2021). Renewable mycelium based composite-sustainable approach for lignocellulose waste recovery and alternative to synthetic materials-a review. *Zeitschrift fur Naturforsch-Section C Journal of Biosciences, 76*, 431–442.

Angrizani, C. C., Cioffi, M. O. H., Zattera, A. J., & Amico, S. C. (2014). Analysis of curaua/glass hybrid interlayer laminates. *Journal of Reinforced Plastics and Composites, 33*, 472–478.

ANSI A208.1. (2016). *Particleboard*. Composite Panel Association.

Arevalo, R., & Peijs, T. (2016). Binderless all-cellulose fibreboard from microfibrillated lignocellulosic natural fibres. *Composites Part A Applied Science and Manufacturing, 83*, 38–46.

Aseer, J. R., Sankaranarayanasamy, K., Jayabalan, P., Natarajan, R., & Dasan, K. P. (2015). Mechanical and water absorption properties of municipal solid waste and banana fiber-reinforced urea formaldehyde composites. *Environmental Progress & Sustainable Energy, 34*(1), 211–221.

Asim, M., Jawaid, M., Abdan, K., & Ishak, M. R. (2017). Effect of pine apple leaf fiber and kenaf fiber treatment on mechanical performance of phenolic hybrid composites. *Fibers and Polymers, 18*(5), 940–947.

Atalla, R. H., & Isogai, A. (2010). 6.16-Celluloses. In H. W. Liu & L. Mander (Eds.), *In comprehensive natural products II* (pp. 493–539). Elsevier.

Attias, N., Reid, M., Mijowska, S. C., Dobrydenm, I., Isakssonm, M., Pokroym, B., Grobman, Y. J., & Abitbol, T. (2021). Biofabrication of nanocellulose-mycelium hybrid materials. *Advanced Sustainable Systems, 5*, 2000196.
Boopalan, M., Niranjanaa, M., & Umapathy, M. J. (2013). Study on the mechanical properties and thermal properties of jute and banana fiber reinforced epoxy hybrid composites. *Composites Part B: Engineering, 51*, 54–57.
Broxterman, S. E., & Schols, H. A. (2018). Interactions between pectin and cellulose in primary plant cell walls. *Carbohydrate Polymers, 192*, 263–272.
Bunsell, A., & Harris, B. (1974). Hybrid carbon and glass fibre composites. *Composites, 5*, 157–164.
Campana, J. M., & Arias, M. (2020). Nanofibers as a delivery system for arbuscular mycorrhizal fungi. *ACS Applied Polymer Materials, 2*, 5033–5038.
Cerimi, K., Akkaya, K. C., Pohl, C., Schmidt, B., & Neubauer, P. (2019). Fungi as source for new bio-based materials: A patent review. *Fungal Biology and Biotechnology, 6*, 1–10.
Chang, J., Chan, P. L., Xie, Y., Ma, K. L., Cheung, M. K., & Kwan, H. S. (2019). Modified recipe to inhibit fruiting body formation for living fungal biomaterial manufacture. *PLoS One, 14*, e0209812.
Chen, P., Lu, C., Yu, Q., et al. (2006). Influence of fiber wettability on the interfacial adhesion of continuous fiber-reinforced PPESK composite. *Journal of Applied Polymer Science, 102*(3), 2544–2551.
Devi, U. L., Bhagawan, S. S., & Thomas, S. (2010). Dynamic mechanical analysis of pineapple leaf/glass hybrid fiber reinforced polyester composites. *Polymer Composites, 31*(6), 956–965.
Diop, C. I. K., Tajvidi, M., Bilodeau, M. A., Bousfield, D. W., & Hunt, J. F. (2017). Evaluation of the incorporation of lignocellulose nanofibrils as sustainable adhesive replacement in medium density fiberboards. *Industrial Crops and Products, 109*, 27–36.
Edhirej, A., Sapuan, S. M., Jawaid, M., & Zahari, N. I. (2017). Cassava/sugar palm fiber reinforced cassava starch hybrid composites: Physical, thermal and structural properties. *International Journal of Biological Macromolecules, 101*, 75–83.
Eero, S. (1993). *Wood chemistry* (pp. 15–112). Elsevier BV.
Elsacker, E., Vandelook, S., Van Wylick, A., Ruytinx, J., De Laet, L., & Peeters, E. (2020). A comprehensive framework for the production of mycelium-based lignocellulosic composites. *Science of the Total Environment, 725*, 138431. https://doi.org/10.1016/j.scitotenv.2020.138431
EN 13986:2004+A1:2015. (2015). *Wood-based panels for use in construction – characteristics, evaluation of conformity and marking*. European Committee for Standardization.
Farooq, U., Wajid, M., Yang, Q., Aziz, A., Xu, J., Zhou, L., & Wang, S. (2020). Biosensors and bioelectronics high-density phage particles immobilization in surface-modified bacterial cellulose for ultra-sensitive and selective electrochemical detection of Staphylococcus aureus. *Biosensors Bioelectronics, 157*, 112–163.
Faruk, O., Bledzki, A. K., Fink, H.-P., & Sain, M. (2014). Progress report on natural fiber reinforced composites. *Macromolecular Materials and Engineering, 299*(1), 9–26.
Ferdosian, F., Pan, Z. H., Gao, G. C., & Zhao, B. X. (2017). Bio-based adhesives and evaluation for wood composites application. *Polymers, 9*, 29.
Gujjala, R., Ojha, S., Acharya, S., & Pal, S. (2014). Mechanical properties of woven jute–glass hybrid-reinforced epoxy composite. *Journal of Composite Materials, 48*(28), 3445–3455.
Haneef, M., Ceseracciu, L., Canale, C., Bayer, I. S., Heredia-Guerrero, J. A., & Athanassiou, A. (2017). Advanced materials from fungal mycelium: Fabrication and tuning of physical properties. *Science Reports, 7*, 1–11.
Haneefa, A., Bindu, P., & Aravind, I. (2008). Thomas S. Studies on tensile and flexural properties of Short Banana/glass hybrid fiber reinforced polystyrene composites. *Journal of Composite Materials, 42*(15), 1471–1489.
Haq, M., Burgueno, R., Mohanty, A. K., & Misra. M. (2008). Hybrid bio-based composites from blends of unsaturated polyester and soybean oil reinforced with nanoclay and natural fibers. *Composites Science and Technology, 68*, (15–16), 3344–3351.

Hawksworth, D. L. (2001). The magnitude of fungal diversity: The 1.5 million species estimate revisited. *Mycological Research, 105*, 1422–1432.

Hemmilä, V., Adamopoulos, S., Karlsson, O., & Kumar, A. (2017). Development of sustainable bio-adhesives for engineered wood panels – A review. *RSC Advances, 7*, 38604–38630.

Herzog, B., Goodell, B., Lopez Anido, R., & Gardner, D. J. (2005). Durability of fiber-reinforced polymer (FRP) composite-wood hybrid products fabricated using the composites pressure resin infusion system (ComPRIS). *Forest Products Journal, 55*(11), 54–60.

Holt, G. A., Mcintyre, G., Flagg, D., Bayer, E., Wanjura, J. D., & Pelletier, M. G. (2012). Fungal mycelium and cotton plant materials in the manufacture of biodegradable molded packaging material: evaluation study of select blends of cotton by products. *Journal of Biobased Materials and Bioenergy, 6*, 431–439.

Javadian, A., Le Ferrand, H., Hebel, D. E., & Saeidi, N. (2020). Application of mycelium-bound composite materials in construction industry: A short review. *SOJ Materials Science & Energy, 7*, 1–9.

Jayamania, E., Hamdan, S., Rahman, M. R., & Bakri, M. K. B. (2014). Comparative study of dielectric properties of hybrid natural fiber composites. *Procedia Engineering, 97*, 536–544.

Jiang, L., Walczyk, D., McIntyre, G., Bucinell, R., & Tudryn, G. (2017). Manufacturing of biocomposite sandwich structures using mycelium bound cores and preforms. *Journal of Manufacturing Processes, 28*, 50–59.

John, K., & Naidu, S. V. (2007). Chemical resistance of sisal/glass reinforced unsaturated polyester hybrid composites. *Journal of Reinforced Plastics and Composites, 26*, 373–376.

Jone, M., Bhat, T., Wang, C. H., Moinuddin, K., & John, S. (2017). Thermal degradation and fire reaction properties of mycelium composites. In *21st international conference on composite materials*. (Xi'an, 2017).

Jones, M., Weiland, K., Kujundzic, M., Theiner, J., Kahlig, H., Kontturi, E., John, S., Bismarck, A., & Mautner, A. (2019). Waste-derived low-cost mycelium nanopapers with tunable mechanical and surface properties. *Biomacromolecules, 20*, 3513–3523.

Jones, M., Mautner, A., Luenco, S., Bismarck, A., & John, S. (2020). Engineered mycelium composite construction materials from fungal biorefineries: A critical review. *Materials & Design, 187*, 108–397.

Karaduman, Y., Onal, L., & Rawal, A. (2015). Effect of stacking sequence on mechanical properties of hybrid flax/jute fibers reinforced thermoplastic composites. *Polymer Composites, 36*, 2167–2173.

Karana, E., Blauwhoff, D., Hultink, E. J., & Camere, S. (2018). When the material grows: A case study on designing (with) mycelium-based materials. *International Journal of Design, 12*, 119–136.

Khalil, H. P. S. A., Hanida, S., Kang, C. W., & Fuaad, N. A. N. (2007). Agro-hybrid composite: The effects on mechanical and physical properties of oil palm fiber (EFB)/glass hybrid reinforced polyester composites. *Journal of Reinforced Plastics and Composites, 26*, 203–218.

Kharazipour, A., Bergmann, K., Nonninger, K., & Huttermann, A. (1998). Properties of fibre boards obtained by activation of the middle lamella lignin of wood fibres with peroxidase and H_2O_2 before conventional pressing. *Journal of Adhesion Science and Technology, 12*, 1045–1053.

Kistaiah, N., Kiran, C. U., Reddy, G. R., & Rao, M. S. (2014). Mechanical characterization of hybrid composites: A review. *Journal of Reinforced Plastics and Composites, 33*(14), 1364–1372.

Kumar, N. M., Reddy, G. V., Naidu, S. V., Rani, T. S., & Subha, M. C. S. (2009). Mechanical properties of coir/glass fiber phenolic resin based composites. *Journal of Reinforced Plastics and Composites, 28*, 2605–2613.

Kumar, V., Sharma, N. K., & Kumar, R. (2013). Dielectric, mechanical, and thermal properties of bamboo-polylactic acid bionanocomposites. *Journal of Reinforced Plastics and Composites, 32*, 42–51.

Lelivelt, R. J. J., Lindner, G., Teuffel, P., & Lamers, H. (2015). The production process and compressive strength of mycelium-based materials. In *First international conference on bio-based building materials. Clermont-Ferrand, France* (pp. 1–6).

Leng, W., Hunt, J. F., & Tajvidi, M. (2017). Effects of density, cellulose nanofibrils addition ratio, pressing method, and particle size on the bending properties of wet-formed particleboard. *Bio Resources, 12*, 4986–5000.

Lu, Z., Hauschild, M., Ottosen, L. M., Ambaye, T. G., Zerbino, P., Aloini, D., & Lima, A. T. (2024). Climate mitigation potential of biobased insulation materials: A comprehensive review and categorization. *Journal of Cleaner Production, 470*, 143356.

Luo, H., Yue, L., Wang, N., Zhang, H., & Lu, X. (2014). Manufacture of binderless fiber board made from bamboo processing residues by steam explosion pretreatment. *Wood Research, 59*, 861–870.

Majumder, S., Paul, A., Kumar, A., Verma, C. K., Divekar, P. A., Rani, V., Rani, A. T., Halder, J., Pandey, K. K., & Singh, J. (2021). Impact of pesticides on microbial population. In P. Bhatt, S. Gangola, D. Udayanga, & G. Kumar (Eds.), *Microbial technology for sustainable environment*. Springer.

Manan, S., Ullah, M. W., Ul-Islam, M., Atta, O. M., & Yang, G. (2021). Synthesis and applications of fungal mycelium-based advanced functional materials. *Journal of Bioresources and Bioproducts, 6*, 1–10.

Mantanis, G. I., Athanassiadou, E. T., Barbu, M. C., & Wijnendaele, K. (2018). Adhesive systems used in the European particleboard, MDF and OSB industries. *Wood Material Science and Engineering, 13*, 104–116.

Marzluf, G. A. (2014). Introduction to mycology. *Methods Gen Mol Microbiol. 6*, 925-928. https://doi.org/10.1128/9781555817497.Part6.

Meng, D., Mukhitov, N., Neitzey, D., Lucht, M., Schaak, D. D., & Voigt, C. A. (2021). Rapid and simultaneous screening of pathway designs and chassis organisms, applied to engineered living materials. *Metabolic Engineering, 66*, 308–318.

Mishra, M., & Yagci, Y. (2009). *Handbook of vinyl polymers: Radical polymerization, process, and technology*. CRC Press.

Mishra, P. K., Gregor, T., & Wimmer, R. (2017). Utilising brewer's spent grain as a source of cellulose nanofibers following separation of protein-based biomass. *BioResources, 12*, 107–116.

Narendar, R., Dasan, K. P., & Jayachandran, J. (2016). Effect of hybridization and chemical treatment on the mechanical properties of coir pith/nylon/epoxy hybrid composites. *Polymer Composites, 37*, 649–657.

Nayak, S. K. (2009). Degradation and flammability behavior of PP/banana and glass fiber-based hybrid composites. *International Journal of Plastic Technology, 13*(1), 47–67.

Parisi, S., Rognoli, V., & Ayala-garcia, C. (2016). Designing materials experiences through passing of time-material driven design method applied to mycelium-based composites. In *Proceedings of the 10th international conference on design and emotion* (pp. 239–255).

Pei, J. C. (2014). *Phytofibrochemistry* (3rd ed., pp. 22–69). Light Industry Press.

Pelletier, M. G., Holt, G. A., Wanjura, J. D., Bayer, E., & McIntyre, G. (2013). An evaluation study of mycelium based acoustic absorbers grown on agricultural by-product substrates. *Industrial Crops and Products, 51*, 480–485.

Perez-Fonseca, A. A., Robledo-Ortiz, J. R., Moscoso-Sánchez, F. J., Rodrigue, D., & González-Nunez, R. (2014). Injection molded self-hybrid composites based on polypropylene and natural fibers. *Polymer Composites, 35*(9), 1798–1806.

Pothan, L. A., Cherian, B. M., Anandakutty, B., & Thomas, S. (2007). Effect of layering pattern on the water absorption behavior of banana glass hybrid composites. *Journal of Applied Polymer Science, 105*, 2540–2548.

Raghavendra, G., Ojha, S., Acharya, S. K., & Pal, S. K. (2014). A comparative analysis of woven jute/glass hybrid polymer composite with and without reinforcing of fly ash particles. *Polymer Composites, 37*(3), 658–665.

Ramamoorthy, S. K., Di, Q., Adekunle, K. F., & Skrifvars, M. (2012). Effect of water absorption on mechanical properties of Soybean oil thermosets reinforced with natural fibers. *Journal of Reinforced Plastics and Composites, 31*, 1191–1200.

Raman, J., Kim, D. S., Kim, H. S., Oh, D. S., & Shin, H. J. (2022). Mycofabrication of mycelium-based leather from brown-rot fungi. *Journal of Fungi, 8*, 317.

Rathika, S., Palanikumar, K., & Raghavan, P. S. (2014). Physical performance of Sisal-PALF-Banana/glass fiber reinforced polyester hybrid composites. *Asian Journal of Chemistry, 26* (14), 4157–4161.

Reddy, N., & Yang, Y. (2007). Natural cellulose fibers from switchgrass with tensile properties similar to cotton and linen. *Biotechnology and Bioengineering, 97*, 1021–1027.

Reddy, G., Venkata, S., Naidu, V., & Rani, T. S. (2008). Impact properties of Kapok based unsaturated polyester hybrid composites. *Journal of Reinforced Plastics and Composites, 27*, 1789–1804.

Robertson, O. (2020). Fungal future: A review of mycelium biocomposites as an ecological alternative insulation material. In *DS 101: Proceedings of Nord Design 2020, Lyngby, Denmark, 12th–14th August 2020* (pp. 1–13).

Robles, E., Czubak, E., Kowaluk, G., & Labidi, J. (2016). Lignocellulosic-based multilayer self-bonded composites with modified cellulose nanoparticles. *Composites Part B Engineering, 106*, 300–307.

Romanzini, D., Ornaghi Junior, H. L., Amico, S. C., & Zattera, A. J. (2012). Preparation and characterization of ramie-glass fiber reinforced polymer matrix hybrid composites. *Materials Research, 15*, 415–420.

Romanzini, D., Lavoratti, A., Ornaghi, H. L., Amico, S. C., & Zattera, A. J. (2013). Influence of fiber content on the mechanical and dynamic mechanical properties of glass/ramie polymer composites. *Materials and Design, 47*, 9–15.

Sadashiva, K., & Purushothama, K. M. (2023). Physical and mechanical properties of bio based natural hybrid composites. *Journal of Materials and Environmental Science, 14*(01), 131–140.

Saha, B., Hoque, M. A., Tahsin, S. H., Rimi, K. N., Prova, A. T., Dey, D., Rahman, S., & Rahman, M. Z. (2024). Biobased composites for advanced applications: Possibilities and difficulties on the path to circularity. *Comprehensive Materials Processing (Second Edition), 12*, 573–588.

Samal, S. K., Mohanty, S., & Nayak, S. K. (2009). Polypropylene-bamboo/glass fiber hybrid composites: Fabrication and analysis of mechanical, morphological, thermal, and dynamic mechanical behavior. *Journal of Reinforced Plastics and Composites, 28*, 2729–2747.

Saw, S. K., & Datta, C. (2009). Thermomechanical properties of jute/bagasse hybrid reinforced epoxy thermoset composites. *BioResources, 4*(4), 1455–1476.

Silva, L. J., Panzera, T. H., Christoforo, A. L., Rubio, J. C. C., & Scarpa, F. (2012). Micromechanical analysis of hybrid composites reinforced with unidirectional natural fibres, silica micro particles and maleic anhydride. *Materials Research, 15*, 1003–1012.

Summerscales, J., & Short, D. (1978). Carbon fibre and glass fibre hybrid reinforced plastics. *Composites, 9*(3), 157–166.

Sun, W., Tajvidi, M., Hunt, C. G., Mcintyre, G., & Gardner, D. J. (2019). Fully bio-based hybrid composites made of wood, fungal mycelium and cellulose nanofibrils. *Scientific Reports, 9*, 37–66.

Sutkowska, M., Stefanska, A., Vaverkova, M. D., Dixit, S., & Thakur, A. (2024). Recent advances in prefabrication techniques for biobased materials towards a low-carbon future: From modules to sustainability. *Journal of Building Engineering, 91*, 109558.

Sydor, M., Bonenberg, A., Doczekalska, B., & Cofta, G. (2022a). Mycelium-based composites in art, architecture, and interior design: A review. *Polymers, 14*, 145.

Sydor, M., Cofta, G., Doczekalska, B., & Bonenberg, A. (2022b). Fungi in mycelium-based composites: Usage and recommendations. *Materials, 15*, 6283.

Tajvidi, M., Gardner, D. J., & Bousfield, D. W. (2016). Cellulose nanomaterials as binders: Laminate and particulate systems. *Journal of Renewable Materials, 4*, 365–376.

Tan, Y., Hanyu, Z., Qi, M., Donglei, P., Yang, L., Lijie, H., Hao, X., Bao, H., & Hainong, S. (2020). From cellulose to cellulose nanofibrils-A comprehensive review of the preparation and modification of cellulose nanofibrils. *Materials, 13*, 5062.

Trabelsi, M., Mamun, A., Klocker, M., Brockhagen, B., Kinzel, F., Kapanadze, D., & Sabantina, L. (2021). Polyacrylonitrile (PAN) nanofiber mats for mushroom mycelium growth investigations and formation of mycelium-reinforced nanocomposites. *Journal of Engineered Fibers and Fabrics, 16*, 1–12.

Tudryn, G. J., Smith, L. C., Freitag, J., Bucinell, R., & Schadler, L. S. (2017). Processing and morphology impacts on mechanical properties of fungal based biopolymer composites. *Journal and Polymers the Environment, 26*, 1–11.

Udayanga, D., & Miriyagalla, S. D. (2021). Fungal mycelium-based biocomposites: An emerging source of renewable materials. In P. Bhatt, S. Gangola, D. Udayanga, & G. Kumar (Eds.), *Microbial technology for sustainable environment* (pp. 529–550). Springer.

Vandelook, S., Elsacker, E., Van Wylick, A., De Laet, L., & Peeters, E. (2021). Current state and future prospects of pure mycelium materials. *Fungal Biology and Biotechnology, 8*, 1–10. https://doi.org/10.1186/s40694-021-00128-1

Vasatko H., Gosch, L., Jauk, J. & Stavric M. (2022). Basic research of material properties of mycelium-based composites. *Biomimetics, 7* (2), 1–16.

Venkateshwaran, N., & Elaya Perumal, A. (2012). Mechanical and water absorption properties of woven jute/banana hybrid composites. *Fibers Polymers, 13*, 907–914.

Wang, J. H. (1999). *Fiber characteristics and fiber atlas of China papermaking raw materials* (1st ed., pp. 3–126). China Light Industry Press.

Widsten, P., & Kandelbauer, A. (2008). Adhesion improvement of lignocellulosic products by enzymatic pre-treatment. *Biotechnology Advances, 26*, 379–386.

Widyorini, R., Xu, J., Watanabe, T., & Kawai, S. (2005). Chemical changes in steam-pressed kenaf core binderless particleboard. *Journal Wood Science, 51*, 26–32.

Winandy, J. E. (2017). Relating wood chemistry and strength: Part II. Fundamental relationships between changes in wood chemistry and strength of wood. *Wood Fiber Science, 49*, 2–11.

Yang, Z., Zhang, F., Still, B., White, M., & Amstislavski, P. (2017). Physical and mechanical properties of fungal mycelium-based biofoam. *Journal of Materials in Civil Engineering, 29*, 04017030.

Zhao, S., Kong, D., Lian, X., Zhang, Y., Xiang, S., Fu, F., & Liu, X. (2024). A review of recent developments for applicational performances in soybean protein-bonded wood composites. *International Journal of Adhesion and Adhesives, 132*, 103676.

Ziegler, A. R., Bajwa, S. G., Holt, G. A., McIntyre, G., & Bajwa, D. S. (2016). Evaluation of physico-mechanical properties of mycelium reinforced green biocomposites made from cellulosic fibers. *Applied Engineering and Agriculture, 32*, 931–938.

Fungal Waste Biomass as a Feedstock for Polylactic Acid (PLA) Production

Apoorva Anshu Jha, Rajkamal Kushwaha, Manisha Trivedi, and Vaibhav Singh

Abstract The rising demand for sustainable materials has led to the exploration of novel sources for biopolymer production, such as fungal waste biomass for polylactic acid (PLA). Fungal waste, comprising agricultural, industrial, and municipal residues, represents a substantial resource for lactic acid production, the key precursor to PLA. Recent studies have demonstrated that enzymatic hydrolysis of fungal biomass can yield up to 80% fermentable sugars, which are subsequently converted into lactic acid through microbial fermentation with efficiencies reaching 90%. Optimization of fermentation techniques, including continuous processes and integrated bioreactor designs, has further enhanced lactic acid yields, surpassing 100 g/L in some cases. Purification methods, such as crystallization, distillation, and, prominently, membrane separation, have been employed to achieve lactic acid purities exceeding 99%, ensuring high-quality PLA production. Challenges remain in scaling up production due to biomass supply variability and economic feasibility. However, recent advancements in genetic engineering have shown promising results in increasing fungal biomass productivity and lactic acid yield. Utilizing fungal waste biomass for PLA production not only mitigates environmental impacts but also addresses waste management concerns, contributing to a more sustainable circular economy. This comprehensive overview emphasizes the necessity of continued research and development to unlock the full potential of the fungal biomass in bioplastic production, thereby fostering a greener future for generations to come.

Keywords Fungal waste biomass · Polylactic acid · Fermentation · Bioreactor · Purification

A. A. Jha · R. Kushwaha (✉) · M. Trivedi · V. Singh
Department of Biochemical Engineering, School of Chemical Technology, HBTU, Kanpur, Uttar Pradesh, India

D. B. Pal, S. L. Bansal (eds.), *Fungal Waste Biomass Management for Energy, Environment and Value-Added Products*, Fungal Biology,
https://doi.org/10.1007/978-3-031-82599-6_2

1 Introduction

The global shift toward sustainable development and the urgent need to combat environmental challenges posed by conventional petrochemical-based plastics have intensified efforts to explore renewable and biodegradable alternatives (Kabir et al., 2020). Polylactic acid (PLA) has garnered significant attention for its biodegradability and ability to serve as a sustainable substitute for traditional plastics. One promising avenue for PLA production involves utilizing fungal waste biomass as a feedstock. Fungal biomass, generated as a by-product in various industrial processes, represents a plentiful and currently underutilized resource rich in organic materials like polysaccharides, proteins, and lipids. These biomolecules can be efficiently converted into lactic acid (LA), a pivotal precursor in PLA synthesis (Abdel-Rahman et al., 2013).

The process begins with the fermentation of fungal biomass, where engineered strains of *Saccharomyces cerevisiae*, widely utilized in industrial settings, play a crucial role. These strains are adept at enhancing lactic acid production, while exhibiting resilience to organic acids, thereby improving overall yield and efficiency (Jang et al., 2021). By leveraging fungal waste biomass, this approach not only addresses waste disposal challenges but also establishes a sustainable pathway for PLA production. It reduces environmental impact, promotes circular economy principles by utilizing waste as a resource, and diminishes reliance on finite fossil-based feedstocks. This holistic approach underscores the potential of fungal waste biomass in advancing sustainable materials and contributes to the broader goal of mitigating plastic pollution and fostering environmental stewardship (Jang et al., 2021).

1.1 Overview of Polylactic Acid (PLA)

Polylactic acid (PLA) is a biodegradable and biocompatible thermoplastic polyester derived from renewable resources such as cornstarch or sugarcane (Xiao et al., 2012). It has gained significant attention as a sustainable alternative to conventional petroleum-based plastics, due to its attractive properties and potential applications in various fields as shown in Fig. 1 PLA can be synthesized using different methods, with the most common being direct polycondensation of lactic acid and ring-opening polymerization (ROP) of lactide (Xiao et al., 2012). The ROP method is preferred as it allows for the production of higher molecular weight (Mw) PLA (Xiao et al., 2012). Produced by polymerizing lactic acid from sugar fermentation, PLA is known for its mechanical properties, biocompatibility, and compostable nature, making it a green alternative to petroleum-based plastics. PLA breaks down into carbon dioxide (CO_2), water, and biomass under industrial composting, reducing its environmental impact. Applications include packaging, textiles, biomedical devices, and 3D printing. Recent advances in microbial fermentation have improved lactic acid yield and purity, enhancing PLA production (Abdel-Rahman et al.,

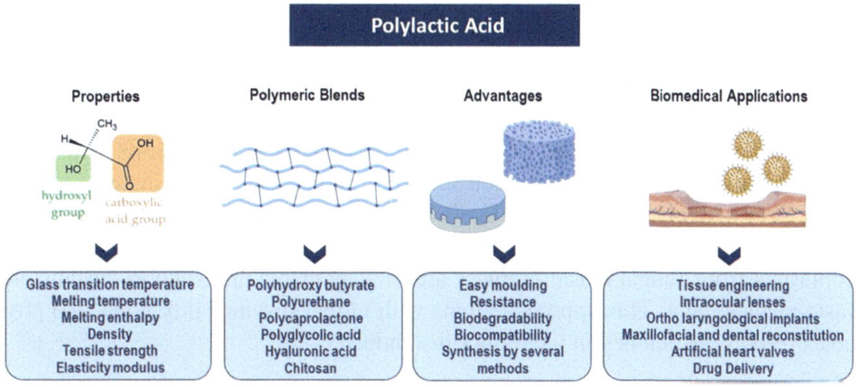

Fig. 1 Overview of polylactic acid (PLA)

2013). Versatile processing techniques further increase PLA's commercial viability (Castro-Aguirre et al., 2016).

In recent years, research and development efforts have focused on optimizing the microbial fermentation processes that produce lactic acid, a crucial precursor for PLA. These advancements have led to higher yields and purities of lactic acid, thereby improving the overall efficiency and sustainability of PLA production. Additionally, innovative processing techniques, such as melt blending and extrusion, have expanded PLA's applications and enhanced its performance in various industrial sectors. As the demand for eco-friendly alternatives to traditional plastics grows, PLA derived from renewable sources continues to play a pivotal role in advancing sustainable material solutions and promoting circular economy practices globally.

Figure 1 is a visual summary of polylactic acid (PLA), focusing on its structure, properties, polymeric blends, advantages, and biomedical applications. PLA's molecular structure is depicted with hydroxyl and carboxylic acid groups. The properties highlighted include glass transition temperature, melting temperature, melting enthalpy, density, tensile strength, and elasticity modulus. Various polymeric blends are listed, such as polyhydroxybutyrate, polyurethane, polycaprolactone, polyglycolic acid, hyaluronic acid, and chitosan. The advantages of PLA are shown as easy molding, resistance, biodegradability, biocompatibility, and synthesis by several methods. The biomedical applications illustrated include tissue engineering, intraocular lenses, ortholaryngological implants, maxillofacial and dental reconstruction, artificial heart valves, and drug delivery systems (Khouri et al., 2024).

1.2 Importance of Sustainable Feedstocks

Fungal waste biomass is a sustainable alternative to traditional feedstocks like corn or sugarcane, which often compete directly with food production (Smith & Johnson, 2020). Using fungal waste for PLA production reduces environmental impact by repurposing waste, lowering greenhouse gas emissions, and conserving energy. It is also cost-effective, especially as it is often available at low or no cost. Unlike conventional plastics, which contribute significantly to pollution (Geyer et al., 2017), bioplastics from fungal waste biomass are biodegradable and support sustainable waste management. This approach aligns with global sustainability goals and promotes resource efficiency in the bioplastics industry.

Importance of Fungal Waste Biomass for PLA Production

(i) *Renewable Resource*

Fungal waste biomass originates from fungal fermentation processes that utilize various organic waste substrates, such as agricultural residues and food industry by-products (Souza Filho, 2018). This feature makes fungal biomass a renewable resource, aiding in the sustainable production of PLA. By employing waste materials, the production process lessens the dependence on fossil fuels and reduces the environmental impact linked to conventional plastic manufacturing (Cottet et al., 2022).

(ii) *High Nutritional Value*

Fungal biomass is abundant in proteins and other essential nutrients, enhancing the overall quality of PLA products. For example, research indicates that fungal species like *Rhizopus oryzae* can produce biomass with up to 43% protein content (Souza Filho, 2018). This high nutritional value not only supports microorganism growth during fermentation but also contributes to the creation of high-performance biocomposites when combined with PLA (Cottet et al., 2022).

(iii) *Biodegradability and Environmental Benefits*

PLA is naturally biodegradable, and when produced from fungal waste biomass, the environmental advantages are even greater. Utilizing organic waste in PLA production helps to create a closed-loop waste management system by transforming waste into valuable products, thereby reducing landfill use and greenhouse gas emissions from waste decomposition (Cottet et al., 2022). Additionally, incorporating fungal biomass into PLA can improve its biodegradability, making it a more eco-friendly material (Cottet et al., 2022).

(iv) *Economic Viability*

Incorporating fungal waste biomass into PLA production can lead to cost savings. Using inexpensive organic waste materials as feedstock can significantly lower raw material costs compared to traditional petroleum-based feedstocks (Talukder et al., 2014). Moreover, the valorization of waste materials through biorefinery approaches can create new economic opportunities in rural areas, bolstering local economies and promoting sustainable practices (Souza Filho, 2018).

(v) *Versatility in Applications*

PLA derived from fungal waste biomass can be customized for various uses, including biodegradable packaging, agricultural films, and medical devices. The versatility of PLA, combined with the sustainable sourcing of its feedstock, positions it as a key player in the move toward more sustainable materials across multiple industries (Cottet et al., 2022).

1.3 Potential of Fungal Waste Biomass

Fungal waste biomass is a promising sustainable feedstock for producing polylactic acid (PLA). Derived from industrial processes such as pharmaceutical and enzyme fermentation, this biomass is rich in polysaccharides, proteins, lipids, and other organic compounds. It provides a valuable source of carbon and nitrogen for biochemical conversion into lactic acid, a PLA precursor (Ren et al., 2022). Utilizing fungal waste biomass addresses sustainability challenges by reducing waste disposal, minimizing environmental impact, and ensuring a stable supply. The conversion process involves enzymatic hydrolysis and microbial fermentation using lactobacillus species to produce efficiently lactic acid. Studies highlight the economic and technical feasibility of using fungal waste biomass in PLA production, enhancing sustainability in bioprocessing industries (Ren et al., 2022).

2 Fungal Waste Biomass: A Renewable Resource

Fungal waste biomass is a versatile resource to enhance polylactic acid (PLA)-based biocomposites. Fungal waste biomass serves as a valuable renewable resource for producing polylactic acid (PLA) due to its high starch content, which can be effectively utilized in microbial fermentation processes (de Eugenio et al., 2023). Derived from industrial by-products and agricultural residues, fungal biomass offers a sustainable alternative to traditional fillers and reinforcements in biopolymer applications. Enhancement is done in the mechanical, thermal, and biodegradation properties of PLA-based materials by incorporating fungal biomass. This approach aims to create biocomposites with improved performance suitable for the applications in packaging, biomedical devices, and environmental remediation. Fungal biomass in reinforcing PLA can reduce dependence on petroleum-based additives and enhance the eco-friendly characteristics of these materials.

The integration of fungal biomass into PLA-based biocomposites not only enhances material properties but also expands the scope of sustainable manufacturing practices. By utilizing fungal waste biomass as a reinforcement agent, biocomposites exhibit improved strength, durability, and biodegradability compared to conventional materials. This approach not only mitigates environmental impact by utilizing agricultural residues and industrial by-products but also promotes circular economy principles by reducing waste and maximizing resource efficiency. Biocomposites reinforced with fungal biomass are poised to play a pivotal role in advancing green technologies across industries, offering viable alternatives to petroleum-based products while aligning with global sustainability objectives (de Eugenio et al., 2023).

In the study, a consolidated bioprocess was utilized, where agro-industrial waste rich in starch was fermented by microbes to produce lactic acid (LA), a precursor to PLA. The research examined the capability of lactic acid bacteria (LAB) and the fungus *Rhizopus oryzae* to produce naturally lactic acid, either in pure cultures or cocultures, highlighting their potential in PLA production. A synergistic interaction was identified between the fungus *Talaromyces amestolkiae* and *Lactiplantibacillus plantarum* M9MG6-B2 in degrading polysaccharides into sugars, indicating the efficiency of fungal–lactobacteria consortia in PLA production (de Eugenio et al., 2023).

The introduction of the quorum-sensing molecule dodecanol enhanced the lactic acid metabolism in the consortium, resulting in improved cooperation and higher yields of lactic acid from both glucose and starch sources. The study also demonstrated the synthesis of lactic acid oligomers and polymers through LA polycondensation and enzymatic ring-opening polymerization catalyzed by a noncommercial lipase OPEr, derived from the fungus *Ophiostoma piceae*, providing a novel enzymatic approach for PLA production (de Eugenio et al., 2023).

Researchers focus on modifying fungal waste biomass to enhance the properties of PLA while still maintaining its biodegradability, making it an environmental-friendly option. By blending modified biomass with PLA, the aim is to address the limitations of PLA, such as its limited toughness, thermal stability, and barrier properties. Modified cellulose, lignin, and starch obtained from fungal waste biomass are used to improve various properties of PLA, including its mechanical strength, thermal stability, and barrier properties (Zhu et al., 2024). Chemical polymerization techniques applied to fungal waste biomass play a crucial role in creating green composites with significant developmental prospects for PLA applications. Therefore, fungal waste biomass presents a promising renewable resource that, when modified and combined with PLA, can lead to the production of environmental-friendly and sustainable composite materials. The combination of fungal waste biomass and lactic acid bacteria in coculture systems offers a promising and sustainable method for PLA production, illustrating the potential of clean technologies in the bioplastic industry (Zhu et al., 2024).

2.1 Sources and Types of Fungal Waste Biomass

(i) *Industrial Processes*: Fungal waste biomass sourced from industrial processes plays a pivotal role in sustainable biopolymer production, particularly for polylactic acid (PLA). Industries such as pharmaceuticals and food processing generate significant fungal biomass through fermentation processes, resulting in residues like brewery and distillery by-products.

(ii) *Research Laboratories*: Research laboratories contribute to fungal waste biomass by producing surplus fungal cultures. These cultures are often utilized in various biotechnological applications, such as enzyme extraction and yeast fermentation, further contributing to biomass production.

Types of Biomasses

(i) *Filamentous Fungi Residues (e.g., Aspergillus)*: Residues from filamentous fungi, such as *Aspergillus*, are commonly utilized in PLA synthesis. These fungi generate substantial biomass during industrial processes, offering a sustainable source for biopolymer production (Asadollahzadeh et al., 2022).

(ii) *Yeast Cells Post-Fermentation*: After fermentation processes, yeast cells contribute to fungal waste biomass. These cells can be repurposed for PLA production, enhancing the eco-friendly nature of biopolymer manufacturing (Asadollahzadeh et al., 2022).

(iii) *Fungal Mycelium on Agricultural Substrates (e.g., Straw)*: Fungal mycelium grown on agricultural substrates, such as straw, represents another valuable type of biomass. This substrate utilization not only utilizes agricultural residues effectively but also underscores the role of fungi in sustainable polymer manufacturing (Asadollahzadeh et al., 2022).

The exploration of these diverse sources and types of fungal waste biomass underscores their critical role in advancing eco-friendly polymer manufacturing practices, contributing to sustainable development goals (Asadollahzadeh et al., 2022).

2.2 Composition and Characteristics

Fungal waste biomass, derived from fungal fermentation processes, is increasingly recognized for its potential in various industries, especially bioremediation and animal feed. The composition and properties of this biomass are influenced by the fungal species and the substrates used during cultivation.

Composition of Fungal Waste Biomass

Fungal biomass mainly consists of proteins, carbohydrates, and lipids. The general chemical composition per kilogram of dry weight for different fungal species includes:

(i) *Proteins*: Protein content ranges from 40% to 60%. For instance, *Rhizopus oryzae* produces biomass with about 43% protein when cultivated on food industry waste substrates (Souza Filho, 2018).

(ii) *Carbohydrates*: Carbohydrates form a substantial part of the biomass, contributing to its energy value.

(iii) *Lipids*: Lipid content is usually lower, around 5–10%, but some fungal species can produce biomass with up to 20% lipids when grown on specific substrates like orange peel extract (Souza Filho, 2018).

Fungal waste biomass, a by-product of industrial processes such as the production of antibiotics and enzymes, is rich in valuable components like chitin, glucans, proteins, and lipids. This composition makes it a promising renewable resource for various applications, including the production of bioplastics such as polylactic acid (PLA). The high carbohydrate content, particularly glucans, is essential for microbial fermentation processes that convert biomass into lactic acid, the monomer for PLA production.

Different fungi have unique compositions based on their metabolic pathways and the substrates they utilize. For example, *Aspergillus niger* is known for its high protein content and biosorption properties, making it suitable for environmental applications such as heavy metal removal (Spanělova et al., 2003).

Moreover, the protein and lipid content can provide supplementary nutrients required for optimal microbial growth and productivity.

Characteristics of Fungal Waste Biomass Functional Properties

(i) *Biodegradability*: Fungal biomass is biodegradable, which makes it an environment-friendly option for waste management and material production.
(ii) *Nutritional Value*: With its high protein content and balanced amino acid profile, fungal biomass is a viable alternative protein source in animal feeds, particularly in aquaculture (Souza, 2018).
(iii) *Biosorption Capacity*: The biomass has demonstrated significant potential in biosorption applications, particularly for heavy metals. *Aspergillus niger*, for instance, has shown a maximum lead biosorption capacity of 93 mg/g dry biomass (Spanělova et al., 2003).

Beyond its biochemical composition, fungal waste biomass offers additional characteristics that enhance its suitability as a feedstock for bioplastic production (Maroua et al., 2023). The presence of chitin, a nitrogen-containing polysaccharide, not only contributes to the structural integrity of fungal cells but also holds potential for applications in biocomposite materials due to its biodegradability and antimicrobial properties. Glucans provide fermentable sugars essential for microbial metabolism during lactic acid fermentation, facilitating efficient conversion processes. Furthermore, the protein and lipid fractions support microbial growth and offer opportunities for value-added applications in nutritional supplements and biofuels (Maroua et al., 2023).

2.3 Current Uses and Disposal Methods

Fungal waste biomass, a by-product of various industrial processes such as antibiotic production and food fermentation, is increasingly recognized as a valuable renewable resource. Originating from the fermentation processes of various fungi, fungal waste biomass holds significant promise for sustainable applications due to its diverse current uses and disposal methods. This biomass is gaining recognition for its role in biotechnological processes, environmental remediation, and as a resource for producing biofuels and biobased materials.

Current Uses of Fungal Waste Biomass

(i) *Microalgae Harvesting*: An innovative application of fungal waste biomass is in the harvesting of microalgae. Research has shown that fungal mycelium can effectively bind to microalgae, aiding their collection for biofuel production and other biotechnological applications. This method offers a cost-effective and environment-friendly alternative to traditional harvesting techniques, such as centrifugation and chemical flocculation, which often have significant drawbacks (Talukder et al., 2014).

(ii) *Biobased Materials*: Fungal biomass is being investigated for the development of biobased materials, such as biodegradable plastics and composites. The complex structure of fungal mycelium allows for the creation of materials that can replace conventional plastics sustainably. For instance, studies have indicated that polylactic acid (PLA)-based composites can be produced using fungal biomass, enhancing the material's properties while promoting environmental sustainability (Cottet et al., 2022).

(iii) *Biosorption*: Fungal waste biomass is also used for biosorption, where it acts as a natural adsorbent to remove pollutants, such as heavy metals and pesticides, from wastewater. Fungi like *Aspergillus niger* have demonstrated effective biosorption capabilities, making them valuable in environmental cleanup efforts Cabrera et al., 2020).

(iv) *Animal Feed*: The high protein content of fungal biomass makes it a viable option for animal feed, particularly in aquaculture. It provides a sustainable protein source that can supplement traditional feed ingredients, reducing reliance on fishmeal and promoting more sustainable livestock practices (Souza Filho, 2018).

Currently, fungal waste biomass is utilized in various ways, such as in enzyme production, biofuels, and animal feed. For instance, fungal waste from *Penicillium chrysogenum* has been repurposed for enzyme production, highlighting its potential for biotechnological applications.

Disposal Methods of Fungal Waste Biomass

(i) *Composting*: A common disposal method for fungal waste biomass is composting. This process not only reduces waste volume but also converts the biomass into nutrient-rich compost that can enhance soil fertility. Composting fungal biomass contributes to nutrient cycling and improves soil health (Bai et al., 2022).

(ii) *Anaerobic Digestion*: Fungal waste biomass can also be processed through anaerobic digestion, a method that transforms organic matter into biogas. This process aids in waste management while generating renewable energy, supporting a circular economy by utilizing waste to produce energy (Bai et al., 2022).

(iii) *Incineration*: In some instances, incineration is used as a disposal method for fungal waste biomass. This method reduces waste volume and can generate energy. However, it is less preferred due to environmental concerns related to emissions and the loss of potential nutrients that could be recovered through composting or anaerobic digestion (Bai et al., 2022).

So, fungal waste biomass is employed in various innovative applications, including microalgae harvesting, biobased material production, and environmental remediation. Disposal methods like composting and anaerobic digestion not only address waste management but also support sustainable practices in agriculture and energy production.

3 Polylactic Acid (PLA): Properties and Applications

With increasing environmental resource constraints, there is a heightened focus on biodegradable materials. Polylactic acid (PLA) has gained prominence due to its excellent mechanical properties, ease of processing, and ability to degrade efficiently after use. However, the production and purification of its precursor, lactide, pose challenges, leading to high costs and limited scalability (Xue, 2023).

Addressing these challenges, carbon dioxide (CO_2) has emerged as a promising resource for both synthesizing and modifying PLA, using innovative approaches such as cell factory technology. This advancement not only expands PLA's application potential but also resolves critical issues in lactide production and purification. Additionally, supercritical carbon dioxide (sc-CO_2) shows promise in PLA production, influencing its properties and broadening its applications (Xue, 2023).

3.1 Chemical Structure and Properties of PLA

Polylactic acid (PLA) is gaining prominence as an eco-friendly alternative to fossil-based polymers due to its minimal environmental impact. PLA, an aliphatic polyester, is typically synthesized from lactic acid through ring-opening polymerization or polycondensation. Commonly processed using techniques like injection molding, hot pressing, and 3D printing, PLA exhibits superior thermal workability among bio-based polymers.

Despite its comparable tensile strength and elasticity modulus to conventional polyesters, PLA's lower toughness can be improved through composite development. Recognized as safe for food contact by the European Food Safety Authority (EFSA), Polylactic Acid (PLA) offers barrier properties comparable to Low-Density Polyethylene (LDPE) and Polystyrene (PS) but with limited inherent antibacterial capabilities, which can be enhanced with additives. While higher costs currently limit PLA's use in packaging, its biodegradability makes it suitable for diverse applications including medical implants, controlled drug release systems, and tissue engineering supports. Further development of processing methods such as blending, nanocomposite creation, and polymer modification is essential to broaden PLA's industrial applications (Hajdek et al., 2023).

3.2 Industrial Applications of PLA

PLA finds versatile applications across several industries. In packaging, PLA's biodegradability and biocompatibility make it a preferred choice for food packaging, significantly lowering environmental impact. In textiles, PLA fibers serve as a sustainable alternative to traditional synthetics, enhancing the eco-friendliness of textile manufacturing. Its biocompatibility extends PLA's utility to medical devices, facilitating the development of implants and drug delivery systems that advance healthcare solutions. PLA's attributes also make it ideal for 3D printing filaments, providing ease of processing and biodegradability crucial for prototyping and production across various sectors. Moreover, PLA is gaining traction in the automotive industry for interior components, leveraging its lightweight properties to potentially reduce carbon emissions in vehicle-manufacturing processes (Paul et al., 2023).

3.3 Environmental Impact of PLA Production

Polylactic acid (PLA) represents a significant advancement in environmental-friendly polymers, derived from renewable resources like cornstarch or sugarcane. Its production process consumes CO_2, resulting in a lower carbon footprint compared to traditional petroleum-based plastics. This not only reduces greenhouse gas emissions but also contributes to a cleaner environment. PLA is compostable and biodegradable, breaking down into harmless natural compounds without leaving toxic residues. By promoting the use of PLA-based materials, we can effectively reduce the accumulation of nonbiodegradable plastics in the environment, fostering a sustainable ecosystem (Deeraj et al., 2022).

4 Conversion of Fungal Biomass to Lactic Acid

Lactic acid, which is a monocarboxylic acid that is both colorless and odorless, has garnered significant interest due to its wide range of uses in both food and nonfood sectors. Notably, there is a growing demand for lactic acid in the production of biodegradable polylactic acid materials, which serve as an eco-friendly substitute for plastics derived from petroleum. This acid can be generated through either chemical synthesis or microbial fermentation processes as depicted in Fig. 2. During chemical synthesis, two optical isomers, L (+)-lactic acid and D (−)-lactic acid, are formed (Lin et al., 2020). The fungal genus *Rhizopus*, particularly *R. oryzae*, has garnered significant interest for its ability to produce high-purity L-lactic acid (LA) efficiently. Unlike L-lactic acid-producing bacteria, *Rhizopus* requires fewer nutrients and can ferment a diverse range of substrates, including lignocellulosic biomass, directly into L-lactic acid. Moreover, it can convert starches into L-lactic acid without prior saccharification. *Rhizopus* filamentous structure facilitates straightforward separation from the fermentation broth, reducing production costs (Ren et al., 2022). Recently, a heat-resistant strain of *Rhizopus* was isolated, achieving L-lactic acid yields of 84.3 $g{\cdot}L^{-1}$ and 105–119 $g{\cdot}L^{-1}$ in batch and fed-batch fermentations, respectively, using liquefied cassava starch at pH 5.5. However, challenges such as the production of by-products like ethanol and fumaric acid during heterofermentation increased energy consumption due to oxygen requirements, and potential interference from filamentous bacteria in mass transfer and L-lactic acid recovery processes remain significant (Ren et al., 2022).

In the recent study, the fungal isolate *Aspergillus japonicus* was identified as having the capability to transform plant biomass residues into various organic acids, including lactic acid. The research employed solid-state fermentation techniques using a blend of sugarcane bagasse and faba bean straw with *Aspergillus japonicus*, which led to the generation of lactic acid along with other organic acids (Mahgoub et al., 2022).

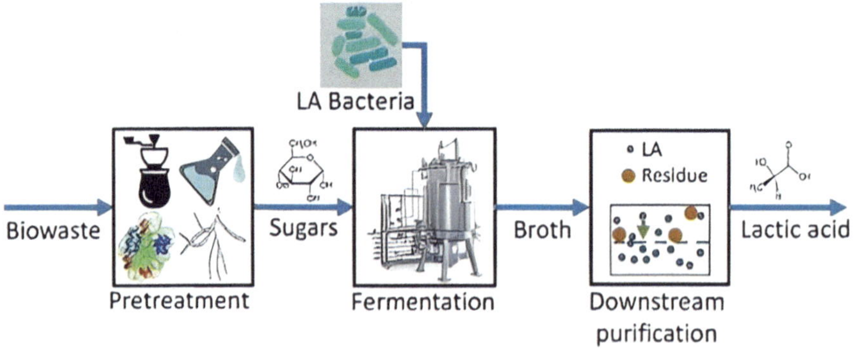

Fig. 2 The process of converting biowaste into lactic acid

During the fermentation process, lactic acid was one of the seven organic acids detected, with their relative abundances ranked as follows: ascorbic acid, oxalic acid, formic acid, malic acid, succinic acid, lactic acid, and citric acid. These findings indicated that *Aspergillus japonicus* showed promise as a biocatalyst for converting sugarcane bagasse and faba bean straw residues into lactic acid. This indicated its potential role in the bioconversion of fungal biomass into valuable organic acids, showcasing its utility in sustainable bioprocessing (Mahgoub et al., 2022). A study was done in which fungal biomass was transformed into lactic acid using a method called electro-fermentation, which involved the application of electrons to microorganisms, thereby modifying their metabolic pathways. Lactic acid bacteria (LAB) were essential in this conversion process due to their metabolic characteristics and their capability to produce lactic acid as a primary end product (Winder et al., 2022).

LAB are notable for their anaerobic metabolism, which allows them to efficiently convert biomass to lactic acid even in the presence of oxygen. These bacteria possess genetic malleability, making it possible to engineer their genetic makeup to regulate product output and enhance lactic acid yields by boosting their electroactivity. Utilizing LAB in electro-fermentation processes can increase the economic value of products derived from fungal biomass by improving lactic acid production (Winder et al., 2022).

The sequence begins with the pretreatment of biowaste, breaking down complex biomass into simpler sugars. These sugars are then subjected to fermentation using lactic acid bacteria, resulting in the production of a fermentation broth. The broth undergoes downstream purification to separate lactic acid from residual components, yielding purified lactic acid as the final product (Thygesen et al., 2021).

4.1 Pretreatment of Fungal Biomass

Preparation of fungal waste biomass through pretreatment a vital step in improving the properties of polylactic acid (PLA) composites is crucial for enhancing its suitability for anaerobic digestion processes. Various methods such as mechanical, chemical, and biological pretreatments are employed to disrupt the fungal cell walls, facilitating easier access to the biomass's internal components during digestion. Mechanical methods involve grinding or milling to reduce particle size, while chemical treatments use acids, alkalis, or organic solvents to break down complex structures and enhance enzymatic accessibility. Biological pretreatments utilize fungi or enzymes to degrade lignocellulosic materials enzymatically. Each method aims to improve the efficiency of anaerobic digestion by increasing substrate availability and reducing digestion time, ultimately enhancing biogas production from fungal waste biomass (Zhang et al., 2020).

Research efforts are geared toward modifying biomass to address PLA's limitations while maintaining its biodegradability, resulting in green composites with significant developmental potential. These pretreatment methods focus on enhancing the mechanical properties, thermal stability, and barrier properties of PLA by utilizing modified cellulose, lignin, and starch from biomass. When modified biomass is blended with PLA, the resulting composite material can overcome the inherent limitations of PLA, making it a more viable alternative to petroleum-based polymers (Zhu et al., 2024).

Furthermore, plasma modification of biomass presents a promising approach for developing reinforced PLA composite materials, suggesting significant future applications. This method aims to create composites with improved properties, thereby addressing both environmental sustainability and performance requirements.

Fungal biomass pretreatment is also essential for the efficient bioconversion of lignocellulose into fermentable sugars, a crucial step in producing bio-based products like PLA. Fungal pretreatment, such as using *Phanerochaete chrysosporium*, is environmental-friendly and has low energy consumption, making it a favorable option for biomass processing. Near-infrared spectroscopy (NIR) has been explored as a rapid and straightforward method to assess lignin content in fungal pretreated wood, offering a high-throughput and user-friendly monitoring approach. The study emphasized the importance of monitoring lignin degradation during fungal pretreatment, highlighting the potential of spectroscopic methods like NIR for assessing lignin content in biomass, which could be beneficial for subsequent PLA production processes (Wittner et al., 2023).

In addition, pretreatment of fungal biomass for PLA production involves processes that enhance the degradation of organic polymers like lignin and hemicellulose, which are essential for biofuel production. Fungi such as *Trichoderma*, *Aspergillus*, *Tinea versicolor*, *Dichomitus squalens*, *Phlebia floridensis*, *Daedalea flavida*, and *Phlebia radiata* are known for their lignin- and hemicellulose-degrading capabilities, making them potential candidates for pretreatment processes. These fungi produce multi-enzymatic complexes like cellulosomes, lignin-degrading auxiliary enzymes, and lignin-modifying enzymes such as laccase and heme-containing peroxidase, which aid in the breakdown of lignocellulosic biomass (Rana Chhetri et al., 2022). Various factors, including strain type, inoculum load, pH, temperature, fatty acids, carbon-to-nitrogen (C/N) ratio, time, aeration, grindability, surface area, particle size, and supplements added, play crucial roles in the pretreatment process to enhance efficiency and yield of the final product. Combining microbial pretreatment with physical, chemical, and mechanical methods can lead to a synergistic effect, improving the overall efficiency of the process and contributing to the sustainable production of biofuels (Rana Chhetri et al., 2022).

4.2 Enzymatic Hydrolysis and Fermentation Processes

Enzymatic hydrolysis and microbial fermentation are pivotal biotechnological approaches for liberating bioactive peptides from lignocellulosic biomass. Lignocellulosic biomass, despite being a renewable energy source, poses challenges in hydrolysis, limiting its use in anaerobic digestion processes. To enhance microbial access to carbohydrates, various pretreatment methods such as physical, chemical, thermochemical, and biological techniques, including biological pretreatment using white rot fungi have been explored due to their cost-effectiveness and eco-friendliness.

Enzymatic hydrolysis utilizes enzymes to break down complex proteins into smaller peptides and amino acids, enhancing their bioavailability and functionality, particularly beneficial in food and pharmaceutical industries for producing peptides with antioxidant or antimicrobial properties. Meanwhile, microbial fermentation employs microorganisms like bacteria, yeast, or fungi to ferment substrates, converting them into bioactive compounds, including peptides. These methods not only facilitate efficient peptide production but also promote sustainable practices by leveraging renewable resources and minimizing waste (Kainthola et al., 2021). Fungal biomass can be transformed into lactic acid through a fermentation process utilizing specific microorganisms. Here is an explanation of how this conversion happens (Asadollahzadeh et al., 2022):

(i) *Fermentation Process*: The conversion of fungal biomass to lactic acid is achieved through a carefully controlled fermentation process. This process involves the breakdown of biomass components by specific microorganisms, such as certain strains of fungi or bacteria. These microorganisms initiate the fermentation by decomposing the complex organic materials within the biomass, ultimately resulting in the production of lactic acid.

(ii) *Microbial Activity*: Within the fermentation environment, microorganisms actively metabolize the sugars, cellulose, and other organic compounds present in the fungal biomass. These microorganisms, through their metabolic processes, convert these compounds into lactic acid. The efficiency of this conversion depends on the type of microorganisms used and the conditions maintained during fermentation, such as temperature, pH, and nutrient availability.

(iii) *Bioprocessing Techniques*: The production of lactic acid from fungal biomass is facilitated by advanced bioprocessing techniques. These techniques involve the use of microorganisms to enzymatically break down the biomass. The metabolic pathways of these microorganisms are harnessed to optimize the yield of lactic acid. This bioprocessing involves a series of steps, including pretreatment of the biomass, inoculation with microorganisms, and maintenance of optimal fermentation conditions to maximize lactic acid production.

(iv) *Biorefinery Concept*: The conversion of fungal biomass to lactic acid is part of the broader biorefinery concept. This approach emphasizes the sustainable and efficient use of biomass resources to produce valuable biochemicals. In a biorefinery, different components of the biomass are fractionated and converted into various products, minimizing waste and enhancing resource efficiency. Lactic acid production from fungal biomass exemplifies this approach, as it transforms renewable biomass into a high-value biochemical with diverse applications.

(v) *Sustainable Production*: The process of converting fungal biomass to lactic acid represents a sustainable and eco-friendly production method. This method reduces dependence on fossil fuels and lowers greenhouse gas emissions associated with conventional chemical production. By utilizing renewable fungal biomass, the process contributes to a circular economy and promotes the development of biodegradable and bio-based materials, such as polylactic acid (PLA). PLA is a versatile polymer used in biocomposites, packaging, and medical devices, highlighting the environmental and economic benefits of this sustainable production pathway.

4.3 Optimization of Lactic Acid Production

Optimizing lactic acid production involves leveraging various microbial and substrate sources. Lactic acid, a key organic compound, is primarily produced through fermentation by diverse microorganisms capable of utilizing different carbohydrate substrates. Lactic acid bacteria, notably *Lactobacillus* spp., demonstrate robust fermentation capabilities, while *Bacillus* spp. offer cost-effective production potential. Genetic engineering has enhanced productivity in *Corynebacterium glutamicum* and *Escherichia coli*. Fungi such as *Rhizopus* spp. utilize renewable carbon sources, particularly with advantageous amylolytic properties, for lactic acid synthesis. Yeasts, tolerant to environmental stresses like acidity, are also manipulated genetically for improved lactic acid yields. Additionally, microalgae and cyanobacteria, functioning as photosynthetic organisms, present alternative methods for lactic acid production without relying on carbohydrate feedstocks. Various substrates—from plant residues like molasses and agricultural waste to dairy by-products supplemented with nitrogen sources—are essential components in the fermentation process (Abedi & Hashemi, 2020).

The production of lactic acid is fundamental for synthesizing polylactic acid (PLA), a polymer that is both biodegradable and biocompatible. One study concluded that the optimization process required concentrating the lactic acid solution to eliminate free water before oligomerization and lactide synthesis can proceed. The synthesis process involved using varying concentrations of tin octoate catalyst at temperatures between 150 °C and 210 °C to maximize the conversion of lactic acid to lactide. Following synthesis and purification, the lactide yield, an essential intermediate for PLA production, ranged from 67% to 69%. The raw lactide undergoes recrystallization in butyl acetate, which was identified as the most effective solvent for laboratory purification, achieving a yield of 41.4%. Further optimization steps involved polymerizing lactide in a reactor unit with a tin octoate catalyst concentration of 500 ppm, resulting in a high conversion rate of 95% and the production of PLA pellets with specific properties (Aliev et al., 2024).

In another study, the researchers investigated that the synthesis of PLA from lactic acid encompassed several critical steps, including the pretreatment of lignocellulosic biomass, fermentation, concentration, and purification. Pressurized hot water pretreatment at specific conditions—such as 180 °C for 30 min at 10 MPa—was essential for efficient carbohydrate separation, facilitating effective lactic acid production. The enzymatic hydrolysis process utilized cellulase and beta-glucosidase enzymes, alongside fermentation with bacteria such as *Lacticaseibacillus rhamnosus* ATCC 7469, to convert biomass into lactic acid. Concentration and purification steps involved extraction with ammonium sulfate and n-butanol and were crucial for obtaining high-purity lactic acid for subsequent PLA processing. Azeotropic dehydration of lactic acid at 140 °C for 24 h in the presence of xylene and $SnCl_2$ catalyst resulted in the formation of lactide (CPLA), a precursor for PLA synthesis. Microwave-assisted polymerization under specific conditions, like 140 °C for 30 min with $SnCl_2$ catalyst, was vital for transforming lactide into PLA with high yield and purity, suitable for various applications including the packaging industry (Senila et al., 2023).

Other study, which aimed at optimizing lactic acid production, utilized overripe plantain as feedstock through response surface methodology (RSM). Variables such as substrate concentration, initial pH, fermentation temperature, and duration were analyzed to maximize lactic acid concentration, which reached up to 163.00 g/l. The study identified the optimal conditions for lactic acid production through numerical optimization: a glucose concentration of 159.10 g/l, pH of 7.0, temperature of 42.3 °C, and fermentation time of 60 h, resulting in 157.53 g/l of lactic acid. The research concluded that overripe plantain is an effective biochemical feedstock for lactic acid production, which can be further utilized in the creation of PLA, promoting a sustainable and eco-friendly polymer (Alao et al., 2023).

5 Polymerization of Lactic Acid to Polylactic Acid (PLA)

Polylactic acid (PLA) is recognized as a leading biodegradable polymer, celebrated for its eco-friendly nature and sustainable production methods. Its popularity in biomedical fields is attributed to its biocompatibility and inherent biodegradability, making it suitable for diverse applications such as medical implants, drug delivery systems, and tissue engineering scaffolds. PLA is derived from lactic acid, a compound that can be efficiently produced through the fermentation process. This fermentation primarily utilizes sugars sourced from renewable materials, such as cornstarch or sugarcane, converting them into lactic acid as shown in Fig. 3.

The lactic acid serves as the precursor for PLA polymerization, offering a sustainable pathway for manufacturing biodegradable plastics without relying on nonrenewable resources like petroleum. This renewable origin not only reduces dependence on fossil fuels but also aligns with global efforts toward sustainable manufacturing practices. As a result, PLA has emerged as a promising alternative to traditional petroleum-based plastics, offering a viable solution to environmental challenges without compromising on performance or functionality in various applications (Chatterjee & Nair, 2021).

The figure illustrates the sustainable life cycle of polylactic acid (PLA), showcasing its production from corn through photosynthesis, fermentation, and polymerization into lactic acid. The PLA is then molded into plastic products, which

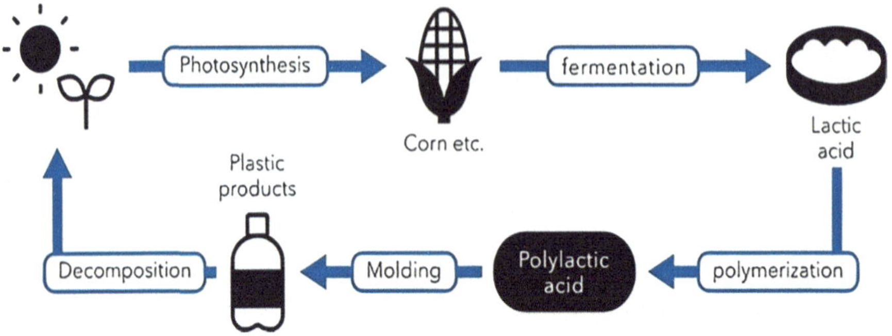

Fig. 3 Sustainable life cycle of polylactic acid (PLA)

decompose back into natural elements, completing the cycle and ensuring minimal environmental impact (HIGHCHEM CO., LTD).

One study investigated that the production of polylactic acid (PLA) from lactic acid involves several intricate steps, commencing with the synthesis of lactide, a pivotal precursor in PLA manufacturing. The process began with the oligomerization and subsequent synthesis of lactic acid under controlled conditions, facilitated by a tin octoate catalyst at elevated temperatures ranging from 150 °C to 210 °C (Aliev et al., 2024). Following synthesis, the crude lactide undergoes purification through recrystallization in butyl acetate, resulting in a yield of 41.4%. Subsequently, polymerization of the purified lactide occurred within a reactor unit using a tin octoate catalyst concentration of 500 ppm, achieving an impressive conversion rate of 95%. The resulting PLA exhibited a molecular weight (Mw) of 228 kDa and a polydispersity index (PDI) of 1.94, indicating the uniformity and quality of the polymer (Aliev et al., 2024). Comprehensive characterization employed techniques such as differential scanning calorimetry, nuclear magnetic resonance (NMR) spectroscopy, and gel permeation chromatography ensuring the thorough analysis of PLA properties, validating its suitability for diverse applications (Aliev et al., 2024).

5.1 Chemical Polymerization Techniques

Chemical polymerization techniques for converting lactic acid into polylactic acid (PLA) involve several methods, each tailored to optimize PLA's properties for specific applications, particularly in biomedicine. PLA is renowned as a biodegradable polymer derived from renewable resources, making it environmental-friendly. It finds extensive use in biomedical applications due to its biocompatibility and ability to degrade safely within the human body.

Common synthesis methods include ring-opening polymerization, melt polycondensation, and microwave irradiation. Lactic acid, the precursor to PLA, can be obtained through fermentation of sugars and various renewable raw materials.

Chemical synthesis of PLA typically begins with the production of a racemic mixture of lactic acid, where acetaldehyde is used as a starting material. This process involves reacting acetaldehyde with hydrogen cyanide, followed by hydrolysis of the resulting lactonitrile, yielding lactic acid along with ammonium chloride as a by-product. Alternative starting materials such as vinyl acetate and glycerol can also be employed in this synthetic pathway (Chatterjee & Nair, 2021).

Fungal biomass residues offer a viable source for producing lactic acid, a crucial precursor in PLA synthesis. Chemical methods are pivotal in converting lactic acid derived from fungal biomass into PLA. Ring-opening polymerization stands out as a widely adopted technique, involving the transformation of lactic acid molecules into PLA chains (Das et al., 2023). Alternatively, catalysts are employed to expedite polymerization reactions, facilitating the conversion of fungal-derived lactic acid into PLA. Optimization of PLA production from lactic acid involves meticulous control of reaction parameters such as temperature, pressure, and catalyst concentration, ensuring enhanced yield and purity of the final product (Das et al., 2023).

5.2 *Role of Catalysts in PLA Synthesis*

As shown in Table 1, catalysts play a pivotal role in achieving controlled and efficient PLA production during the polymerization of lactic acid. Metal catalysts such as zinc 2-ethylhexanoate (ZnOct2) have shown high activity in lactide polymerization experiments, while calcium 2-ethylhexanoate (CaOct2) exhibited lower activity. Optimization of the ring-opening polymerization (ROP) process using these catalysts has yielded polylactides with desirable properties, including molecular weights up to 82.3 kg/mol, lactide conversions up to 85%, and low D-lactic acid

Table 1 Role of catalysts in PLA synthesis

Catalyst	Synthesis Condition	Remark	References
$SnCl_2$	140 °C, 24 h with xylene	Crucial for azeotropic dehydration, forming lactide (CPLA)	Garlotta (2001)
Zinc lactate	160–180 °C, 2–3 h	Facilitates efficient lactide synthesis, achieving moderate yield and purity	Aliev et al. (2024)
Stannous octoate	130–150 °C, vacuum conditions	Used for ring-opening polymerization, results in high-molecular-weight PLA	Aliev et al. (2024)
Magnesium oxide	180 °C, 6 h	Alternative catalyst for lactide formation, with lower environmental impact	Aliev et al. (2024)
Titanium butoxide	140 °C, inert atmosphere	Effective for producing low-molecular-weight PLA, used in combination with other catalysts	Aliev et al. (2024)
Tin chloride	150–170 °C, 5 h	Improves polymerization rate and molecular weight of PLA	Fang et al. (2009)
Aluminum isopropoxide	120 °C, solvent-free conditions	Promotes controlled polymerization, resulting in consistent PLA properties	Fang et al. (2009)

content (8%) (Gadomska-Gajadhur & Ruśkowski, 2020). These characteristics make the resulting PLA suitable for medical and pharmaceutical applications. Metal complexes like tin (II) octoate (Sn (Oct)2) have also been extensively utilized due to their solubility in organic solvents, stability, and FDA approval for medical and food applications. The mechanism of tin-catalyzed lactide polymerization involves coordination and insertion mechanisms, facilitating controlled polymer chain growth and relatively low racemization rates compared to other catalytic systems (Masutani et al., 2014).

5.3 Process Optimization and Scale-Up Considerations

Process optimization and scale-up considerations are crucial in the polymerization of lactic acid to polylactic acid (PLA). Achieving efficient and scalable production involves fine-tuning various parameters such as catalyst concentration, reaction temperature, and reaction time to maximize polymer yield and control polymer properties. Optimization often includes experimental design methodologies to systematically explore and optimize these parameters. Scale-up considerations address the transition from laboratory-scale synthesis to industrial production, ensuring that the process maintains efficiency and quality at larger volumes (Ejiohuo, 2023).

Factors such as reactor design, heat and mass transfer considerations, and raw material sourcing become critical in scaling up PLA production while maintaining economic feasibility and sustainability. Successful scale-up ensures consistent product quality and meets the increasing demand for PLA in various applications, including packaging, textiles, and biomedical devices.

Enhancing PLA production efficiency and product quality involves adjusting parameters such as extruder and bed temperatures, layer height, printing speed, travel speed, infill, and shell count (Raja et al., 2022). Key factors influencing macular defects in PLA straw production include cleaning frequency, mold head temperature, mold outer ring size, and mold material. More frequent cleaning can minimize surface defects, and the optimal die head temperature is 195 °C. The best die mouth outer ring size is 26 mm, and PEEK material is preferred for mold construction (Tian et al., 2023).

Scaling up PLA production requires careful evaluation of equipment capacity, material availability, and production costs to ensure a smooth transition to higher production volumes. Maintaining optimized parameters identified during process analysis is crucial. Consistent cleaning practices help prevent surface defects in larger outputs, and controlling die head temperature becomes even more critical. Proper selection and maintenance of mold materials, such as PEEK, sustain the improved quality in larger production volumes. Optimization of printing parameters, including extrusion rates and layer thickness, can help achieve consistent mechanical properties in PLA products during scale-up. Conducting mechanical tests, such as tensile tests following ASTM D638 standard, can determine the impact of process changes on the mechanical strength of PLA products, guiding optimization efforts for both small- and large-scale productions (Jang et al., 2021).

6 Advantages of Using Fungal Biomass for PLA Production

Using fungal waste biomass for PLA production offers several advantages. First, fungal biomass can be sourced from waste materials, making it a cost-effective and sustainable feedstock. Fungi, such as *Rhizopus* species, are highly efficient in breaking down complex organic materials into simpler sugars, which can then be fermented into lactic acid, the precursor for PLA. This process not only reduces waste but also lowers production costs. Additionally, fungal biomass can be cultivated under mild conditions without the need for expensive inputs, further enhancing its economic viability. Moreover, the use of fungal biomass aligns with the principles of a circular economy, promoting the recycling and repurposing of organic waste into valuable bioproducts like PLA (Asadollahzadeh et al., 2022).

Integrating fungal biomass into PLA can improve the mechanical properties of the resulting biocomposites, such as increased elongation at break and impact strength, effectively addressing the brittleness often seen in PLA. By incorporating fungal biomass, the thermal properties of PLA can be enhanced, as demonstrated by the shifts in glass transition, crystallization, and melting temperatures to lower values, suggesting potential improvements in thermal performance. Fungal mycelium biomass, which is a part of fungal biomass, is obtained from organic waste residues, making it a sustainable option for creating bio-based materials like PLA-based composites. The complex composition of fungal biomass, including fibers and proteins, adds unique characteristics, making it a valuable blending component in biocomposites and enhancing the overall properties of the materials (Asadollahzadeh et al., 2022).

6.1 Sustainability and Environmental Benefits

Using fungal waste biomass for PLA production offers considerable sustainability and environmental benefits. Fungal biomass is biocompatible, biodegradable, and can be sourced from various biomass materials, promoting a circular economy. It provides high stability, flexibility, tunability, hydrophobicity, nonflammability, porosity, and insulating properties. Additionally, fungal biomass is fast-growing, self-repairing, and self-healing, making it an excellent eco-friendly alternative to traditional plastics. The incorporation of fungal biomass in PLA production reduces reliance on nonrenewable resources, lowers carbon footprint, and minimizes waste generation, aligning with sustainable development goals. Moreover, fungal biomass is safe for biological systems and decomposes naturally, reducing environmental impact (Asadollahzadeh et al., 2022).

6.2 Cost-Effectiveness and Economic Viability

Using fungal waste biomass for PLA production is not only environmentally sustainable but also economically viable. Fungal biomass can be sourced from inexpensive and readily available agricultural and industrial waste, significantly reducing raw material costs. The growth and cultivation of fungi require minimal resources and can be carried out under simple conditions, further lowering production expenses.

Additionally, fungal biomass offers high yields of lactic acid, the precursor for PLA, enhancing the efficiency of the production process. This cost-effective feedstock reduces reliance on expensive, nonrenewable resources, contributing to the overall economic viability of PLA production. Moreover, the integration of fungal biomass into existing production systems can lead to operational savings by minimizing waste disposal costs and reducing the need for extensive preprocessing steps. By leveraging fungal biomass, industries can produce PLA more economically, making it a competitive alternative to traditional plastics and promoting sustainable economic growth.

6.3 Comparative Analysis with Other Biomass Sources

Using fungal waste biomass for PLA production offers significant sustainability advantages over other biomass sources. Fungal biomass, sourced from various waste materials, promotes a circular economy by reducing competition for food resources and minimizing environmental impact compared to agricultural biomass. It possesses high stability, flexibility, and hydrophobicity, alongside biocompatibility and biodegradability, making it an eco-friendly alternative to traditional plastics, as shown in Table 2. Incorporating fungal biomass in PLA production lowers carbon footprints, minimizes waste generation, and aligns with sustainable development goals. Additionally, fungal biomass exhibits antibacterial, antioxidant, and skin-brightening properties, suitable for diverse applications in textiles, packaging, medicine, cosmetics, construction, and automotive sectors. This makes fungal biomass a versatile and sustainable feedstock, addressing global challenges while ensuring environmental and human health safety.

7 Challenges and Limitations

Using fungal waste biomass for PLA production presents both opportunities and challenges. While fungal biomass shows promise in producing lactic acid for PLA (Das et al., 2023), incorporating it into PLA composites can decrease tensile strength and Young's modulus while improving elongation at break and impact strength.

Table 2 Comparative analysis of PLA with other fungal-based biomass sources

Aspect	PLA	*Aspergillus niger*	*Saccharomyces Cerevisiae*	References
Source	Cornstarch, sugarcane	Waste agricultural products, industrial by-products	Sugar, molasses, fruit, and vegetable waste	*PLA*: (Mangaraj et al., 2019) *Aspergillus niger:* (Cairns et al., 2018) *Saccharomyces Cerevisiae:* (Hong & Nielsen, 2012)
Properties	Biodegradable, high mechanical strength, stable, moderately flexible	High protein content, enzymatic activity, stable, less flexible	High ethanol production, ease of genetic modification, stable, flexible	*PLA*: (Lim et al., 2008) *Aspergillus niger:* (Meyer et al., 2015) *Saccharomyces cerevisiae:* (Stephanopoulos, 2007)
Applications	Packaging, medical implants, textiles	Enzyme production, biofuels, food industry	Bioethanol production, food additives, pharmaceuticals	*PLA*: (Mangaraj et al., 2019) *Aspergillus niger:* (Cairns et al., 2018) *Saccharomyces Cerevisiae:* (Hong & Nielsen, 2012)
Environmental impact	Renewable, compostable, low carbon footprint	Renewable, compostable, low carbon footprint	Reduces agricultural waste, low greenhouse gas emissions	*PLA*: (Mangaraj et al., 2019) *Aspergillus niger:* (Cairns et al., 2018) *Saccharomyces Cerevisiae:* (Hong & Nielsen, 2012)

Furthermore, the limited production levels of fungal enzymes in conventional host organisms hinder their application in biotechnology. Additionally, there is a need for further exploration and development in using fungal bio-machinery for PLA production.

A significant challenge in utilizing fungal waste biomass for the production of PLA is ensuring the uniformity and quality of the raw materials (Abdelshafy et al., 2023). The composition and characteristics of fungal waste biomass can vary widely based on the fungal species, growth conditions, and processing techniques employed. These variations can result in inconsistencies in the quality and performance of the resulting PLA product, potentially hindering its commercial feasibility. These factors highlight the ongoing research and innovation required to fully exploit the potential of fungal biomass in the PLA manufacturing process.

7.1 Technical and Logistical Barriers

The utilization of fungal waste biomass for PLA production presents several notable technical and logistical barriers. Fungal pretreatment, although a cost-effective method, requires extended incubation periods, leading to longer processing times and reduced yields compared to traditional approaches. This longer incubation period not only increases operational costs but also complicates the logistics of production scheduling and resource management. Moreover, the variability in fungal biomass composition and the need for precise control over growth conditions further add to the complexity. These factors significantly impact the economic feasibility of large-scale production, emphasizing the need for innovative solutions to optimize the integration of fungal biomass in PLA-manufacturing processes while addressing these technical and logistical challenges effectively (Vasco-Correa & Shah, 2019).

7.2 Quality and Consistency of Feedstock

The quality and uniformity of feedstock are critical factors in leveraging fungal waste biomass for polylactic acid (PLA) production. Recent studies have demonstrated that fungal biomass, derived from fermentation residues, can effectively enhance PLA properties, addressing issues such as brittleness. Research has also focused on evaluating fungal biomass sourced from edible potato protein liquor, highlighting its substantial protein content and its potential as a resource for food and feed production due to its rich amino acid and polyunsaturated fatty acid profiles (Sar et al., 2022).

Furthermore, advancements in real-time assessment techniques, such as neural networks and image analysis for evaluating corn stover feedstock, underscore the importance of consistent feedstock quality in optimizing biorefinery processes. Ensuring high-quality fungal waste biomass feedstock is crucial for optimizing the mechanical and thermal properties of PLA biocomposites.

7.3 Regulatory and Market Challenges

The utilization of fungal waste biomass in the production of polylactic acid (PLA) poses significant regulatory and market hurdles. While integrating fungal biomass into PLA composites can improve their properties and address the brittleness typically associated with PLA, challenges such as inconsistencies in production processes, recycling methods, carbon footprint concerns, and pricing complexities impede its widespread adoption, particularly with first-generation feedstocks (Kalinke et al., 2023). Basidiomycete fungi exhibit promise in efficiently degrading plant biomass; however, understanding the regulatory frameworks governing the production of biomass-degrading enzymes in fungi remains a challenge.

Additionally, the complexity of plant biomass degradation by fungi complicates efforts to develop a bio-based economy, highlighting barriers like limited enzyme production and unclear phenotypes during biomass degradation processes (Khosravi et al., 2019). Overcoming these regulatory and market challenges is critical to advancing the utilization of fungal waste biomass for PLA production.

8 Future Perspectives and Research Directions

In the realm of sustainable materials, the future outlooks and avenues for research in utilizing fungal waste biomass for polylactic acid (PLA) production appear promising. Fungal biomass, rich in proteins and carbohydrates, can be cultivated in controlled bioreactors to consistently yield safe biomass, aligning seamlessly with the principles of the bio-based circular economy (Wan Mohtar et al., 2022). Integration of fungal biomass into PLA formulations has the potential to enhance material characteristics, such as improving elongation at break and impact resistance while mitigating brittleness.

Furthermore, harnessing fungal processes for lactic acid production offers a sustainable alternative to petroleum-derived plastics, thereby reducing greenhouse gas emissions and fossil fuel consumption, which plays a pivotal role in environmental conservation efforts. Exploring fungal species like *Ganoderma lucidum* for bioremediation of plastic waste streams and their application in mycelium composites unveils novel opportunities for innovative applications in architecture and waste management. Additionally, symbiotic relationships between fungi and microalgae in wastewater treatment illustrate potential for high-value biomass production and the synthesis of value-added compounds for various industrial sectors, underscoring the versatility and economic viability of fungal applications across diverse fields (Wu et al., 2023).

8.1 Innovations in Fungal Biomass Processing

Innovations in fungal waste biomass processing for PLA production are paving the way for sustainable advancements in biotechnology. Researchers are exploring novel techniques to enhance fungal biomass utilization, focusing on optimizing pretreatment methods to improve efficiency and yield. Innovative approaches include enzymatic hydrolysis and microbial fermentation, which aim to streamline the conversion of fungal waste biomass into lactic acid precursors for PLA synthesis.

Moreover, advancements in bioreactor design and process engineering are facilitating scalable production, addressing technical challenges such as prolonged incubation periods and variable biomass composition. These innovations not only aim to reduce production costs but also enhance the environmental footprint by utilizing renewable resources effectively. As research continues to evolve, integrating these innovations promises to establish fungal biomass as a viable and eco-friendly feedstock in the PLA production landscape.

8.2 Integration with Circular Economy Models

Fungal waste biomass offers a pathway to integrate into circular economy models for PLA production by effectively breaking down lignocellulosic plant biomass and converting sugars into energy-dense molecules, aligning with the principles of a bio-based circular economy as depicted in Fig. 4. Filamentous fungi possess innate capabilities to selectively and efficiently degrade components of wood, making them well suited for aiding in the production of second-generation biofuels and chemicals from wood in biorefinery applications (Meyer et al., 2020).

Fungal biotechnology has significant potential to advance the shift toward a bio-based circular economy by sustainably generating a range of products including food, feed, chemicals, fuels, textiles, and materials for various industries. This information aims to educate scientists, educators, the public, industrial stakeholders, and policymakers about the current fungal biotech advancements, highlighting both the opportunities and challenges in fungal biotechnology. This underscores the critical role of fungi in addressing pressing global challenges and contributing to

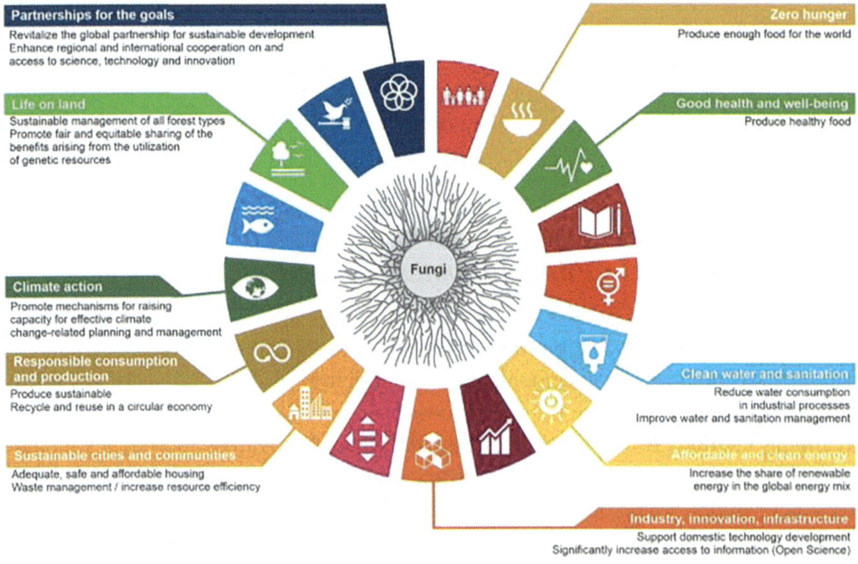

Fig. 4 Visualizing the circular economy with fungal biotechnology

efforts in climate change mitigation and the achievement of sustainable development objectives (Meyer et al., 2020).

The figure highlights the life cycle of filamentous fungi, detailing their ability to decompose organic materials and produce valuable substances such as enzymes, organic acids, and antibiotics. The figure effectively demonstrates how fungi can convert agricultural and forestry by-products into new materials, showcasing their role in a circular bio-economy. This innovative approach not only supports environmental sustainability but also opens up new avenues for industrial applications (Meyer et al., 2020).

8.3 Potential for Genetic and Metabolic Engineering

The recognition of fungal waste biomass as a viable, sustainable source for producing polylactic acid (PLA) is growing, especially within the realm of genetic and metabolic engineering. This strategy not only boosts PLA production efficiency but also aids in advancing a circular economy by repurposing organic waste materials.

Techniques in genetic engineering, such as CRISPR-Cas9, are increasingly used to optimize the metabolic pathways of filamentous fungi, to make them more adept at converting lignocellulosic waste into valuable products like PLA (Chatterjee & Venkata Mohan, 2022). By altering the genetic structure of fungi, scientists can enhance their ability to break down complex carbohydrates from agricultural residues, thus increasing the yield of fermentable sugars for PLA production (Chatterjee & Venkata Mohan, 2022). Advancements in fungal waste biomass for PLA production involve leveraging genetic engineering tools like CRISPR-Cas9 to make precise modifications to fungal genomes. These genetic modifications improve the fungi's ability to efficiently decompose lignocellulosic biomass into fermentable sugars by boosting cellulases, hemicellulases, and lignin-degrading enzymes, which are crucial for optimal biomass conversion.

Metabolic engineering further refines the internal metabolic processes of fungal cells to maximize lactic acid production, essential for PLA. Synthetic biology techniques redesign metabolic networks to enhance the efficiency of substrate conversion from fungal waste biomass (Zhang et al., 2021). This approach may involve boosting the expression of enzymes crucial for lactic acid fermentation and suppressing competing pathways that divert resources away from PLA production (Zhang et al., 2021). Overexpressing key enzymes involved in glycolysis and the pentose phosphate pathway, such as lactate dehydrogenase and pyruvate decarboxylase, enhances yields while minimizing by-products. Concurrently, optimizing fermentation conditions supports fungal growth and maximizes lactic acid output.

Fungal waste biomass, particularly from species such as *Aspergillus oryzae* and *Trichoderma reesei*, can be grown on various organic substrates, including agricultural waste and by-products from the food industry (Cottet et al., 2022). This biomass acts not only as a raw material for PLA production but also provides vital nutrients that can improve the fermentation process. Incorporating fungal biomass into PLA production can lead to the creation of biocomposites with enhanced mechanical properties, broadening their potential applications (Cottet et al., 2022). Downstream processing techniques are essential for extracting and purifying lactic acid from fermentation broths. Methods such as filtration, centrifugation, and advanced separation technologies, including membrane processes, facilitate efficient recovery of high-purity lactic acid.

Using fungal waste biomass for PLA production helps in reducing waste and supports sustainability by converting organic waste into valuable products. This approach aligns with the concept of biorefinery, where biomass is used to produce multiple products, thereby minimizing waste and optimizing resource use (Chatterjee & Venkata Mohan, 2022). Furthermore, utilizing inexpensive organic waste as a feedstock can significantly lower PLA production costs, offering a more competitive alternative to petroleum-based plastics (Zhang et al., 2021). These biotechnological advances not only improve PLA production yields but also support environmental sustainability by minimizing waste and energy consumption. By using fungal biomass as a renewable and eco-friendly feedstock, these strategies contribute to establishing a circular economy model for biopolymer production, aligning with global sustainability goals.

9 Conclusion

In conclusion, the utilization of fungal waste biomass in polylactic acid (PLA) production represents a significant advancement in sustainable bioplastics. By integrating fungal biomass into PLA biocomposites, researchers have explored enhancing the mechanical, thermal, and surface properties through the addition of bio-based plasticizers like triethyl citrate (TEC). This approach not only addresses PLA's inherent brittleness but also demonstrates the potential for developing robust biocomposite materials suitable for diverse applications in packaging, biomedical devices, and beyond (Prendiz et al., 2019).

Furthermore, fungal technology offers a promising avenue for biomass conversion, leveraging its capabilities in lignin degradation, saccharification, fermentation, and lipid accumulation. This eco-friendly and cost-effective solution contributes to reducing environmental impacts associated with conventional plastic production and waste. Studies on composting sisal residue underscore the role of fungal strains such as *Fusarium oxysporum*, *Fusarium* solani, *Penicillium crustosum*, and *Penicillium chrysogenum* in altering biomass chemical characteristics, which further enhances substrate digestibility and bioconversion efficiency (Chai et al., 2022).

In conclusion, the integration of fungal waste biomass into PLA production promotes sustainable practices and aligns with global efforts to reduce plastic pollution. Utilizing fungal biomass as a feedstock for PLA provides a renewable and eco-friendly alternative to traditional petroleum-based plastics, addressing critical environmental issues such as waste management and carbon emissions. Continued research in fungal biotechnology is essential for advancing the bio-economy, offering innovative solutions that mitigate environmental impacts while meeting industrial demands for biodegradable materials.

As technology progresses, optimizing and scaling up these processes will be crucial for realizing the full potential of fungal biomass as a renewable feedstock for bioplastic production. This includes improving the efficiency of fungal biomass conversion, optimizing microbial fermentation techniques, and enhancing PLA properties. Overcoming technical challenges and ensuring the economic feasibility of large-scale production will be vital. By addressing these issues, the full potential of fungal waste biomass can be unlocked, paving the way for a more sustainable and environmental-friendly approach to bioplastic production.

References

Abdel-Rahman, M. A., Tashiro, Y., & Sonomoto, K. (2013). Recent advances in lactic acid production by microbial fermentation processes. *Biotechnology Advances, 31*(6), 877–902. https://doi.org/10.1016/j.biotechadv.2013.04.002

Abdelshafy, A., Hermann, A., Herres-Pawlis, S., & Walther, G. (2023). Opportunities and challenges of establishing a regional bio-based polylactic acid supply chain. *Global Challenges, 7*(7), 2200218. https://doi.org/10.1002/gch2.202200218

Abedi, E., & Hashemi, S. M. B. (2020). Lactic acid production – producing microorganisms and substrates sources-state of art. *Heliyon, 6*(10), 04974. https://doi.org/10.1016/j.heliyon.2020.e04974

Alao, A. I., Agarry, S. E., & Afolabi, T. J. (2023). Optimization of lactic acid fermentation from overriped plantain using response surface methodology (RSM). *Bioprocess Engineering, 7*(1), 1–9. https://doi.org/10.11648/j.be.20230701.11

Aliev, G., Toms, R., Melnikov, P., Gervald, A., Glushchenko, L., Sedush, N., & Chvalun, S. (2024). Synthesis of L-lactide from lactic acid and production of pla pellets: Full-cycle laboratory-scale technology. *Polymers, 16*(5), 624. https://doi.org/10.3390/polym16050624

Asadollahzadeh, M., Mahboubi Soufiani, A., Taherzadeh, M., Åkesson, D., & Lennartsson, P. (2022). Application of fungal biomass for the development of new polylactic acid-based biocomposites. *Polymers, 14*(9), 1738–1738. https://doi.org/10.3390/polym14091738

Bai, Y., Chen, M., Peng, A., Shao, L., Luo, J., Deng, S., Yan, Y., & Peng, B. (2022). Valorization of waste biomass through fungal technology: Advances, challenges, and prospects. *Agronomy and Crop Science, 188*, 115608. https://doi.org/10.1016/j.indcrop.2022.115608

Cabrera, B. G., Gallardo, F., Nesic, A., Taboada, E., Marican, A., Mirabal-Gallardo, Y., Avila-Salas, F., Delgado, N., & de Armas-Ricard, M. (2020). Utilization of industrial by-product fungal biomass from Aspergillus Niger and Fusarium culmorum to obtain biosorbents for removal of pesticide and metal ions from aqueous solutions. *Journal of Environmental Chemical Engineering, 8*(4), 104355. https://doi.org/10.1016/j.jece.2020.104355

Cairns, T. C., Nai, C., & Meyer, V. (2018). How a fungus shapes biotechnology: 100 years of Aspergillus Niger research. *Fungal Biology and Biotechnology, 5*, 13. https://doi.org/10.1186/s40694-018-0054-5

Castro-Aguirre, E., Iniguez-Franco, F., Samsudin, H., Fang, X., & Auras, R. (2016). Poly (lactic acid)-Mass production, processing, industrial applications, and end of life. *Advanced Drug Delivery Reviews, 107*, 333–366. https://doi.org/10.1016/j.addr.2016.03.010

Chai, Y., Bai, M., Chen, A., & Peng, L. (2022). Valorization of waste biomass through fungal technology: Advances, challenges, and prospects. *Industrial Crops and Products, 188*(3), 115608. https://doi.org/10.1016/j.indcrop.2022.115608

Chatterjee, S., & Nair, P. (2021). Extraction and polymerization of lactic acid to polylactic acid. *International Journal of Scientific & Engineering Research, 12*(1), 549. ISSN 2229-5518.

Chatterjee, S., & Venkata Mohan, S. (2022). Fungal biorefinery for sustainable resource recovery from waste. *Waste Management, 132*, 1–12. https://doi.org/10.1016/j.wasman.2021.12.014

Cottet, C., Ramirez-Tapias, Y. A., Delgado, J. F., de la Osa, O., Salvay, A. G., & Peltzer, M. A. (2022). Application of fungal biomass for the development of new polylactic acid-based biocomposites. *Polymers, 14*(9), 1738. https://doi.org/10.3390/polym14091738

Das, M., Santra, S., Rajlakshmi, Saravanabhupathy, S., Dey, S., Banerjee, S., Banerjee, R., Satyanarayana, T., & Deshmukh, S. K. (2023). Lactic acid production from fungal machineries and mechanism of PLA synthesis: Application of AI-based technology for improved productivity. In *Fungi and fungal products in human welfare and biotechnology* (pp. 211–256). Springer. https://doi.org/10.1007/978-981-19-8853-0_8

De Eugenio, L. I., Murguiondo, C., Galea-Outon, S., Prieto, A., & Barriuso, J. (2023). Fungal–Lactobacteria consortia and enzymatic catalysis for polylactic acid production. *Journal of Fungi, 9*(3), 342. https://doi.org/10.3390/jof9030342

Deeraj, B. D. S., Jayan, J. S., Saritha, A., & Joseph, K. (2022). PLA-based blends and composites. In *Biodegradable polymers, blends and composites* (pp. 237–281). https://doi.org/10.1016/B978-0-12-823791-5.00014-4

Ejiohuo, O. (2023). Optimization of the degradation processes for polylactic acid using microbial enzymes: A brief summary. *SustainE, 1*(2), 1–24. https://doi.org/10.55366/suse.v1i2.1

Fang, L., Qi, R., Linbo, L., Gongwen, J., & Suangw, H. (2009). Synthesis of poly(L-lactide) via solvothermal method. *International Journal of Polymer Science, 2009*(7), 1. https://doi.org/10.1155/2009/929732

Gadomska-Gajadhur, A., & Ruśkowski, P. (2020). Synthesis of PLA via ring-opening polymerization. *Organic Process Research and Development, 24*(8), 1435–1442. https://doi.org/10.1021/acs.oprd.0c00149

Garlotta, D. (2001). A literature review of poly (lactic acid). *Journal of Polymers and the Environment, 9*(2), 63–84. https://doi.org/10.1023/A:1020200822435

Geyer, R., Jambeck, J. R., & Law, K. L. (2017). Production, use, and fate of all plastics ever made. *Science Advances, 3*(7), 1700782. https://doi.org/10.1126/sciadv.1700782

Hajdek, K., Smoljan, B., Šarkanj, B., & Sitek, W. (2023). Processing technologies, properties and application of poly (lactic acid) (pla). *International Journal of Modern Manufacturing, 15*(1), 87–97.

Hong, K. K., & Nielsen, J. (2012). Metabolic engineering of Saccharomyces cerevisiae: A key cell factory platform for future biorefineries. *Cellular and Molecular Life Sciences, 69*, 2671–2690. https://doi.org/10.1007/s00018-012-0945-1

Jang, B.-K., Ju, Y., Jeong, D., Jung, S.-K., Kim, C.-K., Chung, Y.-S., & Kim, S.-R. (2021). l-Lactic acid production using engineered Saccharomyces cerevisiae with improved organic acid tolerance. *Journal of Fungi, 7*(11), 928. https://doi.org/10.3390/jof7110928

Kabir, E., Kaur, R., Lee, J., Kim, K.-H., & Kwon, E. E. (2020). Prospects of biopolymer technology as an alternative option for non-degradable plastics and sustainable management of plastic wastes. *Journal of Cleaner Production, 258*, 258. https://doi.org/10.1016/j.jclepro.2020.120536

Kainthola, J., Podder, A., Fechner, M., & Goel, R. (2021). An overview of fungal pretreatment processes for anaerobic digestion: Applications, bottlenecks and future needs. *Bioresource Technology, 321*, 124397. https://doi.org/10.1016/j.biortech.2020.124397

Kalinke, C., Crapnell, R. D., Oliveira, P. R., Janegitz, B. C., Bonacin, J. A., & Banks, C. E. (2023). Opportunities and challenges of establishing a regional bio-based polylactic acid supply chain. *Global Challenges, 7*(7), 2200218. https://doi.org/10.1002/gch2.202200218

Khosravi, C., Kowalczyk, J. E., Chroumpi, T., Battaglia, E., Aguilar Pontes, M.-V., Peng, M., Wiebenga, A., Ng, V., Lipzen, A., He, G., Bauer, D., Grigoriev, I. V., Grigoriev, I. V., & de Vries, R. P. (2019). Transcriptome analysis of Aspergillus Niger xlnR and xkiA mutants grown on corn stover and soybean hulls reveals a highly complex regulatory network. *BMC Genomics, 20*(1), 853. https://doi.org/10.1186/s12864-019-6235-7

Khouri, N. G., Bahú, J. O., Blanco-Llamero, C., Severino, P., Concha, V. O. C., & Souto, E. B. (2024). Polylactic acid (PLA): Properties, synthesis, and biomedical applications – A review of the literature. *Journal of Molecular Structure, 1309*, 138243. https://doi.org/10.1016/j.molstruc.2024.138243

Lim, L. T., Auras, R., & Rubino, M. (2008). Processing technologies for poly (lactic acid). *Progress in Polymer Science, 33*(8), 820–852. https://doi.org/10.1016/j.progpolymsci.2008.05.004

Lin, H. T. V., Huang, M. Y., Kao, T. Y., Lu, W. J., Lin, H. J., & Pan, C. L. (2020). Production of lactic acid from seaweed hydrolysates via lactic acid bacteria fermentation. *Fermentation, 6*(1), 37. https://doi.org/10.3390/fermentation6010037

Mahgoub, S. A., Kedra, E. G. A., Abdelfattah, H. I., Abdelbasit, H. M., Alamoudi, S. A., Al-Quwaie, D. A., Selim, S., Alsharari, S. S., Saber, W. I. A., & El-Mekkawy, R. M. (2022). Bioconversion of some agro-residues into organic acids by cellulolytic rock-phosphate- solubilizing Aspergillus japonicus. *Fermentation, 8*(9), 437. https://doi.org/10.3390/fermentation8090437

Mangaraj, S., Mohanty, S., Swain, S., et al. (2019). Development and Characterization of Commercial Biodegradable Film from PLA and Corn Starch for Fresh Produce Packaging. *Journal of Packaging Technology and Research, 3*, 127–140. https://doi.org/10.1007/s41783-019-00055-y

Maroua, G., Hiligsmann, S., Sayed, M., Alloun, W., & Kacem Chaouche, N. (2023). The study of the industrial aptitude of Aspergillus fumigatus strain for xylanase production. *Egyptian Journal of Chemistry*. https://doi.org/10.21608/ejchem.2023.174799.7195

Masutani, K., Kimura, Y., Jiménez, A., Peltzer, M., & Ruseckaite, R. (2014). Poly (lactic acid) science and technology: Processing, properties, additives and applications. *The Royal Society of Chemistry, 1*, 1–36. https://doi.org/10.1039/9781782624806-00001

Meyer, V., Andersen, M. R., & Brakhage, A. A. (2015). Industrial biotechnology of filamentous fungi. *Applied Microbiology and Biotechnology, 99*(3), 909–917. https://doi.org/10.1007/s00253-014-6148-5

Meyer, V., Basenko, E. Y., Benz, J. P., Braus, G. H., Caddick, M. X., Csukai, M., de Vries, R. P., Endy, D., Frisvad, J. C., Gunde-Cimerman, N., Haarmann, T., Hadar, Y., Hansen, K., Johnson, R. I., Keller, N. P., Kraševec, N., Mortensen, U. H., Perez, R., Ram, A. F. J., Record, E., Ross, P., Shapaval, V., Steiniger, C., van den Brink, H., van Munster, J., Yarden, O., & Wösten, H. A. B. (2020). Growing a circular economy with fungal biotechnology: A white paper. *Fungal Biology and Biotechnology, 7*(5), 5. https://doi.org/10.1186/s40694-020-00095-z

Paul, A., Sreedevi, K., Sharma, S. S., Anjana, V. N., Thomas, S., Ajitha, A. R., Jose Chirayil, C., & Thomas, B. (2023). *Polylactic acid (PLA) handbook of biopolymers* (pp. 1195–1227). Springer. https://doi.org/10.1007/978-981-19-0710-4_44

Prendiz, J., Vega-Baudrit, J. R., & Mena, M. (2019). Polylactic acid (PLA) as a bioplastic and its possible applications in the food industry. *Food Science & Nutrition, 5*(2), 100048. https://doi.org/10.24966/FSN-1076/100048

Raja, S., Agrawal, A. P., Patil, P. P., Thimothy, P., Capangpangan, R. Y., Singhal, P., & Wotango, M. T. (2022). Optimization of 3D printing process parameters of polylactic acid filament based on the mechanical test. *International Journal of Chemical Engineering, 2022*, 1. https://doi.org/10.1155/2022/5830869

Rana Chhetri, B., Acharya, D., Gautam, A., Bajracharya, N., Shrestha, A., & Khadka, S. (2022). Microbial pre-treatment of lignocellulosic biomass for biofuel production: A review. *International Journal of Applied Sciences and Biotechnology, 10*(3), 140–148. https://doi.org/10.3126/ijasbt.v10i3.47510

Ren, Y., Wang, X., Li, Y., Li, Y.-Y., & Wang, Q. (2022). Lactic acid production by fermentation of biomass: Recent achievements and perspectives. *Sustainability, 14*(21), 14434. https://doi.org/10.3390/su142114434

Sar, T., Larsson, K., Fristedt, R., Undeland, I., & Taherzadeh, M. J. (2022). Demo-scale production of protein-rich fungal biomass from potato protein liquor for use as innovative food and feed products. *Food Bioscience, 47*, 101637. https://doi.org/10.1016/j.fbio.2022.101637

Senila, L., Cadar, O., Kovacs, E., Gal, E., Dan, M., Stupar, Z., Simedru, D., Senila, M., & Roman, C. (2023). L-Poly(lactic acid) production by microwave irradiation of lactic acid obtained from lignocellulosic wastes. *International Journal of Molecular Sciences, 24*(12), 9817. https://doi.org/10.3390/ijms24129817

Smith, J. K., & Johnson, L. M. (2020). Sustainable feedstocks for bioplastic production: A review. *Environmental Science and Pollution Research, 27*(4), 3489–3503. https://doi.org/10.1007/s11356-019-07054-2

Souza Filho, P. F. (2018). *Fungi-based biorefinery model for food industry waste: Progress toward a circular economy*. Swedish Centre for Resource Recovery. ISBN 978-91-88838-01-8.

Spaněİova, M., Machovic, V., & Brezina, M. (2003). Characterization and sorption properties of Aspergillus Niger waste biomass. *Central European Journal of Chemistry, 3*(2), 192–200. https://doi.org/10.2478/BF02476223

Stephanopoulos, G. (2007). Challenges in engineering microbes for biofuels production. *Science, 315*(5813), 801–804. https://doi.org/10.1126/science.1139612

Talukder, M. R., Das, P., & Wu, J. C. (2014). Immobilization of microalgae on exogenous fungal mycelium: A promising separation method to harvest both marine and freshwater microalgae. *Biochemical Engineering Journal, 91*, 53–57. https://doi.org/10.1016/j.bej.2014.07.014

Thygesen, A., Tsapekos, P., Alvarado-Morales, M., & Angelidaki, I. (2021). Valorization of municipal organic waste into purified lactic acid. *Bioresource Technology, 342*, 125933. https://doi.org/10.1016/j.biortech.2021.125933

Tian, Y., Wang, J., & Zhang, L. (2023). Manufacturing process analysis and optimization of polylactic acid straw based on improving macular defects. *Materials and Automation Technology*, 12801. https://doi.org/10.1117/12.3007331

Vasco-Correa, J., & Shah, A. (2019). Techno-economic bottlenecks of the fungal pretreatment of lignocellulosic biomass. *Fermentation, 5*(2), 30. https://doi.org/10.3390/fermentation5020030

Wan Mohtar, W. H. M., Wan-Mohtar, W. A. A. Q. I., Zahuri, A. A., Ibrahim, M. F., Show, P. L., Ilham, Z., Jamaludin, A. A., Abdul Patah, M. F., & Rowan, N. J. (2022). Role of ascomycete and basidiomycete fungi in meeting established and emerging sustainability opportunities: A review. *Bioengineered, 13*(7–12), 14903–14935. https://doi.org/10.1080/21655979.2023.2184785

Winder, J. C., Hewlett, M., Liu, P., & Love, J. (2022). Conversion of biomass to chemicals via electro fermentation of lactic acid bacteria. *Energies, 15*(22), 8638. https://doi.org/10.3390/en15228638

Wittner, N., Gergely, S., Slezsák, J., Broos, W., Vlaeminck, S. E., & Cornet, I. (2023). Follow- up of solid-state fungal wood pretreatment by a novel near-infrared spectroscopy-based lignin calibration model. *Social Science Research Network*, 106725–106725. https://doi.org/10.2139/ssrn.4352587

Wu, W., Li, S., Xie, P., Li, X., Chang, H., & Ho, S. H. (2023). Algal-fungal interactions and biomass production in wastewater treatment: Current status and future perspectives. *Algal Research-Biomass Biofuels and Bioproducts, 70*, 103021–103021. https://doi.org/10.1016/j.algal.2023.103021

Xiao, L., Wang, B., Yang, G., & Gauthier, M. (2012). Poly (lactic acid)-based biomaterials: Synthesis, modification and applications. *Polymeric Biomaterials*, 247–282. https://doi.org/10.5772/23927

Xue, J. (2023). Synthesis, modification and application of polylactic acid. *Highlights in Science Engineering and Technology, 52*, 8–16. https://doi.org/10.54097/hset.v52i.8700

Zhang, Z., Liu, J., & Zhang, Y. (2020). An overview of fungal pretreatment processes for anaerobic digestion: Applications, bottlenecks and future needs. *Bioresource Technology, 298*, 122399.

Zhang, Y., Liu, Y., & Wang, Y. (2021). Metabolic engineering of filamentous fungi for the production of biofuels and biochemicals. *Biotechnology Advances, 49*, 107738. https://doi.org/10.1016/j.biotechadv.2021.107738

Zhu, J., Sun, H., Yang, B., & Weng, Y. (2024). Modified biomass-reinforced polylactic acid composites. *Materials, 17*(2), 336. https://doi.org/10.3390/ma17020336

Utilizing Fungal Mycelium in Bio-nanocomposites as a Replacement for Chemically Synthesized Lightweight Polymers

Manisha Trivedi, Rajkamal Kushwaha, Bablu Mordina, Apoorva Anshu Jha, and Vaibhav Singh

Abstract In recent years, the growing attention and demand for environmental-friendly products, combined with the need to reduce carbon emissions, microplastic pollution, and plastic contamination in the global ecosystem, has resulted in the development of biodegradable material innovations in a variety of industries. Manufacturing mycelium-based green composites (MBCs) has emerged as a popular notion and a viable alternative to the current production of eco-friendly materials. MBCs are created by growing the vegetative component of mushroom mycelia on various lignocellulosic substrates, which act as a biological binder that bonds substrate particles through a biotechnological process. This technique has the potential to generate a flexible and biodegradable material, offering an alternative to typical synthetic and other conventional materials. Even though MBCs are renewable and biodegradable, they lack some strength, durability, and other critical attributes when compared to traditional materials. To address these disadvantages and realize the full potential of MBCs, agricultural residual waste, which accounts for a significant portion of global solid waste, has emerged as an appealing option for improving the properties of MBCs because it typically contains hemicellulose acting as a binding agent which has a high ability to improve the mechanical and physical properties of composite materials. This approach can produce MBCs that are

M. Trivedi · R. Kushwaha (✉) · A. A. Jha
Department of Biochemical Engineering, School of Chemical Technology, HBTU, Kanpur, Uttar Pradesh, India

B. Mordina
Nano Science and Coating Division, Defence Materials and Stores Research and Development Establishment-DRDO, Kanpur, Uttar Pradesh, India

V. Singh
Department of Biochemical Engineering, School of Chemical Technology, HBTU, Kanpur, Uttar Pradesh, India

Nano Science and Coating Division, Defence Materials and Stores Research and Development Establishment-DRDO, Kanpur, Uttar Pradesh, India

D. B. Pal, S. L. Bansal (eds.), *Fungal Waste Biomass Management for Energy, Environment and Value-Added Products*, Fungal Biology,
https://doi.org/10.1007/978-3-031-82599-6_3

stronger, more durable, and suitable for a broader range of uses. Furthermore, this research not only adds to the use of eco-friendly materials, but it also helps to achieve the long-term aims of encouraging circular economy principles and minimizing agricultural waste.

Keywords Mycelium-based composites · Lignocellulosic waste · Biodegradable material · Circular economy · Sustainability

1 Introduction

The incorporation of fungal mycelium into bio-nanocomposites is emerging as a game changer in the search for sustainable materials that can replace traditional lightweight polymers. Mycelium, the vegetative structure of fungus, is made up of a network of hyphae that can bind to lignocellulose substrates, resulting in biocomposites with both ecological and mechanical benefits. These mycelium-based composites (MBCs) are distinguished by their biodegradability, low energy requirements during production, and ability to utilize agricultural and industrial waste, which is consistent with circular economy concepts (Meyer et al., 2020a, b).

Studies show that careful selection of fungal species and substrates can greatly improve the properties of MBCs. Research, for example, has shown that certain mixes of mycelium and cellulose nanofibrils (CNFs) can result in enhanced mechanical characteristics, like a higher modulus of elasticity and a lower absorption of water, which makes them appropriate for use in construction and packaging (Sun et al., 2019). Mycelium's fibrous structure is responsible for its capacity to form strong adhesive bonds with a variety of substrates. This feature facilitates efficient colonization and integration with the materials, improving the composites' overall performance (Gandia et al., 2021).

Furthermore, MBCs' environmental benefits are highlighted by their ability to reduce plastic waste and carbon footprints, when compared to standard synthetic materials. MBC production has minimum waste and energy consumption, making it an appealing option for environmental sustainability (Appels et al., 2019; Girometta et al., 2019). Despite these advantages, there are still obstacles in optimizing production processes and standardizing MBC characteristics for industrial applications. Ongoing research efforts are aimed at discovering ideal fungal species and substrates to improve the mechanical and physical properties of these composites, opening the path for their widespread usage in a variety of industries (Sharma & Sumbria, 2022).

Fungal mycelium is a tiny biomass matrix made up of interconnecting elongated cells that play an important role in the natural decomposition and recycling of various biomass sources. Mycelium-based materials are biocompatible, biodegradable, stable, flexible, tunable, hydrophobic, nonflammable, porous, insulating, rapidly growing, self-repairing, self-organizing, and self-healing. They have antimicrobial, antioxidant, and skin-lightening qualities. Fungal mycelium has applications in a variety of industries, including textiles, packaging, pharmaceuticals, cosmetics, insulation, and the automotive sector, providing sustainable alternatives to standard materials. Research in the biosciences discipline focuses on utilization for biomaterial synthesis and environmental remediation to address global concerns sustainably (Verma et al., 2023).

To summarize, the study of fungal mycelium in bio-nanocomposites marks a huge step forward in material science, providing a sustainable alternative to synthetic polymers while addressing crucial environmental concerns. As research progresses, the potential for mycelium-based materials to contribute to a more sustainable future becomes clearer, emphasizing their significance in developing new, eco-friendly solutions for a wide range of applications.

1.1 Background on the Negative Environmental Impact of Chemically Synthesized Lightweight Polymers

The widespread use of chemically synthesized lightweight polymers has transformed several sectors by increasing adaptability, durability, and cost-effectiveness. However, their environmental impact has become a major global problem. These materials, which are predominantly generated from fossil fuels, contribute significantly to pollution, greenhouse gas emissions, and the buildup of nonbiodegradable trash in ecosystems. This study investigates the multiple environmental difficulties faced by these polymers, from production to disposal, and lays the groundwork for researching sustainable alternatives and mitigation techniques.

Chemically synthesized lightweight polymers, such as polyethylene, polypropylene, and polystyrene, have become popular in modern life due to their versatility, durability, and affordability. However, the widespread use of these polymers has resulted in severe environmental costs. Chemically synthesized lightweight polymers have a detrimental influence on the environment due to a variety of variables, including manufacture, use, and disposal.

The technique of producing chemically synthesized lightweight polymers is energy-intensive and significantly reliant on fossil fuels. The extraction, refining, and processing of these raw materials, all contribute considerably to greenhouse gas emissions, which are a key cause of climate change. According to a study published in *Science*, plastic manufacture alone accounts for roughly 4% of global greenhouse gas emissions (Zheng & Suh, 2019). This amount is projected to rise as the demand for these polymers increases.

One of the most alarming aspects of chemically synthesized lightweight polymers is their longevity in the environment. These polymers can break down over hundreds of years, accumulating in many environments. A study published in *Environmental Science & Technology* discovered that microplastics and nanoplastics, which are minute bits of bigger plastic products, have been observed in a variety of marine ecosystems, including the deepest portions of the ocean (Peng et al.,

2018). The existence of these particles in the food chain has sparked worry about their possible influence on marine life and public health.

As chemically synthesized lightweight polymers degrade, harmful substances can be released into the environment. These compounds, such as plasticizers and antioxidants, can be harmful to living beings. A study published in *Environmental Health Perspectives* discovered that exposure to certain plastic additives, such as bisphenol A (BPA), can disrupt endocrine function and cause developmental and reproductive issues in animals (Vandenberg et al., 2013). The release of these substances into the environment may also disturb sensitive ecosystems.

Waste management systems have substantial hurdles when it comes to disposing of lightweight polymers that are chemically synthesized. Since many of these polymers are difficult to recycle and do not biodegrade, they end up piling up in landfills and the ocean. A 2017 analysis from the Ellen MacArthur Foundation states that only 14% of plastic packaging is collected worldwide for recycling, with the remaining 86% ending up in landfills, the ocean, or burnt. As a result of the waste management dilemma, innovative approaches to minimize plastic waste and encourage the use of recyclable and biodegradable materials have been developed.

The negative environmental impact of chemically synthesized lightweight polymers is a critical issue that must be addressed immediately. The manufacturing, use, and disposal of these polymers contribute to climate change, environmental deterioration, and the buildup of hazardous waste. To address these concerns, we must promote the development and use of biodegradable and recyclable alternatives, enhance waste management systems, and raise awareness about the environmental impact of our plastic consumption patterns.

1.2 Brief Overview of Bio-nanocomposites and Their Potential as Sustainable Alternatives

Bio-nanocomposites are a potential class of materials that can serve as long-term alternatives to chemically synthesized lightweight polymers. These materials are made of biopolymer matrices like cellulose, chitosan, and starch, which are supplemented with nanoscale fillers to improve their characteristics.

Bio-nanocomposites, particularly those based on fungal mycelium, are a big step forward in the creation of sustainable materials. These composites are created by combining mycelium, the vegetative portion of fungi, with a variety of organic substrates, typically lignocellulosic material. The resulting mycelium-based composites (MBCs) have unique features that make them appropriate for a variety of applications, including packaging and construction materials.

One of the most significant advantages of bio-nanocomposites is their reduced dependency on fossil fuels. Unlike petroleum-based alternatives, bio-nanocomposites are sourced from renewable and biodegradable sources, making them a more ecologically responsible choice. This is especially crucial when addressing growing concerns about greenhouse gas emissions and plastic pollution.

Mycelium-based composites are distinguished by their biodegradability, low carbon footprint, and low energy consumption during manufacturing. They use waste materials to reduce environmental impact and promote a circular economy. MBC manufacture is fundamentally sustainable because it generates no waste and allows the finished goods to be recycled (Alemu et al., 2022; Appels et al., 2019). Furthermore, MBCs can be developed to improve mechanical qualities by choosing specific fungal species and substrates, as well as optimizing production procedures (Antinori et al., 2020; Li et al., 2022).

Additionally, bio-nanocomposites' special qualities—such as enhanced mechanical strength, thermal stability, and gas barrier performance—make them appropriate for a variety of uses, such as energy storage systems, biomedical devices, and packaging. The prospective uses of bio-nanocomposites can be further expanded by developing materials with customized functionalities, which can be achieved by modifying their composition and structure.

Mycelium-based composites can be used in a variety of industries due to their adaptability. Recent research, for example, has shown that MBCs have potential use in the construction industry as lightweight, fire-resistant materials with acoustic absorption qualities. Furthermore, it has been demonstrated that adding cellulose nanofibrils (CNFs) to mycelium composites increases their mechanical strength and moisture resistance, which qualifies them for usage in packaging and furniture.

Despite its benefits, there are several obstacles in the way of mycelium-based composites'widespread use. These include the requirement for more in-depth knowledge of the interactions between various fungal species and substrates as well as the necessity for standardized production techniques. The best combinations of fungus and substrates to improve the performance features of MBCs are still being researched.

Fungal mycelium-based bio-nanocomposites are a viable substitute for traditional materials that are in line with the objectives of global sustainability. The potential of these materials to contribute to a more sustainable future is becoming more apparent as research advances, opening new avenues for creative uses in a variety of industries.

1.3 Recent Research on Key Areas in the Field of Bio-nanocomposites

(i) *Nanofiller Incorporation*

Currently researchers are exploring various types of nanoscale fillers, like cellulose nanofibrils, clay nanomaterials, and carbon-based nanostructures to improve the properties of biopolymer matrices.

(ii) *Biopolymer Matrix Optimization*

Efforts are being made to optimize the composition and structure of biopolymer matrices, like cellulose, chitosan, and starch to enhance their compatibility with nanoscale fillers and overall performance.

(iii) *Manufacturing Processes*

Researchers are investigating scalable and cost-effective manufacturing methods for bio-nanocomposites, including techniques such as extrusion, injection molding, and 3D printing.

(iv) *Functional Properties*

Studies are focused on developing bio-nanocomposites with tailored properties, such as improved mechanical strength, thermal stability, gas barrier performance, and biodegradability, to fulfill the demands of various applications.

(v) *Sustainability and Life Cycle Assessment*

Researchers are conducting comprehensive life cycle assessments (LCAs) to evaluate the environmental impact of bio-nanocomposites and ensure their long-term sustainability.

These research trends and developments are paving the way for the widespread adoption of bio-nanocomposites as viable and sustainable alternatives to traditional chemically synthesized lightweight polymers.

2 Fungal Mycelium as a Sustainable Material

Fungal mycelium, the root-like structure of fungi, is currently being recognized as a sustainable material with high potential for usage in bio-nanocomposites. This organic network is mostly made of hyphae that grow by decomposing organic matter, such as agricultural by-products and lignocellulosic waste. Mycelium cultivation, which uses these substrates, not only aids waste reduction but also promotes a circular economy by converting discarded materials into valuable resources. Mycelium's sustainability is further boosted by its low energy requirements during production, which allows it to survive in a variety of environmental situations with minimal input. From Table 1, studies have shown a variety of materials which can be formed through fungal mycelium.

Definition and Characteristics of Fungal Mycelium

Fungal mycelium is the complicated, filamentous structure which comprises the vegetative part of mushrooms and plays an important role in their development, reproduction, and ecological functions. Mycelium is made up of a network of hyphae, which are elongated tubular cells that can be septate (with cross-walls) or coenocytic (with multiple nuclei but no cross-walls). This network enables mycelium to span large areas, hence increasing the surface area for nutrition absorption and promoting interactions with diverse substrates. Mycelium is mostly composed of chitin, a polysaccharide that provides stiffness and structural support, making it resistant to environmental stresses such as desiccation and mechanical damage.

Table 1 Fungal mycelium studies for developing biocomposites

Study	Authors	Year	Fungal species	Material type	Key findings
Development of biodegradable foams			*Ganoderma lucidum*	Mycelium-based foam	Demonstrated the potential of mycelium-based foams as a sustainable alternative to synthetic foams
Mycelium bricks for construction	Appels et al.	2019	*Trametes versicolor*	Mycelium bricks	Highlighted the strength and insulation properties of mycelium bricks for construction
Sustainable packaging materials			*Pleurotus ostreatus*	Mycelium packaging	Showed mycelium packaging as a biodegradable option for reducing plastic waste
Composite materials for automotive	Haneef et al.	2017	*Schizophyllum commune*	Mycelium composites	Investigated mycelium composites as lightweight, sustainable materials for automotive applications
Insulation properties of mycelium	Pelletier et al.	2013	*Pleurotus eryngii*	Mycelium insulation panels	Demonstrated excellent thermal insulation properties of mycelium-based panels
Acoustic properties of mycelium	Islam et al.	2018	*Ganoderma applanatum*	Mycelium acoustic panels	Evaluated the sound absorption capabilities of mycelium-based acoustic panels
Biodegradable mycelium composites	Holt et al.	2012	*Pleurotus ostreatus*	Mycelium biocomposites	Assessed biodegradability and mechanical properties of mycelium composites
Mycelium-based furniture	Vandeloo et al.	2015	*Fomes fomentarius*	Mycelium furniture	Explored the use of mycelium for creating sustainable, biodegradable furniture

Mycelium growth is characterized by apical extension, in which hyphal ends elongate and branch out, generating a dense mat capable of penetrating various substrates such as soil, decaying organic debris, and even living creatures in the case of parasitic fungus. This ability to colonize varied settings is a major feature of mycelium, allowing fungi to play important roles in nutrient cycling and breakdown in ecosystems. Mycelium can also respond dynamically to environmental factors like moisture, temperature, and nutrient availability, allowing it to adapt and optimize resource acquisition (Lücking et al., 2021).

In addition to its ecological importance, mycelium has attracted interest for its potential applications in biotechnology and materials science. Due to its biodegradability and renewability, recent studies have investigated using mycelium as a sustainable alternative to traditional materials like plastics and foams. Mycelium-based composites have been created for a wide range of applications, including packaging, insulation, and even construction materials, demonstrating their versatility and promise to reduce environmental impact (Haneef et al., 2021). Furthermore, mycelium's ability to develop symbiotic interactions with plants, such as mycorrhizal associations, improves nutrient intake and plant health, highlighting its value in agricultural techniques (Smith & Read, 2008).

In short, fungal mycelium is a complex and versatile structure that acts as the basis for fungal life. Its distinctive traits, such as its filamentous form, chitin composition, and ecological roles, have drawn the attention of researchers and practitioners alike.

2.1 Advantages of Using Fungal Mycelium in Bio-nanocomposites

Fungal mycelium integration into bio-nanocomposites offers several benefits that make it a competitive substitute for lightweight polymers made through chemical synthesis. One of the biggest advantages is that mycelium-based materials are biodegradable; they may naturally break down and reduce the environmental damage that comes with synthetic composites and traditional plastics. This quality is especially crucial given the current state of plastic waste accumulation and the pressing demand for sustainable materials. In addition, mycelium has exceptional mechanical qualities that may be customized by choosing specific fungal species and growth environments, such as high tensile strength, flexibility, and durability. Because of their versatility, materials can be designed to fulfil needs for a range of applications, such as building, packaging, and biomedical (Haneef et al., 2021).

Furthermore, mycelium can successfully use agricultural and forestry waste as substrates for development, fostering a circular economy by converting low-value biomass into high-value products without producing more trash. Mycelium composites' lightweight nature adds to their appeal, making them ideal for applications requiring weight reduction, such as those in the automotive and aerospace industries (Danninger et al., 2022). Furthermore, mycelium's natural porous structure offers good insulation qualities, making it an appealing choice for thermal and acoustic applications (Attias et al., 2020).

Another important advantage of mycelium-based bio-nanocomposites is their capacity for carbon sequestration. As mycelium grows, it can absorb and store carbon dioxide from the environment, aiding in climate change mitigation efforts. The ability of mycelium to merge with nanotechnology provides another layer of functionality, allowing the production of innovative bio-nanocomposites with better features such as barrier performance and mechanical robustness. Incorporating nanoparticles, such as silver or zinc oxide, into mycelium composites can impart antibacterial qualities, making them ideal for applications.

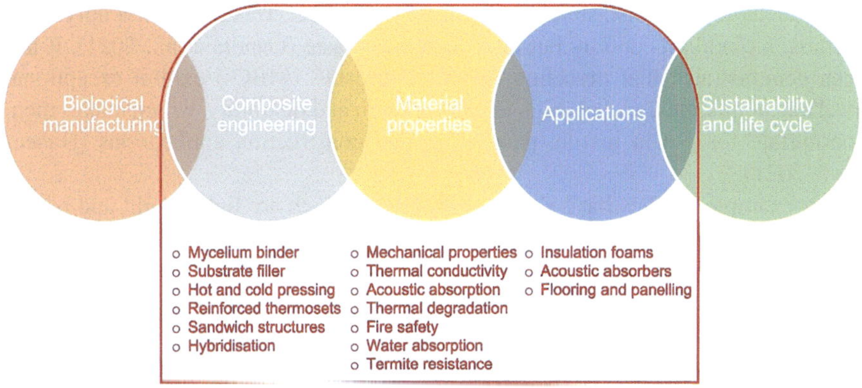

Fig. 1 Properties of fungal mycelium-based composites (Jones et al., 2020)

Mycelium composites have customizable material properties based on their composition and manufacturing process and can replace foams, timber, and plastics for applications, such as insulation, door cores, paneling, flooring, cabinetry, and other furnishings. Due to their low thermal conductivity, high acoustic absorption, and fire safety properties, outperforming traditional construction materials, such as synthetic foams and engineered woods, they show promise as thermal and acoustic insulation foams. From Fig. 1, it can be inferred that fungal mycelium-based composites have a blend of good material properties to be used for variety of applications (Jones et al., 2020).

The negative side makes mycelium an appealing candidate for sustainable material development, in line with global sustainability goals. Additionally, mycelium's ability to develop symbiotic interactions with plants, such as mycorrhizal associations, improves soil health and biodiversity, bolstering its role in sustainable agriculture (Smith & Read, 2008).

The benefits of employing fungal mycelium in bio-nanocomposites not only address major environmental issues but also provide a road to manufacturing high-performance, sustainable materials for a wide range of sectors. Mycelium's unique features, combined with its ecological benefits and potential for creative uses, make it a major participant in the shift to more sustainable material solutions.

2.2 *Previous Studies on the Use of Fungal Mycelium in Sustainable Materials*

Recent years have seen a major increase in research on the use of fungal mycelium in sustainable materials, with the potential for this material to displace conventional, chemically synthesized lightweight polymers. Several studies have shown that mycelium can grow on a variety of substrates made from industrial and agricultural waste, including sawdust, straw, and cotton by-products. This feature not only helps promote a circular economy but also decreases waste (Gandia et al., 2021). It has been demonstrated that mycelium-based composites (MBCs) exhibit exceptional mechanical qualities, such as high tensile strength and flexibility, which makes them appropriate for use in textile, packaging, and construction applications (Haneef et al., 2021).

For example, studies have shown that MBCs, which are lightweight and biodegradable, can successfully replace synthetic foams in building materials and offer comparable insulating qualities (Attias et al., 2020). Because of its porous structure, which may be designed to improve thermal and acoustic insulation, MBCs, in addition to their mechanical qualities, show variety in their applications (Danninger et al., 2022). Expanding the range of possible applications, the integration of nanomaterials into mycelium composites has also been investigated to enhance their functional characteristics, such as mechanical strength and antibacterial activity. Additionally, research has shown that mycelium has the capacity to absorb carbon during growth, which could aid in attempts to mitigate climate change and increase its appeal as a sustainable material option. The adaptability of mycelium in the creation of bio-based hybrid composites has been established by recent investigations. The study of Livne et al. (2022), which carried out a life cycle assessment of mycelium bio-composites and showed their potential as low-embodied energy, CO_2-sink building materials, is a noteworthy example. According to the study, mycelium composites have the potential to drastically lower carbon emissions when compared to traditional building materials, supporting green building techniques. Additionally, research has been done to improve the mechanical qualities and moisture resistance of mycelium composites by including cellulose nanofibrils, hence expanding their range of applications.

The environmental benefits of mycelium-based materials have received attention in the literature, with studies highlighting their significance in reducing plastic pollution and carbon footprints. To improve the effectiveness of MBCs, research has emphasized the need to select appropriate fungal species and substrates. For example, Li et al. (2022) conducted a comprehensive analysis of mycelium-based materials based on their substrates, emphasizing the benefits of combining organic and inorganic materials to improve functioning. The study revealed over 70 fungal species that can be used to produce MBCs, emphasizing the importance of specialized techniques to maximize their ecological and mechanical qualities. Additionally, the use of mycelium in the production of biodegradable packaging has been investigated, with results indicating that mycelium-based materials can be effective substitutes for traditional plastics in a variety of applications, including food packaging and disposable items.

Recent advances in mycelium technology have resulted in novel applications in electronics, such as the creation of mycelium-based substrates for flexible and biodegradable electronic devices (Danninger et al., 2022). These applications highlight mycelium's potential to not only replace synthetic materials in existing industries but also open up new markets for sustainable solutions. Also, the integration of mycelium with biopolymers has been studied to improve the overall properties of bio-nanocomposites by leveraging the strengths of both materials to create composites with specific functionality (Haneef et al., 2021).

Additionally, Sharma and Sumbria (2022) talked about the incorporation of nanotechnology into mycelium composites, which can result in better qualities, including increased durability and strength. The development of functionalized materials that can satisfy certain application needs has new opportunities because of this integration. Further investigation into the mechanical and biological properties of mycelium is nevertheless required, because of the difficulties in standardizing production processes and guaranteeing uniform quality across various applications (Appels et al., 2019; Girometta et al., 2019).

Further research has focused on mycelium's potential as a natural binder in composite materials, removing the need for synthetic adhesives, which can be detrimental to the environment. This is consistent with the emerging trend of developing entirely biodegradable materials that do not compromise performance. Furthermore, research has investigated the role of mycelium in soil health and ecosystem restoration, showing its capacity to increase soil structure and nutrient cycling, both of which are required for sustainable agriculture (Smith & Read, 2008).

Overall, the research demonstrates the versatility and sustainability of fungal mycelium as a possible alternative to standard materials, encouraging more investigation and development in this creative field.

3 Properties and Performance of Fungal Mycelium Bio-nanocomposite

The use of fungal mycelium in bio-nanocomposites has gained popularity in recent years due to its ability to substitute chemically synthesized lightweight polymers. Mycelium-based composites (MBCs) are a sustainable alternative to standard materials, offering a variety of advantages such as biodegradability, mechanical strength, and thermal insulation. This chapter will investigate the qualities and performance of fungal mycelium bio-nanocomposites, including mechanical, thermal, and environmental aspects.

3.1 Mechanical Properties

Fungal mycelium bio-nanocomposites (MBCs) have a variety of mechanical qualities that make them appropriate for a wide range of applications, particularly as environmental-friendly alternatives to existing materials. Several factors influence the mechanical performance of these composites, including the fungus species employed, the growth substrate type, and the exact processing parameters used during production. According to studies, MBCs can achieve tensile strengths comparable to typical synthetic materials, making them suitable for structural applications.

For example, studies show that depending on the growth circumstances and the fungal species used, mycelium composites formed from agricultural waste substrates, such as straw and sawdust, can have tensile strengths ranging from 1.5 to 4.5 MPa (Haneef et al., 2021). Mycelium's distinct fibrous structure helps to maintain its mechanical integrity due to the hyphal network providing a solid foundation for efficiently distributing loads.

The commonly formed fungal mycelium-based composites have mechanical properties in the range as depicted in Table 2. Furthermore, MBC mechanical properties can be modified using a variety of approaches, including optimizing growth factors such as temperature, humidity, and nutrient availability. For example, including nanomaterials, such as cellulose nanofibers or clay, in mycelium composites has been demonstrated to considerably improve mechanical performance. These nanofillers can increase tensile, flexural, and impact strength in composites, making them more appropriate for demanding applications. Furthermore, using cross-linking agents during the post-processing step might improve the mechanical stability of mycelium-based materials, allowing them to be used in industries such as construction, packaging, and even automotive.

However, there are still issues in standardizing the mechanical properties of MBCs for large-scale applications. Variability in fungal species and growth conditions might cause variances in performance, prompting additional research to develop dependable production techniques. Nonetheless, the promising mechanical properties of fungal mycelium bio-nanocomposites make them a potential alternative to traditional materials, especially in situations where sustainability and environmental impact are critical (Gandia et al., 2021).

3.2 *Thermal Properties*

The thermal properties of fungal mycelium bio-nanocomposites are another important factor that determines their use in a variety of sectors. MBCs typically have good thermal insulation qualities, making them appropriate for use in building and

Table 2 Mechanical properties of fungal mycelium-based biocomposites

Property	Value range	Description
Density	60–200 kg/m^3	Depends on the type of fungal mycelium and composite materials
Compressive strength	1.5–5.5 MPa	Varies with the type of fungal strain and reinforcement materials used
Tensile strength	1.5–4.5 MPa	Influenced by the type and amount of reinforcement
Flexural strength	0.2–1.5 MPa	Depends on the composite structure and materials used
Young's modulus	10–300 MPa	Indicates the stiffness of the material

packaging. Mycelium's porous structure adds to its low thermal conductivity, which has been reported to range between 0.05 and 0.15 W/m•K depending on the density and composition of the composite. This low thermal conductivity is beneficial for insulating materials since it promotes temperature stability in buildings and decreases energy consumption for heating and cooling.

The commonly formed fungal mycelium-based composites have thermal properties in the range as depicted by Table 3. Furthermore, the types of substrates chosen and the particular fungal species employed have an impact on the thermal stability of mycelium-based composites. Research, for example, has demonstrated that MBCs created from lignocellulosic materials have better thermal stability than MBCs made from pure mycelium (Haneef et al., 2021). The thermal characteristics of MBCs can be further improved by adding nanomaterials. For example, it has been shown that adding silica or graphene oxide to mycelium composites increases their thermal stability and fire resistance, making them more appropriate for use in settings where heat exposure is a concern.

Thermogravimetric analysis (TGA) has also been used to study the thermal degradation behavior of MBCs, and the results show that these composites may tolerate temperatures as high as 250 °C before experiencing appreciable degradation. For uses in packaging and construction, where materials may be exposed to temperature fluctuations, this thermal resilience is essential. To attain the intended thermal characteristics while preserving the mechanical integrity of the composites, it is crucial to optimize the formulation and processing conditions.

In the final analysis, the thermal characteristics of fungal mycelium bio-nanocomposites indicate their potential for usage in energy-efficient applications. MBCs are a potential alternative to conventional insulating materials due to their low thermal conductivity and better thermal stability, which contributes to sustainability and energy conservation efforts in a variety of sectors (Gandia et al., 2021).

Table 3 Thermal properties of fungal mycelium-based biocomposites

Property	Value range	Description
Thermal conductivity	0.03–0.04 W/mK	Indicates the material's ability to conduct heat; low values suggest good insulating properties
Specific heat capacity	1.3–2.5 kJ/kg·K	Indicates the amount of heat required to change the temperature of a unit mass by 1 °C
Thermal diffusivity	0.1–0.2 mm^2/s	Indicates a measure of how quickly heat moves through the material
Thermal expansion coefficient	20–50 μm/m·K	Indicates how much the material expands when heated
Decomposition temperature	200–250 °C	Indicates the temperature at which the mycelium composite begins to decompose
Heat of combustion	15–20 MJ/kg	Indicates the amount of energy released during combustion
Flame retardancy	V-0 to V-2 (UL94 rating)	Indicates the material's resistance to catching fire

3.3 Environmental Impact and Biodegradability

The environmental impact and biodegradability of fungal mycelium bio-nanocomposites are among its most notable benefits. Since MBCs are a naturally occurring substance formed by fungi, they can naturally break down in the environment without adding to the problem of plastic pollution.

Fungal mycelium-based composites (MBCs) present a strong substitute for traditional chemically synthesized lightweight polymers due to their biodegradability and environmental effect. The vegetative portion of fungi, known as mycelium, forms a fibrous network that links organic substrates like agricultural waste together as it grows by colonizing these materials. In addition to making use of waste materials, this approach helps to lessen the pollution and environmental damage caused by typical plastic manufacture (Appels et al., 2019; Girometta et al., 2019).

While MBCs are naturally biodegradable and decompose into nontoxic metabolites through microbial action, their long-term environmental impact is greatly reduced. Research has indicated that, in contrast to conventional plastics, which took hundreds of years to disintegrate, mycelium composites can disintegrate in weeks to months when exposed to natural circumstances.

Studies show that fungi's various enzymatic capacities, which can efficiently break down lignocellulosic materials and other organic compounds, aid in the biodegradation of MBCs (Li et al., 2022). Mycelium's natural composition contains polysaccharides such chitin and glucans, which further enhance its biodegradability. This makes it appealing for use in packaging and building applications where environmental considerations are crucial (Gandia et al., 2021). In addition, compared to standard materials, the manufacture of products based on mycelium is associated with a more sustainable lifetime because of its low energy consumption and minimum waste generation (Sharma & Sumbria, 2022).

Mycelium composites' life cycle assessments (LCAs) have shown that they have the capacity to function as carbon sinks, storing carbon dioxide during their growth and assisting in the fight against climate change. Using agricultural by-products helps create a circular economy by turning trash into useful resources and lowering the requirement for virgin materials (Haneef et al., 2017). Haneef et al. (2017) conducted a comparative study that demonstrated the notable reduction in carbon footprint between mycelium-based materials and typical plastics. This finding underscores the ecological benefits of implementing these biocomposites across a range of industries.

Furthermore, MBCs influence the ecosystem that goes beyond their biodegradability. Mycelium cultivation can enhance soil health and encourage biodiversity. Mycelium breaks down into nutrients that enhance the soil, increasing its fertility and promoting plant growth (Gandia et al., 2021). Furthermore, it has been shown that mycelium can absorb heavy metals and other environmental pollutants, suggesting that it may find utility in bioremediation applications (Saravanan et al., 2021). This capacity improves the general health of ecosystems, in addition to helping to clean up dirty areas.

Despite these advantages, there are still limitations in optimizing growing conditions and scaling manufacturing processes for industrial applications. Substrate diversity, fungal strain selection, and growing conditions, all have a major impact on MBC mechanical characteristics and degradation rates (Appels et al., 2019; Girometta et al., 2019). Ongoing research focuses on developing standardized procedures to improve the quality and performance of mycelium-based materials while ensuring that their environmental benefits are fully realized.

Finally, the environmental impact and biodegradability of fungal mycelium-based composites demonstrate their promise as sustainable alternatives to traditional materials. MBCs'capacity to decompose organically, combined with their minimal carbon footprint and beneficial benefits on soil health, puts them as a promising solution to pressing environmental concerns. As research progresses, the incorporation of mycelium-based materials into numerous applications may play an important role in promoting sustainability and lowering dependency on nonbiodegradable plastics.

4 Manufacturing and Applications of Fungal Mycelium Bio-nanocomposites

The development and deployment of fungal mycelium bio-nanocomposites (MBCs) is a ground-breaking step in the search for sustainable materials that may effectively replace traditional, chemically synthesized lightweight polymers. As the environmental impact of plastic waste grows, research into bio-based alternatives becomes more vital. Fungal mycelium, with its distinct features and growth characteristics, is a possible answer. This section will go over the steps required in manufacturing bio-nanocomposites from fungal mycelium, as well as their possible uses in a variety of industries. By studying the approaches and breakthroughs in this field, we can gain a deeper understanding of how mycelium-based products can contribute to a more sustainable future.

4.1 Process of Creating Bio-nanocomposites Using Fungal Mycelium

The synthesis of fungal mycelium for producing bio-nanocomposites is a ground-breaking concept in sustainable materials research. As the world grapples with the environmental issues posed by plastic waste and resource depletion, the demand for innovative and eco-friendly solutions has never been greater. Fungal mycelium, the vegetative portion of fungi, has unique qualities that make it an excellent option for the development of biodegradable composites that can replace traditional lightweight plastics.

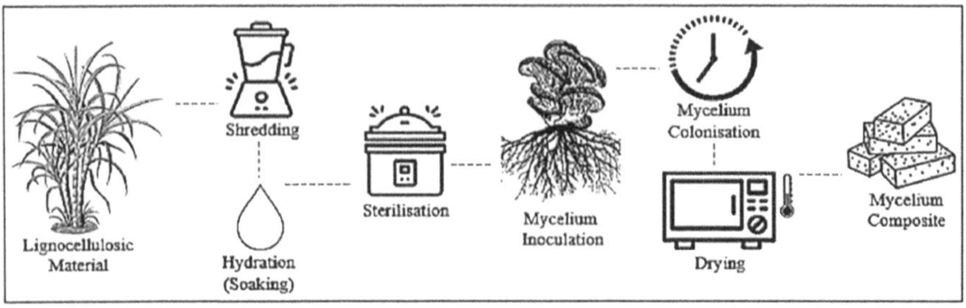

Fig. 2 Synthesis of fungal mycelium-based biocomposites (Lingam et al., 2023)

There are several steps involved in the successful production of mycelium composites as shown in Fig. 2 (Lingam et al., 2023). This section goes into the complex process of creating bio-nanocomposites from mycelium, highlighting the essential phases required, such as substrate selection and preparation, fungal strain injection, and regulated growing conditions. Each phase of the process is designed to maximize mycelium's natural binding capabilities while also integrating nanomaterials to improve the mechanical and physical qualities of the final composites.

Our main goal in investigating the processes used to create mycelium-based bio-nanocomposites is to provide a thorough understanding of how these materials can be effectively generated and modified for a variety of applications. The use of nanotechnology to mycelium composites not only increases their performance but also broadens their potential applications in areas like packaging, building, and textiles. In the end, this investigation of the procedure highlights the potential of fungal mycelium as a sustainable substitute for artificial materials, opening the door for creative fixes that are consistent with the tenets of a circular economy. The knowledge gathered by researching the synthesis of mycelium-based bio-nanocomposites will be invaluable in promoting sustainable material practices and tackling the pressing environmental issues of our day as this field of study develops.

4.1.1 Selection and Preparation of Substrates

The preparation of bio-nanocomposites with fungal mycelium begins with the careful selection of appropriate substrates. Agricultural leftovers, including straw, corn stalks, and sawdust, are common substrates. These substrates function as both a nutrient supply and a structural component of mycelium. Pretreatment of these substrates is frequently required to improve their digestibility and availability for fungal colonization. Mechanical shredding, steam explosion, and chemical treatments are used to break down the lignocellulosic structure, allowing the mycelium to penetrate and develop (Li et al., 2022).

4.1.2 Inoculation with Fungal Strains

Following substrate preparation, the substrates are inoculated with specific fungal strains recognized for their quick growth and binding capabilities, such as *Ganoderma lucidum*, *Pleurotus ostreatus*, or *Trametes versicolor*. To avoid contamination, the inoculation process is normally carried out in a sterile environment. Fungal spores or mycelium are put into the pretreated substrate to ensure equitable dispersion. This is essential for fostering consistent mycelium development across the substrate (Gandia et al., 2021).

4.1.3 Controlled Growth Conditions

After being inoculated, the substrate–fungal mixture is grown in a controlled environment that is dark and humid. These circumstances are necessary for the best possible mycelial growth. The mycelium colonizes the substrate during the incubation period, which can span from days to weeks. During this time, a dense network of hyphae forms, connecting and binding the components together. The mycelium secretes lignocellulolytic enzymes that aid in the process of self-assembly by breaking down the substrate and enabling the hyphae to unite to form a cohesive composite material (Appels et al., 2019).

4.1.4 Post-harvest Treatment

Once adequate colonization is achieved, the mycelium composites are subjected to a post-harvest treatment to stabilize the material and improve its mechanical qualities. This treatment may include drying, heat treatment, or sterilization, all of which prevent future fungal development and eradicate possible pathogens (Sharma & Sumbria, 2022). Drying composites reduces moisture content, which increases mechanical strength and extends the material's shelf life. Heat treatment can improve the thermal stability of composites, making them ideal for a variety of applications such as insulation and packaging.

4.1.5 Incorporation of Nanomaterials

Nanomaterials can be added to mycelium-based composites during the mixing process to improve their qualities even further. Nanofillers such as cellulose nanofibrils, graphene oxide, or silica nanoparticles can be applied to the substrate prior to inoculation. The addition of these nanoparticles has been found to improve the mechanical strength, thermal stability, and barrier properties of the bio-nanocomposites. For example, adding cellulose nanofibrils can improve the tensile strength and flexibility of mycelium-based composites, allowing them to compete with standard plastics.

4.1.6 Customization for Applications

The adaptability of the mycelium-based bio-nanocomposite manufacturing method enables customization based on the required end-use applications. Researchers can tune the mechanical and physical properties of composites to satisfy specific industry requirements by modifying the kind of substrate, fungal strain, and nanofiller content (Gandia et al., 2021; Li et al., 2022). This versatility makes mycelium-based composites a feasible option for meeting the growing need for sustainable materials.

In final analysis, several procedures must be followed to create bio-nanocomposites using fungal mycelium. These procedures include substrate preparation and selection, fungal strain inoculation, controlled growth conditions, post-harvest treatment, incorporation of nanomaterials, and customization for applications. This creative strategy minimizes the negative effects on the environment and encourages the use of waste materials while also supporting sustainable development by lowering the use of nonbiodegradable plastics.

4.2 Potential Applications in Various Industries

The mycelium-based composites having tailored structural, physical, chemical, mechanical, and biological properties are relying on the strain, feeding substrate, and the manufacturing process. The mycelium-based functional materials with tunable properties are synthesized by selecting the desired components and the synthesis method. The pure and composites of stiff, elastic, porous, less dense, fast-growing, and low-cost mycelium-derived materials with efficient antimicrobial, antioxidant, and skin-whitening properties pave their way in various applications such as construction, packaging, medicine, and cosmetics. Figure 3 depicts the applied areas of fungal mycelium-based advanced functional materials (Manan et al., 2021).

4.2.1 Packaging Industry

Fungal mycelium-based composites (MBCs) are developing as novel biodegradable alternatives in the packaging sector. Mycelium's ability to be molded into a variety of shapes enables it to efficiently replace traditional plastic packaging. Research has shown that mycelium composites can provide strong product protection while dissolving organically in composting conditions, greatly decreasing plastic waste (Gandia et al., 2021). Companies such as Evocative Design have successfully commercialized mycelium-based packaging solutions that not only replace Styrofoam but also provide a sustainable end-of-life option, demonstrating the potential for widespread use in the packaging industry (Ecovative, 2023).

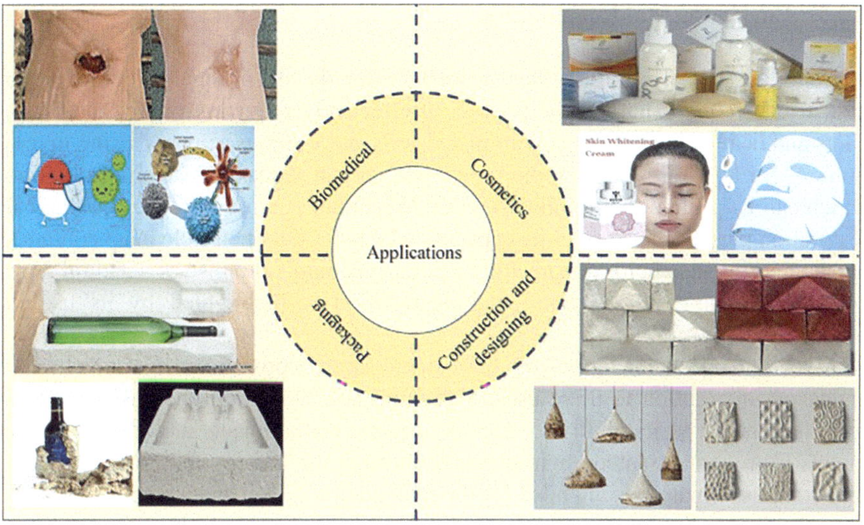

Fig. 3 Application of fungal mycelium in different areas (Manan et al., 2021)

4.2.2 Construction Industry

In the construction work, mycelium composites are being investigated for their potential as insulating materials and structural components. Their low heat conductivity and lightweight nature make them ideal for energy-efficient building strategies. According to research, mycelium-based insulation can surpass traditional materials in terms of thermal performance while remaining completely biodegradable. Furthermore, mycelium bricks made from growing mycelium on agricultural waste have shown promising mechanical qualities, making them a feasible choice for sustainable building. This application not only enhances resource efficiency but also helps reduce the carbon footprint connected with building materials.

Mycelium composites are being investigated in the building industry for their possible use as insulation materials, wall panels, and structural components. Mycelium's natural insulation capabilities, along with its lightweight nature, make it an appealing choice for energy-efficient building materials. Research has shown that mycelium-based insulation can give equivalent thermal performance to traditional materials while being more environmental-friendly (Attias et al., 2020). Furthermore, the ability to form mycelium composites into diverse shapes enables creative building designs that improve both aesthetic appeal and utility.

4.2.3 Environmental Remediation

Mycelium's special qualities have a lot to offer the environmental remediation industry. Mycoremediation, the process of effectively removing organic contaminants and heavy metals from contaminated soils and water, is achieved using mycelium. According to research, some fungal species can sequester lead and cadmium, which can help to detoxify polluted areas. Mycelium is a useful tool in ecological restoration projects because of its capacity, which not only helps clean up contaminated areas but also helps restore soil health. Mycelium's promise as a natural remedy for dealing with environmental degradation is highlighted by its use in bioremediation.

Mycelium bio-nanocomposites have demonstrated promise in bioremediation endeavors in addition to these uses. Mycelium is a useful tool for environmental remediation due to its ability to degrade organic contaminants and absorb heavy metals from contaminated environments. Through the incorporation of mycelium into composite materials intended for the remediation of soil and water, researchers may alleviate pollution issues and generate valuable materials at the same time.

4.2.4 Agriculture

Mycelium-based products are being researched for use in agriculture as biodegradable soil supplements and plant containers. Through the reduction of plastic waste and improvement of soil health, these applications support sustainable farming methods. Plant growth can be supported by mycelium, which can enhance soil structure and nutrient availability (Sharma & Sumbria, 2022). Furthermore, since some fungal species can prevent the formation of dangerous pathogens in the soil, less chemical pesticide is required, the use of mycelium as a natural pest deterrent is being investigated. This application supports organic farming practices and enhances crop productivity.

4.2.5 Biomedical Field

The biomedical industry offers promising potential for mycelium applications, particularly in the production of wound dressings and tissue-engineering scaffolds. Mycelium's biocompatibility and capacity to encourage cell growth make it appropriate for these applications, providing a sustainable alternative to synthetic materials (Appels et al., 2019; Girometta et al., 2019). Studies have demonstrated that mycelium-based materials can promote cell proliferation and differentiation, which are essential for optimal tissue regeneration. Furthermore, mycelium composites can be tailored to include bioactive chemicals, improving their therapeutic potential for a variety of medical applications.

4.2.6 Electronics and Smart Materials

The combination of mycelium composites and nanotechnology creates new opportunities for innovation in electronics and smart materials. Nanofillers, such as graphene or cellulose nanofibrils, can improve the electrical conductivity and tensile properties of mycelium-based composites, making them appropriate for use in flexible electronics, sensors, and biodegradable circuit boards (Li et al., 2022). This combination of mycelium and nanotechnology establishes these composites as a forward-thinking option in the search for sustainable materials in high-tech applications, meeting the growing demand for environmental-friendly electronics.

Additionally, the use of mycelium composites as substrates for flexible and biodegradable electrical devices is being investigated by the electronics industry. In place of conventional materials used in electronics, mycelium can offer a lightweight and environmental-friendly substitute, possibly lowering electronic waste and advancing the ideas of the circular economy (Danninger et al., 2022). Mycelium-based components that can be integrated with current electronic technologies are the subject of continuous research, opening up new possibilities for wearable technology and smart packaging.

4.2.7 Textile Industry

The promise of fungus mycelium as a sustainable material is also being recognized by the textile sector. Mycelium has the potential to be transformed into materials like leather that are both biodegradable and devoid of the toxic chemicals commonly employed in the traditional leather-making process. To provide customers with eco-friendly options without sacrificing quality or style, brands are beginning to experiment with mycelium-based textiles for upholstery and fashion. The trend toward sustainable textiles is in line with the rising need for fashion that is both morally and ecologically sound.

In summary, the numerous possible uses of fungal mycelium-based composites in packaging, construction, environmental remediation, agriculture, biomedicine, and electronics demonstrate their versatility and promise as sustainable alternatives to traditional materials. As research in these subject advances, it is envisaged that the creation and commercialization of mycelium-based goods will increase, paving the path for a more sustainable future.

5 Outlook and Challenges

Fungal mycelium bio-nanocomposites (MBCs) hold great potential as a sustainable substitute for chemically synthesized lightweight polymers; however, there are several obstacles that need to be overcome before they can be widely adopted and commercialized. There will be a huge increase in demand for bio-based alternatives such as mycelium composites as awareness of environmental issues and the need for sustainable materials grows. Mycelium's distinct qualities, such as its mechanical strength, biodegradability, and capacity for thermal insulation, make it a strong contender for a range of uses, from building and textiles to packaging. To fully utilize MBCs, though, a few significant obstacles must be removed.

One of the main challenges is the mycelium production's scalability. Although research conducted on a laboratory scale has shown that it is possible to produce mycelium composites of excellent quality, there are still numerous challenges to overcome before these methods can be used in industrial production. It is challenging to guarantee consistency and dependability in large-scale manufacturing due to variations in fungal species, substrate kinds, and growing settings, which can result in variations in the final products' qualities. Additional research is required to produce standardized techniques that are easily replicable in industrial settings and to optimize growth factors to overcome this issue. Furthermore, to improve the effectiveness and economics of mycelium cultivation, developments in fermentation technologies and bioprocess engineering will be essential (Gandia et al., 2021).

The requirement for a thorough understanding of the underlying biological mechanisms controlling mycelium development and its interaction with different substrates is another difficulty. Although a lot of work has been made in describing the mechanical and thermal properties of mycelium composites, it is still unclear how the fungi's biological traits relate to the emerging material properties. The inability to customize mycelium-based products for uses is a result of this knowledge gap. In addition to investigating the possibility of using genetic engineering to improve desired characteristics like growth rate and mechanical strength, future research should concentrate on clarifying the genetic and metabolic pathways involved in mycelium development.

The economic viability of mycelium composites is also an important consideration that will determine their future use. Currently, mycelium-based materials may have higher manufacturing costs than conventional materials, owing to the expenses of substrates, growing, and processing. To increase the economic viability of MBCs, it is critical to investigate the use of low-cost agricultural waste and by-products as substrates, which can assist reduce production costs while also solving waste management challenges. Furthermore, effective supply chains and market strategies will be critical for increasing the commercialization of mycelium composites.

Despite these hurdles, the future of fungal mycelium bio-nanocomposites looks promising, thanks to a growing community of academics and companies interested in the possibilities of mycelium materials. Interdisciplinary collaboration among experts in biology, materials science, and engineering will be required to improve the area and open new applications. Furthermore, as consumer demand for environmental-friendly products grows, there is a tremendous possibility for mycelium composites to gain market share in a variety of industries, including packaging, construction, and textile.

In short, while the approach going forward for fungal mycelium bio-nanocomposites is laden with problems, the potential benefits in terms of sustainability and environmental impact make them an appealing field of research and development. Mycelium composites can play an important part in the transition to a more sustainable and circular economy by tackling challenges such as scalability, economic viability, and scientific understanding.

5.1 Opportunities for Further Research and Development

The study of fungal mycelium bio-nanocomposites (MBCs) as sustainable alternatives to conventional materials offers numerous prospects for future research and development. As global demand for ecologically benign materials rises, the potential for mycelium-based composites to meet a variety of industrial difficulties becomes more evident. Here are several major areas where additional study and development can dramatically improve our understanding and application of mycelium composites.

(i) *Optimization of Growth Conditions and Substrate Utilization*

One of the key fields of study is optimizing mycelium growth conditions. Understanding how different environmental conditions, such as temperature, humidity, light exposure, and nutrition availability, influence mycelium development rate and qualities can lead to more effective production procedures. Research might concentrate on determining the optimal conditions for various fungus species and substrates, which can differ greatly. Agricultural waste items like straw, sawdust, and coffee grounds can be used as substrates, but their performance is dependent on pretreatment methods and growth conditions. Investigating the connections between substrate composition and fungal metabolism can provide information that improves mycelial development and the mechanical qualities of the composites.

(ii) *Integration of Nanotechnology*

The incorporation of nanotechnology into mycelium composites is another potential area of research. The addition of nanomaterials such as cellulose nanofibers, silica, or metal nanoparticles can improve the mechanical strength, thermal stability, and barrier properties of mycelium-based products. Future research might look at the synergistic effects of these nanomaterials when mixed with mycelium, and how they interact at the molecular level to increase composite performance. Furthermore, the scalability of adding nanomaterials into the manufacturing process must be addressed to ensure that these improvements can be implemented affordably and efficiently.

(iii) *Biological and Genetic Research*

Additional research into the biology and genetic elements of mycelium can provide useful insights on optimizing its qualities. Genetic engineering approaches can be used to improve specific properties in fungal strains, such as growth rate, robustness, and substrate utilization. Understanding the genetic processes that control mycelial growth and development can lead to the production of specialized fungal strains for specific purposes. Furthermore, studying the biochemical interactions between mycelium and its surroundings can provide novel strategies to improve the performance of mycelium composites.

(iv) *Life Cycle Assessment and Environmental Impact Studies*

Conducting extensive life cycle assessments (LCAs) on mycelium bionanocomposites is critical for evaluating their environmental impact when compared to traditional materials. Quantifying the carbon footprint, energy consumption, and trash generation associated with mycelium composite manufacture and disposal can be the subject of future research. These data can assist in identifying areas for improvement in the manufacturing process and providing vital information to stakeholders contemplating the use of mycelium-based materials. Furthermore, studies on the biodegradability and compostability of mycelium composites in different environments can help to substantiate their sustainability claims.

(v) *Market Development and Economic Viability*

To ensure widespread utilization of mycelium composites, market development and economic viability studies are required. This includes investigating the potential of mycelium-based materials in industries such as packaging, construction, textiles, and electronics. Understanding market trends, customer preferences, and regulatory frameworks can assist in discovering commercialization prospects (Gandia et al., 2021). Furthermore, investigations on the cost-effectiveness of mycelium manufacturing processes, such as substrate procurement and processing methods, can shed light on how to make mycelium composites competitive with standard materials.

(vi) *Innovative Applications*

Mycelium composites' adaptability creates an endless number of chances for creative uses. Mycelium can be used to absorb toxins or heavy metals from contaminated settings in developing disciplines like bioremediation; research into these applications is possible. Furthermore, there is much work to be done in developing mycelium-based products for use in healthcare, such as biodegradable implants or wound dressings. More research is necessary to determine whether mycelium may act as a natural glue in composite materials, doing away with the need for artificial adhesives.

(vii) *Interdisciplinary Collaboration*

Finally, developing multidisciplinary collaboration among researchers, industry stakeholders, and policymakers is crucial to furthering the area of mycelium bionanocomposites. Collaboration can help with knowledge-sharing, resource pooling, and the creation of standardized standards for production and testing. By bringing together experts in materials science, biology, engineering, and environmental science, the research community can address the numerous problems of mycelium composites, accelerating their development and adoption.

Ultimately, the potential for future study and development in fungal mycelium bio-nanocomposites are numerous and diverse. Researchers can realize the full potential of mycelium-based materials by optimizing growth conditions, incorporating nanotechnology, investigating genetic alterations, conducting life cycle assessments, and finding creative applications. As the demand for sustainable alternatives to traditional polymers rises, mycelium composites are ready to play an important part in the transition to a more sustainable and circular economy.

5.2 Potential Obstacles and How to Overcome Them

While the development and implementation of fungal mycelium bio-nanocomposites (MBCs) provide tremendous promise for sustainable materials, numerous major barriers must be overcome before they can be widely used. These difficulties transcend the technical, economic, and regulatory sectors, and overcoming them will necessitate collaborative efforts from researchers, industry, and policymakers.

(i) *Scalability of Production*

Scalability of production is one of the biggest barriers to mycelium composites' commercialization. The viability of producing high-quality mycelium-based products has been shown through laboratory-scale research; nevertheless, the translation of these methods to industrial-scale manufacturing remains a considerable issue. It is challenging to guarantee consistency and dependability in large-scale manufacturing due to variations in fungal species, substrate kinds, and growing settings, which can result in variations in the final products' qualities.

To address this issue, additional research is required to optimize growth factors and develop standardized techniques that may be easily replicated in industrial settings. This includes determining the optimum fungal strains and substrates for certain applications, as well as developing ideal growing and processing procedures. Collaborations between university researchers and industrial partners can promote the exchange of knowledge and resources, resulting in the creation of scalable production systems. Furthermore, advances in bioprocess engineering and automation technologies can improve the efficiency and uniformity of mycelium production.

(ii) *Variability in Material Properties*

The qualities of mycelium composites vary greatly depending on the fungus species utilized, the substrate composition, and the growth conditions. This diversity can make it challenging to achieve industrial standards for mechanical, thermal, and barrier qualities in specific applications.

Solution: To address this issue, extensive research should be done to thoroughly study the correlations between growth circumstances, substrate composition, and resultant material qualities. Creating a comprehensive database of fungal species and their performance characteristics might assist in picking the best strains for certain applications. Furthermore, using modern characterization techniques, such as scanning electron microscopy (SEM) and mechanical testing, can provide insights into the microstructural features of the composites, allowing researchers to fine-tune the production process.

(iii) *Economic Viability*

Currently, mycelium-based materials may have higher manufacturing costs than conventional materials, owing to the expenses of substrates, growing, and processing (Gandia et al., 2021). This price barrier may impede the use of mycelium composites in competitive markets.

Solution: To increase the economic feasibility of MBCs, researchers could look at low-cost agricultural waste and by-products as substrates, which can assist cut production costs while also solving waste management challenges. Furthermore, building effective supply chains and market strategies will be critical to advancing the commercialization of mycelium composites. Engaging with industry stakeholders to understand market trends and consumer preferences can help build goods that satisfy specific needs while being cost-effective.

(iv) *Regulatory Challenges*

The introduction of novel materials into the market frequently necessitates navigating complex regulatory systems, especially in areas like food packaging and construction, where safety and performance demands are rigorous. The lack of recognized norms for mycelium-based goods might cause uncertainty for both makers and consumers.

Solution: To overcome regulatory hurdles, it is critical to consult with regulatory organizations early in the development phase to ensure that mycelium composites meet the relevant safety and performance standards. Conducting comprehensive testing and gaining approvals for mycelium-based products can help to increase customer trust and market adoption. Furthermore, working with industry associations and advocacy groups can provide useful insights into regulatory requirements and assist design policies that encourage the use of sustainable materials.

(v) *Limited Awareness and Acceptance*

Despite the potential benefits of mycelium composites, consumers and industry stakeholders may have a limited understanding and adoption of their advantages over traditional materials. This lack of familiarity may impede market penetration and inhibit the adoption of mycelium-based goods.

Solution: To overcome this challenge, specific marketing and educational programs should be launched to create knowledge about the benefits of mycelium composites. Engaging customers through workshops, demonstrations, and informational tools can help to highlight the materials' sustainability and performance benefits. Furthermore, highlighting successful case studies and applications can demonstrate the practical benefits of mycelium composites, leading to increased acceptance and interest in their utilization.

(vi) *Research and Development Funding*

It might be difficult to get financing for mycelium composites research and development, especially in a competitive funding environment. A lack of funding may make it more difficult to carry out in-depth research and create novel solutions. One potential solution to overcome financial issues for researchers is to work with government agencies, non-profit organizations that prioritize sustainability and innovation, and industry partners. Funding sources can be attracted to grant applications that highlight the economic feasibility and possible environmental advantages of mycelium composites. Furthermore, forming alliances with companies that are enthusiastic about sustainable materials might help exchange resources and finance research projects. In short, there exist multiple possible barriers to the extensive implementation of fungal mycelium bio-nanocomposites; yet there exists a plethora of approaches to surmount these impediments. Stakeholders can cooperate to forward the development and commercialization of mycelium-based materials by concentrating on scalability, material consistency, economic viability, regulatory compliance, consumer awareness, and funding opportunities. The effective incorporation of mycelium composites into several industries has the potential to make a substantial impact on sustainability initiatives and the shift toward a circular economy.

6 Conclusion

In conclusion, the study of fungal mycelium as a fundamental component in bio-nanocomposites offers a transformative chance to address the serious environmental concerns faced by chemically synthesized lightweight polymers. This chapter has demonstrated traditional plastics' considerable harmful effects on ecosystems and human health, emphasizing the urgent need for sustainable alternatives. The extensive usage of synthetic polymers has led to pollution, resource depletion, and long-term waste in landfills and oceans, creating an unsustainable loop that endangers both the environment and future generations.

Fungal mycelium, with its own qualities and capacities, appears as a potential alternative, providing benefits such as biodegradability, low environmental effect, and the ability to use agricultural waste as substrates. Mycelium's natural growing process enables the fabrication of composites that are not only lightweight and robust but can also decompose in natural surroundings, reducing the long-term ecological footprint associated with typical plastics. Mycelium-based composite manufacturing technologies are both inventive and adaptable, enabling customization to satisfy unique industry needs in packaging, construction, environmental remediation, agriculture, and biomedicine.

Additionally, the mechanical and thermal qualities of mycelium composites, together with their environmental benefits, position them as viable alternatives to traditional materials. These composites have been proven in studies to match or even outperform synthetic materials in certain applications, making them appropriate for a wide range of uses. Their potential in packaging is particularly remarkable, as they can replace single-use plastics while also contributing to a circular economy by using waste goods as raw materials.

However, the path toward broader adoption is not without hurdles. Issues such as standardizing production procedures, scaling-up processes, and guaranteeing consistent quality must be addressed through ongoing R&D. The use of nanotechnology in mycelium composites opens intriguing opportunities for improving characteristics, but it also necessitates careful consideration of safety and environmental concerns. As the area advances, collaboration among researchers, industry stakeholders, and policymakers will be critical to overcome these challenges and fully realize the potential of mycelium-based bio-nanocomposites.

In closing, the incorporation of fungal mycelium into bio-nanocomposites is a positive development in the direction of a more sustainable future. We can create materials that not only satisfy the needs of contemporary applications but also support ecological balance and resource conservation by utilizing the inherent qualities of fungi. Such creative solutions are what material science will look like in the future, and the continuous study of fungal mycelium will surely be crucial in creating a more sustainable and greener planet. Prioritizing multidisciplinary cooperation and funding for studies that concentrate on the scalability and commercialization of mycelium-based products is essential as we move forward. By doing this, we could open the door to a new era of environmental-friendly materials that support the health of the earth for coming generations and are in line with global environmental goals.

References

Alemu, D., Tafesse, M., & Mondal, A. K. (2022). Mycelium-based composite: The future sustainable biomaterial. *International Journal of Biomaterials, 2022*, 1–12. https://doi.org/10.1155/2022/8401528

Antinori, M. E., et al. (2020). Fine-tuning of physicochemical properties and growth dynamics of mycelium-based materials. *ACS Applied Bio Materials, 3*(2), 1044–1051. https://doi.org/10.1021/acsabm.9b01031

Appels, F. V. W., et al. (2019). Fabrication factors influencing mechanical, moisture- and water-related properties of mycelium-based composites. *Materials & Design, 161*, 64–71. https://doi.org/10.1016/j.matdes.2018.11.027

Attias, N., et al. (2020). Material function of mycelium-based bio-composite: A review. *Frontiers in Materials, 8*, 737377. https://doi.org/10.3389/fmats.2021.737377

Danninger, D., Pruckner, R., Holzinger, L., Koeppe, R., & Kaltenbrunner, M. (2022). MycelioTronics: Fungal mycelium skin for sustainable electronics. *Science Advances, 8*(45). https://doi.org/10.1126/sciadv.add7118

Ecovative Design. (2023). *Mycelium packaging*. Retrieved from [Ecovative] https://ecovative.com

Gandia, M., et al. (2021). Recent advances in the construction of biocomposites based on fungal mycelia. *Materials Today, 20*(5), 254–261. https://doi.org/10.1016/j.mattod.2017.02.004

Girometta, C., et al. (2019). Mycelium-based composites: A review of their potential applications. *Journal of Cleaner Production, 221*, 241–257. https://doi.org/10.1016/j.jclepro.2019.02.025

Haneef, M., et al. (2017). The potential of fungal mycelium in sustainable materials: A review. *Materials Today: Proceedings, 4*, 10456–10461. https://doi.org/10.1016/j.matpr.2017.09.039

Haneef, M., et al. (2021). Mycelium-based composites: A sustainable alternative to synthetic materials. *Sustainable Materials and Technologies, 28*, e00283. https://doi.org/10.1016/j.susmat.2021.e00283

Jones, M., Mautner, A., Luenco, S., Bismarck, A., & John, S. (2020). Engineered mycelium composite construction materials from fungal biorefineries: A critical review. *Materials & Design, 187*(108397), 108397. https://doi.org/10.1016/j.matdes.2019.108397

Li, K., Jia, J., Wu, N., & Xu, Q. (2022). Recent advances in the construction of biocomposites based on fungal mycelia. *Frontiers in Bioengineering and Biotechnology, 10*, 1067869. https://doi.org/10.3389/fbioe.2022.1067869

Lingam, D., Narayan, S., Mamun, K.A. and Charan, D. (2023). Engineered mycelium-based composite materials: Comprehensive study of various properties and applications. Construction and Building Materials, 391, pp. 131841–131841. https://doi.org/10.1016/j.conbuildmat.2023.131841.

Livne, A., Wösten, H. A. B., Pearlmutter, D., & Gal, E. (2022). Fungal mycelium bio-composite acts as a CO_2-sink building material with low embodied energy. *ACS Sustainable Chemistry & Engineering, 10*, 12099. https://doi.org/10.1021/acssuschemeng.2c01314

Lücking, R., Aime, M. C., Robbertse, B., Miller, A. N., Aoki, T., Ariyawansa, H. A., Cardinali, G., Crous, P. W., Druzhinina, I. S., Geiser, D. M., Hawksworth, D. L., Hyde, K. D., Irinyi, L., Jeewon, R., Johnston, P. R., Kirk, P. M., Malosso, E., May, T. W., Meyer, W., et al. (2021). Fungal taxonomy and sequence-based nomenclature. *Nature Microbiology, 6*(5), 540–548. https://doi.org/10.1038/s41564-021-00888-x

Manan, S., Ullah, M. W., Ul-Islam, M., Atta, O. M., & Yang, G. (2021). Synthesis and applications of fungal mycelium-based advanced functional materials. *Journal of Bioresources and Bioproducts, 6*(1), 1–10. https://doi.org/10.1016/j.jobab.2021.01.001

Meyer, M., et al. (2020a). Recent advances in the construction of biocomposites based on fungal mycelia. *Materials Today, 20*(5), 254–261. https://doi.org/10.1016/j.mattod.2017.02.004

Meyer, H., et al. (2020b). Mycelium-based composites: A review of their properties and applications. *Advanced Materials, 32*(18), 1905636. https://doi.org/10.1002/adma.201905636

Peng, X., Chen, M., Chen, S., Dasgupta, S., Xu, H., Ta, K., Du, M., Li, J., Guo, Z., & Bai, S. (2018). Microplastics contaminate the deepest part of the world's ocean. *Geochemical Perspectives Letters, 9*, 1–5.

Saravanan, P., et al. (2021). Mycoremediation: Fungal remediation of heavy metals and organic pollutants. *Environmental Science and Pollution Research, 28*(45), 64371–64385. https://doi.org/10.1007/s11356-021-13607-5

Sharma, R., & Sumbria, R. (2022). Mycelium bricks and composites for sustainable construction industry: A state-of-the-art review. *Innovative Infrastructure Solutions, 7*(5). https://doi.org/10.1007/s41062-022-00903-y

Smith, S. E., & Read, D. (2008). *Mycorrhizal symbiosis*. Elsevier. https://doi.org/10.1016/b978-0-12-370526-6.x5001-6

Sun, W., Tajvidi, M., Hunt, C. G., McIntyre, G., & Gardner, D. J. (2019). Fully bio-based hybrid composites made of wood, fungal mycelium and cellulose Nanofibrils. *Scientific Reports, 9*(1). https://doi.org/10.1038/s41598-019-40442-8

Vandenberg, L. N., Ehrlich, S., Belcher, S. M., Ben-Jonathan, N., Dolinoy, D. C., Hugo, E. R., Hunt, P. A., Newbold, R. R., Rubin, B. S., Saili, K. S., Soto, A. M., Wang, H.-S., & vom Saal, F. S. (2013). Low dose effects of bisphenol A. *Endocrine Disruptors, 1*(1), e26490. https://doi.org/10.4161/endo.26490

Verma, N., Satya Eswari, J., & Mahapatra, C. (2023). Green sustainable biocomposites: Substitute to plastics with innovative fungal mycelium based biomaterial. *Journal of Environmental Chemical Engineering, 11*(5), 110396–110396. https://doi.org/10.1016/j.jece.2023.110396

Zheng, J., & Suh, S. (2019). Strategies to reduce the global carbon footprint of plastics. *Nature Climate Change, 9*(5), 374–378. https://doi.org/10.1038/s41558-019-0459-z

Life Cycle Assessment and Techno-Economic Analysis on Setting Up of Pilot Scale Plants

Magesh Kumar Marichamy

Abstract In recent times, technocrats and scientists employ advanced technologies in industrial processes to mitigate greenhouse gas emissions, which subsequently abate climate change. Momentum is increasing among humans to combat unprecedented climate change and environmental degradation for sustainable livelihood. Biomass utilization for green fuel production is a viable alternate for fossil fuels to sort out the issues related with gas emissions, land pollution, and inflated economy. The major challenge in biomass usage is the difficulty in handling voluminous quantity and poor profit margin of valorized products. Owing to high energy density and abundance in availability, the agricultural residues and woody biomass are considered as excellent raw materials to produce biofuels and other carbon-deprived fuels. Before switching over from fossil fuel energy source to biomaterial energy source, it is necessary to evaluate the environmental impacts, cradle-to-grave carbon footprint size in biomass usage, capital costs, operating costs, product yield, and scale-up costs of bioconversion process. In order to assess the environmental impacts of a process and products, the life cycle assessment (LCA) is used, whereas the techno-economic analysis (TEA) evaluates the performance of a process in terms of monetary benefits. Raw materials procurement and processing is the first step in LCA, while the energy emission terms related to raw materials are embodied into produced goods. LCA has become the reliable tools to monitor environmental qualities and forecasts the strategies to follow in biomass conversion processes. The manufacturers, policymakers, and deciding authorities choose the process with least environmental consequences based on the output from LCA. These approaches upskill the end users to understand polluting potential of products and processes and educate public for meticulous use of products. The guidelines to implement and practice LCA in industries are defined in ISO 14040 and ISO 14044 standards. The environmental consequences of a product or a process are addressed in ISO 14040, while ISO 14044 standard defines the four phases, namely, goal and scope, life

M. K. Marichamy (✉)
Department of Chemical Engineering, SRM Institute of Science and Technology, Kattankulathur, Tamil Nadu, India

D. B. Pal, S. L. Bansal (eds.), *Fungal Waste Biomass Management for Energy, Environment and Value-Added Products*, Fungal Biology,
https://doi.org/10.1007/978-3-031-82599-6_4

cycle inventory (LCI) analysis, life cycle impact assessment (LCIA), and interpretation, in LCA. The economic and technical performance of a process and product is analyzed using TEA parameters such as net present value (NPV), payback period, and discounted cash flow rate of return (DCFR). This chapter discusses the scope of LCA and TEA in biomass valorization.

Keywords Biomass · Carbon footprint · Life cycle assessment · Climate change · Techno economic analysis · Economic indicators

1 Introduction

Globalization, urbanization, industrialization, and open trade agreements have been the buzz words, which strongly correlated to environmental pollution and climate change. The ruthless exploration and unfair utilization of fossil fuels are the primary reasons for greenhouse gas emissions in an uncontrolled manner. The socioeconomic status of human beings and economic-growth indicator of a nation depend mainly on energy production and utilization rate. There is a need to meet global energy demand without compromising the quality of environment. It can be achieved by shifting the dependency from conventional fossil fuels to biomass residues. Being an organic material obtained from plants, the energy content in biomass is synthesized from sunlight and water by a process called photosynthesis. It is a highly versatile feedstock because it can be combusted with air for direct heat production or indirectly converted to liquid/gaseous fuels by different processes. Wood-processing wastes, agricultural residues, food wastes, and biogenic materials in household wastes are the typical sources of biomass. The energy from these sources is obtained by the processes namely direct combustion, thermochemical reformation, biological transformation, and chemical conversion. Biomass accounts for 14% of the total energy supply globally and primary energy source for 45% of rural population (Gomathi et al., 2024). Thermochemical conversion process deals with heating, dehydrating, and stabilizing of biomass feedstocks segregated from municipal solid wastes. For efficient burning of biomass, it is converted to bone-dry material by heating in the temperature range of 200–320 °C by torrefaction process. The dried, blackened product is briquetted to form high-energy density pellets used for direct heating or co-firing purpose. Biomass co-fired along with coal minimizes the demand for coal and quantity of greenhouse gases produced. Pyrolysis is an alternate way of handling of biomass in anaerobic condition to alter the chemical structure of biomass. The products obtained by pyrolysis of biomass are pyrolysis oil, syngas, and biochar. The pyrolysis oil also called as bio-oil can be used to generate electricity and is used as additional compound in other fuels. Syngas can be converted to methane and is considered a suitable replacement for natural gas. It is a gas mixture consisting primarily of hydrogen and carbon monoxide. Biomass is an integral part of the earth's carbon cycle, which replenish the carbon amount in different spherical zones of earth namely atmosphere, hydrosphere, biosphere, and lithosphere. The intensity of solar radiation entering into earth's atmosphere is controlled by the amount of carbon dioxide (CO_2) present. The exchange of carbon occurs via photosynthesis, respiration, and human activities. Fossil fuels are produced from plants and, animal sources and they are stored deep beneath the earth crust for millions of years. Due to industrialization and technology development, the fossil fuels are explored and burned for energy, thereby releasing the sequestered carbon into atmosphere, whereas biomass is produced from contemporary carbon source. It is feasible to maintain dynamic equilibrium between the rate of carbon consumption in biomass growth and rate of carbon produced from biomass conversion. Biomass energy has the potential to reduce greenhouse gas emissions to greater extent. Burning of fossil fuels produce fresh leash of carbon dioxide gas from carbon compounds accumulated due to photosynthesis millions of years ago. The carbon dioxide gas produced during energy generation from fossil fuels is literally considered as newcomer to atmospheric air. On the contrary, the carbon emitted from biomass burning is equivalent to carbon utilized in photosynthesis during biomass growth. Hence, the net quantity of carbon compound emitted is zero.

The types of biomass feedstocks used in energy production are purposely grown crops, agricultural crop residues, forest and wood-processed residues, and municipal wastes. Purposely grown energy crops are categorized as herbaceous and woody crops. Switch grass, miscanthus, bamboo, sweet sorghum, wheat grass, etc., are some examples of herbaceous crops, which are perennial grasses. Short-rotation woody crops are rapidly growing hardwood trees which can be harvested after maturation between 5 and 8 years of planting. Hybrid poplar, hybrid willow, silver maple, eastern cottonwood, green ash, black walnut, and sweet gum are examples of woody crops. Agricultural residues are abundant, diversified, and distributed widely across the world. Secondary income is generated for the farmers when these residues are sold to local industries to cater to their energy needs. Activated carbon prepared from biomass is used in supercapacitors and batteries, owing to its excellent energy storage ability (Chaudhary et al., 2024). Mitigation of atmospheric CO_2 level and sustainable life is achieved by the proper management of biomass wastes and afforestation of degraded landscapes. Energy conservation via biomass is achieved in two ways: planting trees for shading, windbreaks, avenues, green corridors, green building, and increasing the use of biomass feedstocks for heat production and electricity generation. The purposely grown energy crops have the potential to reduce carbon emission to the tune of 0.6×10^9 ton carbon/year (Sampson et al., 1993). Quality of life indicator and sustainability depend on proper management of domestic biomass wastes because they are an integral part of human activities and a viable alternate for fossil fuels. Hydrothermal processing methods are mostly used to convert these hydrated wastes to energy and fuels (Wang & Tester, 2023). A holistic and exemplified approach is needed to understand monetary benefits and environmental advantages if biomass is properly managed. There are challenges associated with biomass handling, managing, and in conversion technologies. Emphasis must be given on safe handling and technology upgradation for valorization of biomass wastes. The common occupational hazards in biomass management include exposure to biological airborne pollutants (bacterial, fungal spores), respiratory tract infection, mold inhalation, emission of soot particles, organic dusts, and volatile organic compounds (Singh & Kalia, 2017). The successful biomass waste management strategies depend on how effectively the energy is liberated without affecting the ambient qualities and health of personnel. Biomass can be considered as energy storage device like a capacitor, because it stores the energy produced during photosynthesis process. The quantum of energy stored in biomass depends on the intensity of solar radiation, rate of photosynthesis process, and competitive growth with other trees. Biomass is a plant-based material consisting mainly of cellulose, hemicellulose, lignin, and organic volatiles. On dry weight basis, a sample of wood contains 45% cellulose, 20% hemicellulose, 30% lignin, and the remaining 5% is inert ash, polymeric gums, sugars, starches, volatile materials, etc. Hemicellulose is easy to hydrolyze and degrade because it is having amorphous structure, whereas cellulose resists hydrolysis and degradation owing to its crystalline nature. Lignin plays the role of binding agent to produce intimate contact between cellulose and hemicellulose fiber for providing firmness to trees and plants.

2 Biomass Management and Conversion Technologies

The biomass conversion technologies are categorized as thermochemical conversion methods, biochemical conversion methods, and physicochemical conversion techniques. Figure 1 outlines the types of biomass conversion technologies.

Thermochemical conversion process involves the utilization of direct heat to produce energy from biomass and it is designed to combust woody residues having low moisture content. In direct combustion process, the fuel is oxidized at high temperature to produce hot gases which in turn is utilized to produce high pressure steam for electricity generation. Carbon and hydrogen present in woody biomass is converted into carbon dioxide and water by exothermic reaction. General stoichiometry for this reaction is proposed as:

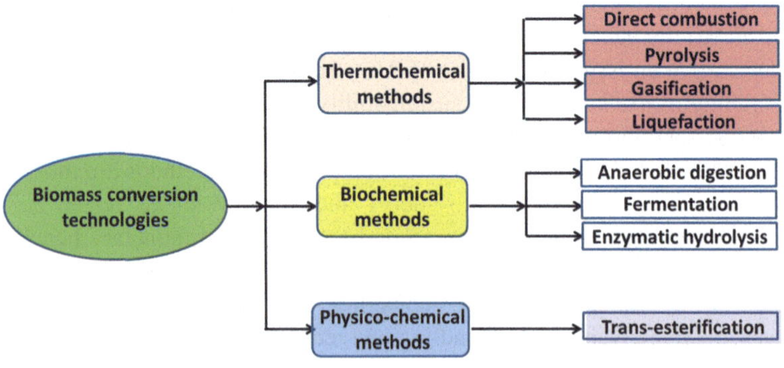

Fig. 1 Types of biomass conversion technologies

$$\text{Biomass} + O_2\,(\text{from } air) \rightarrow CO_2 + H_2O. \tag{1}$$

If air is supplied in excess than stoichiometric requirement, then complete combustion takes place to form carbon dioxide; otherwise, incomplete combustion occurs to form carbon monoxide. Incomplete combustion is not desirable, owing to its tendency to produce low enthalpy content gases. The efficiency of combustion process depends on biomass size, its moisture content, actual flame temperature and moles of air supplied. Pyrolysis is another process involve with heating of biomass without air. It is classified as slow and fast pyrolysis based on the rate of heating. The operating temperature in slow process ranges from 400 °C to 600 °C, whereas for fast pyrolysis it ranges between 400 °C and 800 °C. The heating rate in slow pyrolysis lies in the range 0.1–1 C/s, whereas in fast pyrolysis the heating rate of biomass ranges from 10 °C to 200 °C for a time interval of 1–10 s. The energy in biomass is released at an elevated temperature of 500–1400 °C to form voluminous gaseous products with low char content in gasification process. The products are reduced to form combustible gases primarily consisting of carbon monoxide, hydrogen, and methane. The purpose of liquefaction process is to convert biomass into bio-oil product by using a liquid medium. The product yield reaches maximum when the process is carried out in water solvent at a temperature and pressure of 250–400 °C and 5–25 MPa, respectively (Huang et al., 2017). Biochemical methods use microorganisms to break down complex cellulosic biomass to useful end products. In anaerobic digestion process, the biomass is broken in the absence of oxygen to form biogas. The typical yield of biogas in this process is 36 m^3 methane/1000 kg biomass. The efficiency of microbial production depends on substrate concentration, solution pH, feed temperature, and dilution factor. The fermentation process of biomass is facilitated using yeast to form ethanol and carbon dioxide by degradation of grains and decomposition of ethyl alcohol to form acetic acid. Enzymatic hydrolysis process is used to convert lignocellulosic biomass (polysaccharides) to monosaccharides, which is consequently converted to biofuels. The rate of hydrolysis is hampered due to the resistance offered by lignin surrounding the cellulose. Pretreatment method is mandatory to loosen the cellulose from lingo-cellulose matrix followed by enzymatic hydrolysis to form simple sugars from breaking of cellulose. The simple sugar is subjected to microbial fermentation process to yield bioethanol. Cellulosic biofuels can reduce fossil fuel usage and assist in balancing CO_2 gas emissions. Moreover, cellulosic biomasses do not contribute to the release of additional carbon to the atmosphere when it is grown on deforested landscapes. All the biomass conversion techniques are either costly or time-consuming or technically challenging. In recent times, the biomass conversion techniques focus on alternate ways including deconstruction of biomass, fractionation of deconstructed chemicals, synthesis of intermediates from distilled products, and upgradation of biofuels. The sequence of process flow is depicted in Fig. 2.

The longevity and sustainability of life on earth depends on the "exploration and utilization" rate of natural resources. If the resources are exploited blindly, the wrath of nature will ultimately affect the human's life quality index. The well-being of

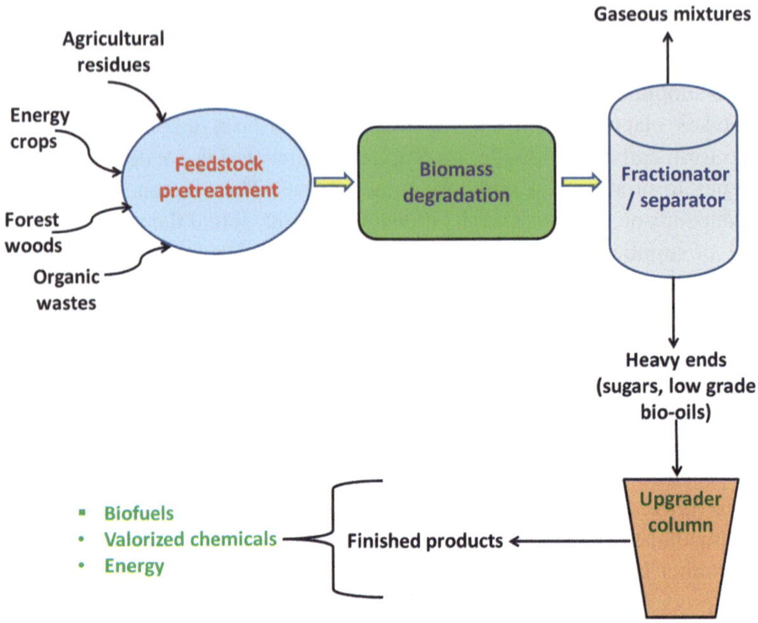

Fig. 2 Sequence of process flow in biomass conversion to energy

earth is determined by five elements in nature viz. land, water, air, space, and fire. Since they are considered as the anatomy of earth, the prosperity of living things decrease when the quality of elements is not up to the standard. The population explosion and unwarranted activities of humans have posed severe stress on natural elements. Land resources are polluted due to synthetic agrochemicals, plastic wastes, solid wastes, and liquid wastes. The similar fate ensued for water (due to industrial effluents, domestic wastewater), air (owing to industrial emissions, vehicle emissions, biomass incineration, airline emissions), and space (expired satellites, telecommunication gadgets). Except fire, almost 80% of natural elements are degraded, purely because of human activities. It can be curtailed by carrying out life cycle assessment (LCA) for all manmade products and activities.

3 Life Cycle Assessment (LCA) and Major Ingredients

Life cycle assessment (LCA) is a process that evaluates the impacts of manmade products on environment and energy from its production to disposal. It assist the consumers to decide whether the product is eco-friendly and to follow disposal practices with proper understanding of environmental concerns. It is associated with identifying and quantifying the raw material requirement, amount of energy consumed, and wastes generated when products are made from natural resources. Each and every step in product-making is assessed to minimize the pollution load and look for eco-friendly options to make products. The general block diagram explaining the sequence of steps in LCA is elucidated in Fig. 3.

LCA will contribute for sustainable life as it provides an insight into overall environmental implications of synthetic products. It enlightens the end users about power consumption, fate mechanism of used products, and detrimental consequences of expired goods. The challenges associated with LCA includes difficulty in defining the scope and boundary of problems, availability of reliable information from product manufacturers, and data logger and finding the line of similarity between material flow and environmental impacts. To analyze the environmental impacts of biomass conversion processes, it is necessary to perform LCA. For pyrolysis of biomass, three life cycle stages are assessed: characterization of woody wastes, biofuel production by pyrolysis, and combustion of bio-oil in steam generating unit. The properties of feedstock determine the type of size reduction equipment and the quantum of drying required. The total energy utilized for drying is the sum of electricity needed for crushing, grinding, and drying operations. The products obtained in pyrolysis process are biofuel, char, syngas, and ash. Biofuel is the major product which is used as an alternate for fossil fuels. The char is a coproduct which is recovered after pyrolysis process for use in furnace toward energy generation. Syngas can be transported to a power integration unit inside the plant and burned along with char for process heat. LCA is an ideal tool to analyze the environmental impacts of each steps in biological conversion process. The major product obtained in anaerobic digestion process is methane and secondary product produced is digested wastes which can be used as organic fertilizer. This method has the potential to handle wet biomass without any pretreatment, definitive gas production, and reduced emissions while burning the product gases (Evangelisti et al., 2014). For anaerobic digestion process, LCA is conducted based on the guidelines framed in

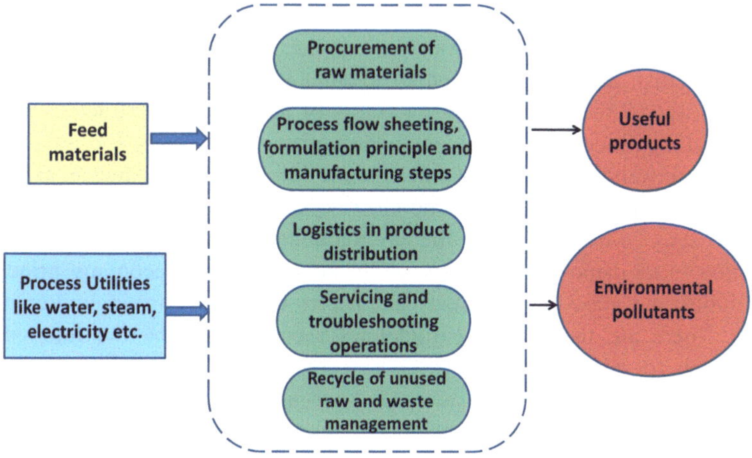

Fig. 3 Block diagram of the sequence of steps in LCA

ISO 14040 and ISO14044, which comprises four sequences of steps: goal and scope definition, life cycle inventory, life cycle impact assessment, and life cycle interpretation. Software such as OpenLCA version 2.2.0 and ASPEN plus are used to perform LCA biological process because they have necessary data in database. The first step is to demarcate system boundary from input and output variables in a process. It is necessary to define the timeline for achieving designated goal and predicted outcome from LCA. The quantity of material and energy entering and leaving the system boundary and its relation with environment is analyzed in life cycle inventory step. The impact assessment stage details the normalization, giving weightage to every impact category (Product carbon footprint, 2024). Since LCA evaluates the environmental impacts of each step in product-making at microscopic level, it assists end users to compare the energy and material used for different varieties of products. It provides insight to make decisions that could minimize the damage to environment.

3.1 Objectives and Steps in LCA

Life cycle assessment is a versatile approach used in environmental pollution remediation with the aim of developing cleaner production technologies. The protocol to be followed is well defined with the objective of examining the attributes of input/output materials, energy consumption, and environmental impacts of the process and products. LCA is now an integral part of product-manufacturing companies striving to achieve net zero emissions, more sustainable production, and mitigated consumption. The assessment starts with studying the detrimental effects of raw material extraction on environmental parameters. It is followed by analyzing the influences of manufacturing steps, transportation steps, usage pattern, and waste disposal on the quality of environment. Three major approaches are followed to carry out LCA analysis: cradle-to-gate step measures the environmental influences of various unit operations, unit processes carried out to extract the raw materials from earth crust, and transportation to manufacturer's site. The cradle-to-grave approach assesses the environmental hazards of raw material extraction process, product-making process, and usage methodology of end products. It is an extensive method that covers all impacts at microscopic level till the product is exhausted without environmental hindrance. The most detailed assessment is called "cradle-to-cradle" stage, which measures the environmental impacts of products from its production to its disposal. This stage advocates the circular economy principles, recyclability, and reuse of products to attain zero tolerance limit in environmental impacts. There are two standard guidelines recommended by the International Organization for Standardization (ISO) to carry out LCA. In ISO 14040: 2006 method, the following items are covered: environmental domain to analyze, time frame required, methods adopted, and delineation of environmental quality parameters to make it transparent to common public. ISO 14044: 2006 is used as a supportive standard in concurrence with ISO 14040. It determines the environmental impacts of a product by analyzing the model of a product.

3.2 Opportunities and Tools of LCA

The quantification of impact on climate change by a product is one of the outputs of LCA study. As of now, LCA focuses on evaluating the carbon emissions linked with raw materials utilization, product-making, and other processes. But the scope of LCA can be extended to other impacts such as eutrophication, ozone depletion, eco-toxicity, photochemical smog formation, land degradation, and resource depletion. It is necessary for a product manufacturer to identify "hotspots" in the life cycle of a product where the product makers can take appropriate action to mitigate environmental impact. This holistic approach of LCA can make valuable contributions to protect hydrosphere, atmosphere, lithosphere, and stratosphere regions of earth. It also can identify new use value for waste products, which ultimately leads to boom in circular economy. LCA permits the users to understand the environmental implications of different products used for the similar work. For example, the electric vehicles are eco-friendly rather than IC engine vehicles powered by fossil fuels. Steel is usually produced using iron ore, but the same type of steel can be made using electric arc furnace method using recycled materials. Based on LCA output, the user can decide the category of product to use without affecting the qualities of environment. It also elucidates the environmental hazards of products after they are disposed of improperly. LCA tools are either software or analytical instruments used to assess the environmental footprint left by products, process, and other accessorial activities. Energy consumption, greenhouse gas emissions, water utilization, and waste generation are assessed systematically and quantitatively using these tools. The outputs of these tools are used by the product makers to design an efficient process via suitable alternative ways, to identify the key areas for improvement, and to make informed choices. Some of the software tools to conduct LCA and assessment of sustainable practices in different manufacturing units are listed below:

(a) SimaPro9: It is used to measure the sustainability index and find the metrics to move toward positive direction of mitigating environmental degradation and is apt for manufacturing sectors, energy conservation, agriculture, transportation, and consumer goods.
(b) GaBi10: It is another tool that supports the decarbonization journey of a company and fulfills their product's sustainability needs. It is applicable to diverse sectors such as manufacturing units, chemical industries, energy production, electronic goods fabrication, automotive, and packaging.
(c) OpenLCA11: It is a versatile open-source tool that is suitable to different sectors and applications, including product manufacturing, green building construction, waste management, and renewable energy production.
(d) Umberto12: It is suitable to assess sustainability index in chemical-manufacturing industries, energy production units, waste-handling sites, and circular economy initiatives.
(e) EarthSmart13: The primary use of this tool is to assess products sustainability and packaging across various sectors, including consumer goods, electronics.

(f) Sustainable Minds14: It is suitable for product designers and engineers working in multiple domains, like consumer goods, electronics items, furniture, and building materials.

3.3 Role of LCA in Environmental Policies

In recent times, the awareness among public, stakeholders, and customers is increasing to know the environmental policies, societal services, and governance of a company. Transparency in data sharing, disclosure of environment-related policies, and carbon footprint size should be posted in official website and newspapers. There is a steady growth in LCA strategies gradually progressing from predefined standards to rules and guideline documents. Product environment footprint (PEF) and environmental product declarations (EPDs) can be used to test the consistency of data received using LCA studies. It is the right time to expand the boundary of LCA from industrial environmental impacts to new domains like societal LCA, organizational LCA, educational institutions LCA, and more. In order to attain better output value from LCA, it is necessary to compare and align the output in relevance to environmental, societal, and governance. The success of LCA study depends on the support rendered by LCA for transition from linear economy to circular economy. Instead of using LCA to optimize the system performance within a limited boundary, it is important to envisage the end state of a product, environmental consequences, attainment of circular economy, and innovations to achieve it. LCA can be a pivot element for successful implementation of environmental policies. It can target any specific stages, within the life cycle of a product, which is the prime contributor to environmental impacts. Moreover, the weak areas for improvement in a product's life cycle are addressed by LCA, and this includes resource, energy intensive stages, and highly polluting phases in product-making. The output from LCA should be resilient against external factors, which may vary with manufacturing locations or change with respect to time. Process methods, enhancement of waste recycling system, and adaption to new technologies may vary from one company to another. The effects of these external factors on LCA output can be assessed by manipulating the input variables of LCA. When dealing with similar systems, LCA is more efficient to compare their behavior and responses. The efficacy of LCA is maximized when the input data is authentic and less sensitivity. It is very effective when the material flow and energy flow through the system is obvious. LCA should be applied with caution during the earlier stages of innovation process, whereas applying LCA at later phases of innovation need not have restrictions. Environmental impacts due to crazy activities of humans are evaluated from affected system's perspective. The benefits from sustainable practices and trade-off convenience can be analyzed using LCA (Ciacci & Passarini, 2020).

3.4 Need for LCA of Biomass Conversion Process Plants

During the process of biomass conversion, pollutants like particulate matter, oxides of nitrogen, oxides of sulfur, lead, vaporized mercury, and other pollutants are emitted into the air. If the pollutants are not mitigated or captured, then the pollutants can cause severe damage to soil, water, and air. The manner in which biofuels are produced can be used to frame policy guidelines for sustainability in production. It is mandatory to perform the LCA of biofuel production chain because the process involves many pathways and different varieties of products and co-products are formed (Osman et al., 2021). LCA analysis is needed to understand the similarities, differences, feasibilities, and technical information in all biomass conversion process. Moreover, a major thrust and vital information is given by the LCA study to combat climate change and unusual weather pattern (Yu et al., 2022). LCA study determines the success of transferring the pilot-scale operation to commercial-scale process, taking care of environmental impacts of all unit operations, unit processes, feedstocks, and products (Hussin et al., 2023).

3.5 Benefits and Challenges of LCA

Life cycle assessment is used to support sustainable growth and production. It is used to assess environmental impacts of a product during its entire lifetime: from extraction of raw material through processing, product manufacturing, distribution and use of products, and disposal or recycling of wastes. LCA allows consumers to make a smart choice by comparing two similar categories of products and to select the product which has less negative impact on environment. The major advantage is that it offers a holistic view on all factors affecting the environment rather than paying attention to only one variable affecting the environmental standards. It assists the end users to select the best values, best ways, and best methods to produce eco-friendly products.

Each LCA study can have different competence so that the impacts analyzed in one process may be ignored in another. The assumptions made in one study may be different from other studies and hence the results are entirely different. Consequently, the results are confusing and very difficult to compare and interpret, resulting in skepticism and criticism. Also, this approach needs more data to get solid results and concrete conclusions, otherwise it is not worthy as the data collection is laborious and time-consuming. The problem is also associated with communicating the results via reliable sources to the stakeholders and public.

4 Techno-Economic Analysis (TEA) of Biomass-Processing Plants

The techno-economic analysis (TEA) method is used for evaluating the economic performance of a technology. It allows the evaluator to analyze the relationship between invested costs and benefits. This is done to assess competitiveness between new technology and an existing technology. The successful commercialization of new technology depends on expense and cost incurred. The manufacturing costs to make a product can be segregated as capital expenses (CapEx) and operating expenses (OpEx). The money spent on building construction, infrastructure development, and equipment purchase is known as capital costs or nonrecurring costs. In TEA, the capital expenditures invested on intangible assets are amortized over a productive lifetime of machineries and other investments. The operating expenses include raw materials procurement costs, energy utilization cost, and labor-related costs. In cost benchmark approach, the economic feasibility of new technology is compared to commercial benchmark technology. The cost-wise performance of each technology is compared by adding the CapEx and OpEx components. For example, consider a new catalyst is developed to manufacture a product. The manufacturing cost of new catalyst is 20% more than the manufacturing cost of existing catalyst. Since the catalyst manufacturing cost is increased, it will be reflected in inflated total cost per unit mass of catalyst used. Meanwhile, if the active life period of the new catalyst is 30% longer than that of exiting catalyst lifetime, then there is no need to replace new catalyst frequently. This accounts for reduced consumption of new catalyst to produce certain volume of product, which is equivalent to volume produced by benchmark catalyst.

The maximum achievable efficiency of biomass-processing plants is achieved by techno-economic analysis phase. Single-stage biomass-processing units struggle to achieve the optimum output because of the high degree of uncertainty in agricultural products availability; erratic rainfall pattern; and extreme global calamities like landslides, floods, and droughts. The raw material demand and uncertainty in supply classifies the biomass treatment units as high-risk investment zone. Money-lending organizations and financial institutions charge higher interest rates for the investments made in biomass process industries. This is like an additional burden on the overall capital investment of the company and stakeholders and well-wishers would have a second thought to make an investment. To overcome these circumstances, the individual units are integrated to attain maximum economic benefit, outweighing individual investment costs and operations costs (Lamers et al., 2015).

4.1 *Techno-Economic Analysis Concepts and Estimation Methods*

Techno-economic analysis (TEA) is considered as an important tool to evaluate the economic performance and index of manufacturing processes. TEA is performed during inception phase of a project keeping the objective as to maximize the profit, yield, purity and minimize the total costs, energy utilization, and environmental impacts. The applications of TEA in industrial sector are phenomenal to meet the competition from other companies. It is conducted in a holistic manner through sequential steps including process flow sheeting, process design and scale up, equipment sizing and fabrication, and cost estimation and cash flow analysis. The results obtained from TEA can be used to evaluate the process performance in terms of product quality and economic indicators. The economic parameters like net present value (NPV), payback period, internal rate of return (IRR), return of investment, and discounted cash flow rate of return and profit–loss balance sheet are used to analyze the profit-yielding capability of process and products. Various techno-economic parameters such as process yield, extent of reaction, installed and actual production capacity, guaranteed selling prices, initial investment cost, and environmental implications are addressed clearly in TEA before scaling-up of pilot-scale plant to industrial scale. Biomass conversion suffers from inherent limitations like poor conversion, unwarranted side reactions, feeble microbe strain, and difficulty in isolating the desired product by downstream process. All these hurdles are wiped out, when the volatility in feedstock price and demand–supply rate is addressed in TEA (Hakeem et al., 2022). A brief TEA is necessary to attract more investors and pioneers in power generation sector for investing in biomass industries. Moreover, the process feasibility can be suggested keeping in mind the sustained feedstock availability and fluctuations in demand and selling cost. The major operating conditions like molar flow rate of inlet and outlet streams and energy consumed or heat produced in process are provided by mass balance and energy balance of process. This information's are pivotal to determine the standard size of equipment using CHEMCAD sizing database, Perry's Chemical Engineers' Handbook and John J. McKetta authored "Encyclopedia of Chemical Processing and Design". The actual equipment cost is a function of base cost, size of equipment, and scaling factor. It is defined by

$$\begin{pmatrix}\text{equipment}\\ \text{cost}\\ \text{in \$}\end{pmatrix} = \begin{pmatrix}\text{base}\\ \text{cost}\\ \text{in \$}\end{pmatrix} * \left[(\text{equipment capacity})/(\text{base capacity})\right]^{(\text{scaling factor})}. \quad (2)$$

In biomass-processing plants, the escalation of equipment cost is calculated using Chemical Engineering Plant Cost Index (CEPCI). The total capital investment for a typical process industry is reported in Table 1 (Peters et al., 2003).

Table 1 Summary of total capital investment for a typical process industry

Component	Average value of fixed capital investment (%)
Direct costs	
Purchased equipment	27.5
Installation of purchased equipment	10
Installation of measuring instruments and control system	7
Pipe material and laying	10.5
Installation of electrical systems	6
Building materials and construction	10
Yard improvements and maintenance	3.5
Service facilities	19
Land	1.5
Indirect costs	
Engineering and supervision	12
Plant construction	10.5
Expenses toward legal check	2
Contractor's service fee	4
Contingency	10

The four steps that are applied to TEA are the development of technology analysis plan (TAP), performance model creation, cost estimation, and report-making. A TAP is like a roadmap for implementing TEA. It gives shrewd ideas about the approach and methodology to be followed for conducting TEA. The process performance model is developed using material balance, energy balance, and momentum balance principles. The proposed models are subjected to constraints and assumptions that reduce complexity of equation and minimize the mathematical steps needed. Cost estimation is based on techno-feasibility report, environmental compliances to meet, and demand for the product. The report consists of detailed flow diagram, simulation model, CapEx, OpEx, and sensitivity study.

4.2 Analysis of Capital Expenses and Operating Expenses

In a biomass conversion industry, the profit margin can be increased when the process equations and economic models are integrated instead of handling them separately. The augmented model is called "techno-economic model" (TEM), which connect process parameters to finance metrics. This approach can help increase business output and understand the factors that significantly affect profit margin. TEM is useful for making progress in technology development lifestyle. When new ideas and technology is conceived, the innovators can find the economic feasibility and potential of the process using this technique. Chemical engineers are familiar with process modeling, simulation, equipment sizing, and economic calculations.

Instead of analyzing the technical performance and economic parameters of a bio-process separately, it is better to build a holistic model by integrating and leveraging technical and economical features. To develop TEM, a process model based on material and energy balance is developed first, followed by calculating equipment sizing variables. Once the technical phase is completed, the capital and operating costs are estimated and a user interface is created. The simulated results from process model are used to compute equipment size and operating condition. The parameters obtained in equipment sizing calculations are used to estimate capital and operating costs. A network of algorithms is used to link process variables and economic parameters to financial metrics. If the connections are automated, then the calculations are less time-consuming and models are user-friendly and viable tools for process analysis. Sensitivity analysis data obtained from model can completely eliminate or minimize operator intervention to control the process (AIChE, 2018). Capital expenses are the present investments made to harvest future monetary benefits and attain sustainability in production. The net CapEx is the product of gross CapEx and working interest (Belyadi et al., 2019)

Renewable energy sources like solar, wind, and geothermal are available abundantly in nature and there is no need to collect it manually and transport by vehicles. This is not the case for biomass, which requires collection, storage, and transportation. The investment cost depends on guaranteed long-term supply of biomass throughout the year at cheaper cost. Biomass feedstocks occupy 40% to 50% of the total investment cost. In India, the price for bagasse range between ₹1007 to ₹1175/ton, and the price of rice husks is around ₹1847/ton (International Renewable Energy Agency, 2012). There are some economic benefits if the conversion technology and output is scaled up. The capital cost of circulating fluidized bed (CFB) boiler combined with steam turbine is ₹ 12.42 lakh for producing 0.5 MW power. But when the power generation unit is scaled up to produce 8.8 MW power, the investment cost drastically reduces to ₹ 3.36 lakh. Operation and overhauling refer to the fixed and variable costs linked with the operation of biomass conversion plants. Fixed costs are generally expressed as percentage of capital costs, and in case of biomass power generation plant, the range of costs lies from 1% to 6% of the initial CapEx per year. The fixed operating cost is lower when the plant size is larger because of the influence of pay economies specifically for the labor required.

4.3 *Assessment of Economic Viability of Pilot Scale*

For an industrial process, economic viability study is important to decide whether the invested amount can be recovered without depreciation in a specific period. A project is considered as viable economically, if the monetary benefits and returns exceed the initial investments in a short duration. Viability study should cover production cost and should provide additional rate of return more than the invested capital amount. The initial success of biomass depends primarily on the support given by the government by announcing minimum support price (MSP) for biomass cultivators, biofuel producers, and other personnel involved in biomass conversion. The economic viability function depends on factors like cost price and quality of feedstocks, revenue from sales, cultivation and maintenance costs, market demand for the product, and stiff competition from other organizations.

5 Conclusion

Biomass, a viable and suitable alternative for fossil fuels, can reduce the carbon-related contaminants vented to the atmosphere, land, and ocean. The use of biomass for sustained power generation is eco-friendly and economically beneficial. The challenges associated with biomass production, procurement, conversion to power, and waste disposal can be met successfully if LCA and TEA is conducted meticulously well before the initiation of the project. The environmental impacts of a product and process can be minimized when LCA is performed in each and every phase of a process. The data obtained from LCA analysis can be used to develop process models, recommend suitable alternate strategies, and facilitate speedy decisions. Harvesting the maximum economic benefits, fulfilling environmental specifications, and meeting the market demand are achieved when LCA and TEA is performed scrupulously.

References

American Institute of Chemical Engineers (AIChE). (2018). https://www.aiche.org/resources/publications/cep/2018/january/techno-economic-modeling-new-technology-development

Belyadi, H., Fathi, E., & Belyadi, F. (2019). *Hydraulic fracturing in unconventional reservoirs: Theories, operations, and economic analysis* (2nd ed.). Gulf Professional Publishing. https://doi.org/10.1016/C2018-0-01643-6. ISBN 978-0-12-817665-8.

Chaudhary, P., Bansal, S., Sharma, B. B., Saini, S., & Joshi, A. (2024). Waste biomass-derived activated carbons for various energy storage device applications: A review. *Journal of Energy Storage, 78*, 109996. https://doi.org/10.1016/j.est.2023.109996

Ciacci, L., & Passarini, F. (2020). Life Cycle Assessment (LCA) of environmental and energy systems. *Energies, 13*(22), 5892. https://doi.org/10.3390/en13225892

Evangelisti, S., Lettieri, P., Borello, D., & Clift, R. (2014). Life cycle assessment of energy from waste via anaerobic digestion: A UK case study. *Waste Management, 34*(1), 226–237. https://doi.org/10.1016/j.wasman.2013.09.013

Gomathi, R., Priya, A. K., Gnanasekaran, L., Rajendran, S., & Hoang, T. K. A. (2024). Biomass and waste derived silica, activated carbon and ammonia-based materials for energy-related applications—A review. *Fuel, 355*, 129490. https://doi.org/10.1016/j.fuel.2023.129490

Hakeem, I. G., Sharma, A., Sharma, T., Sharma, A., Joshi, J. B., Shah, K., Ball, A. S., & Surapaneni, A. (2022). *Techno-economic analysis of biochemical conversion of biomass to biofuels and platform chemicals*. Society of Chemical Industry. https://doi.org/10.1002/bbb.2463

Huang, H., Yuan, X., & Wu, G. (2017). *Liquefaction of biomass for bio-oil products* (pp. 231–250). Springer. https://doi.org/10.1007/978-3-319-49595-8_11

Hussin, F., Hazani, N. N., Khalil, M., & Aroua, M. K. (2023). Environmental life cycle assessment of biomass conversion using hydrothermal technology: A review. *Fuel Processing Technology, 246*, 107747. https://doi.org/10.1016/j.fuproc.2023.107747

International Renewable Energy Agency (IRENA). (2012). *Biomass for power generation* (p. 39). https://www.irena.org/-/media/Files/IRENA/Agency/Publication/2012/RE_Technologies_Cost_Analysis-BIOMASS.pdf

Lamers, P., Roni, M. S., Tumuluru, J. S., Jacobson, J. J., et al. (2015). Techno-economic analysis of decentralized biomass processing depots. *Bioresource Technology, 194*, 205–213. https://doi.org/10.1016/j.biortech.2015.07.009

Osman, A. I., Mehta, N., Elgarahy, A. M., Al-Hinai, A., Al-Muhtaseb, A.'a. H., & Rooney, D. W. (2021). Conversion of biomass to biofuels and life cycle assessment: a review. *Environmental Chemistry Letters, 19*, 4075–4118. https://doi.org/10.1007/s10311-021-01273-0

Peters, M. S., Timmerhaus, K. D., & West, R. E. (2003). *Plant design and economics for chemical engineers* (5th ed., p. 240). McGraw-Hill Chemical Engineering Series, ISBN 0-07-119872-5.

Product carbon footprint and life cycle assessment services, TÜV SÜD. (2024). https://www.tuv-sud.com/en-in/landing/asmea/product-carbon-footprint-and-life-cycle-assessment

Sampson, R. N., Wright, L. L., Winjum, J. K., Kinsman, J. D., Benneman, J., Kürsten, E., & Scurlock, J. M. O. (1993). Biomass management and energy. *Water, Air, and Soil Pollution, 70*, 139–159. https://doi.org/10.1007/BF01104993

Singh, L., & Kalia, V. C. (2017). *Waste biomass management—a holistic approach*. Springer. https://doi.org/10.1007/978-3-319-49595-8

Wang, K., & Tester, J. W. (2023). Sustainable management of unavoidable biomass wastes. *Green Energy and Resources, 1*, 100005. https://doi.org/10.1016/j.gerr.2023.100005

Yu, Z., Ma, H., Liu, X., Wang, M., & Wang, J. (2022). Review in life cycle assessment of biomass conversion through pyrolysis-issues and recommendations. *Green Chemical Engineering, 3*(4), 304–312. https://doi.org/10.1016/j.gce.2022.08.002

Importance of Utilizing Fungus Rather Than Bacteria for Biomass Valorization

Shraddha Awasthi, Mohammad Izhar Alam, and Dan Bahadur Pal

Abstract Fungi are a novel and promising method for biomass valorization when compared to conventional bacterial methods. Fungi possess distinct physiological and biochemical properties. For example, they may produce a wide variety of extracellular enzymes, including hemicelluloses, ligninases, and cellulases, which are necessary for the efficient breakdown of complex organic molecules like lignocellulose. Fungi are valuable in various industrial contexts because, in contrast to bacteria, they can thrive in a wide range of conditions. Fungi are incredibly diversified in substrate utilization due to their wide metabolic pathways, enabling them to produce high-value compounds like medications, biofuels, and bioplastics.

More biomass is converted more quickly and thoroughly when fungi are used instead of bacteria because they are superior in substrate diversity, harsh condition tolerance, and enzyme production efficiency. The main topics of this chapter are the biochemical mechanisms underlying the degradation of fungal biomass, the production of valuable chemicals, and the sustainability of fungal activity over the long term. It also covers current technological applications, issues, and the discipline's future directions. Fungi become an important component of biomass valorization when higher yields, shorter processing times, and less environmental impact are obtained through the use of fungal capacity.

Keywords Biomass Valorization · Fungi · Enzyme production · Lignocellulose degradation · Cellulases · Ligninases · Hemicelluloses · Biofuels · Bioplastics · Pharmaceuticals · Metabolic pathways · Industrial applications · Sustainability

S. Awasthi (✉)
Department of Environmental Studies, MLNCE, University of Delhi, Delhi, India

M. I. Alam
Department of Environmental Studies, University of Delhi, Delhi, India

D. B. Pal
Department of Chemical Engineering, Harcourt Butler Technical University, Kanpur, Uttar Pradesh, India

D. B. Pal, S. L. Bansal (eds.), *Fungal Waste Biomass Management for Energy, Environment and Value-Added Products*, Fungal Biology,
https://doi.org/10.1007/978-3-031-82599-6_5

1 Introduction

1.1 Overview of Biomass Valorization

Biomass valorization is the process of converting organic waste products into commodities such as materials, chemicals, and biofuels. This strategy is crucial to enhancing sustainable development since it reduces waste, reduces greenhouse gas emissions, and lessens reliance on fossil fuels (Bozell & Petersen, 2010). As a renewable resource, biomass can be utilized to produce high-value chemicals and energy, thereby promoting a circular economy. It consists of agricultural wastes, forestry debris, and industrial by-products (Okolie et al., 2022). Pretreatment, fermentation, hydrolysis, and product recovery are the several steps in the multistep process of biomass valorization (De Bhowmick et al., 2018; Chandel et al., 2018). Pretreatment is necessary to break down the complex structure of lignocellulosic biomass, which is composed of cellulose, hemicellulose, and lignin. This step is necessary to make the carbohydrates accessible for hydrolysis and fermentation (Sarkar et al., 2012).

Enzymatic hydrolysis, often facilitated by cellulases and *hemicellulases*, transforms these carbohydrates into fermentable sugars. However, these carbohydrates are converted into fermentable sugars by enzymatic hydrolysis, which is frequently aided by *hemicellulases* and cellulases. Microorganisms ferment these sugars to create useful compounds or biofuels like ethanol (Mankar et al., 2021). Biomass valorization technologies can be broadly classified as thermochemical and biochemical processes, according to Sarvanan et al. (2023). Biochemical processes include microbial and enzymatic transformations, whereas thermochemical processes include pyrolysis, gasification, and combustion. The ability to convert microbial biomass into valuable products under mild and environmentally friendly conditions has gained significant attention (Hassan et al., 2019; Kumar et al., 2009).

1.2 Historical Context: Bacterial Versus Fungal Approaches

In the past, biomass valorization has been dominated by bacterial methods. The capacity of bacteria like *Clostridium thermocellum* and *Escherichia coli* to convert biomass into various compounds and biofuels has been the subject of much research. The quick growth rates, well-characterized genetics, and accessibility of sophisticated genetic engineering techniques are the reasons behind the favoring of these organisms (Demain, 2000).

It is especially efficient for bacteria to ferment simple sugars that come from biomass. For example, *E. coli* has been modified to make succinic acid and ethanol from glucose (Lee et al., 2007). On the other hand, it is commonly known that *Clostridium* species can ferment acetone, butanol, and ethanol (ABE) to produce solvents like butanol (Jones & Woods, 1986). However, when working with complex lignocellulosic substrates, bacterial systems often encounter challenges due to their relatively narrow substrate range and reduced tolerance to inhibitory chemicals commonly found in biomass hydrolysates.

However, fungi have emerged as a competitive alternative to biomass valorization due to their unique physiological and biochemical properties; it is commonly recognized that lignocellulosic biomass can be broken down by a wide range of extracellular enzymes produced by fungi such as *Aspergillus niger*, *Phanerochaete chrysosporium*, and *Trichoderma reesei*. *Hemicellulases*, *cellulases*, and *ligninases* are among the complex polymeric structures of biomass that these enzymes collectively break down into sugars that can be fermented. Table 1 shows the comparison of fungal and bacterial biomass valorization capabilities (Singhania et al., 2013a, b; Bayer et al., 2007).

The use of fungi in biomass valorization has several advantages. First, fungi are useful for a variety of industrial uses because they can adapt to a wide range of environmental factors, such as extreme pH and temperature (Howard et al., 2003). Second, because fungi have a variety of metabolic pathways, they can produce a wide range of high-value compounds, including medicines, biofuels, and bioplastics (Berka et al., 2011). Because of their adaptable metabolism, fungi can utilize a wide variety of substrates, including forestry waste, industrial by-products, and agricultural residues.

Moreover, fungi are widely recognized for their potent enzyme-producing capabilities. For example, *T. reesei* possesses a large amount of *cellulases* in its secretory capacity, which is essential for the efficient conversion of cellulose to glucose (Cherry & Fidantsef, 2003). This high enzyme-production efficiency translates to faster and more comprehensive biomass conversion processes with shorter processing times and higher yields.

The capacity to incorporate fungal processes into current industrial frameworks is another factor driving the move in biomass valorization toward fungal approaches. Conventional fermentation equipment may be used to quickly scale up fungal fermentation, and the finished products can be readily included in the existing supply chains (Ghorai et al., 2009). Moreover, fungal strains can now be modified in unique ways for improved performance and product specificity because of developments in fungal biotechnology, including genetic engineering and synthetic biology (Schindler, 2020).

Table 1 Comparison of fungal and bacterial biomass valorization capabilities (Singhania et al., 2013a, b; Bayer et al., 2007)

Feature	Fungi	Bacteria
Enzyme production	High (extracellular)	Moderate (intracellular)
Substrate diversity	Wide range	Limited
Environmental tolerance	Broad (extreme conditions)	Narrow
Metabolic pathways	Extensive	Limited
Industrial applications	Diverse	Specific
Energy efficiency	High	Moderate

2 Biochemical and Physiological Capabilities of Fungi

2.1 *Extracellular Enzyme Production* (Cellulases, Ligninases, Hemicelluloses)

Fungi are recognized for producing a variety of extracellular enzymes, which are essential for breaking down complex chemical molecules like lignocellulose. The three different extracellular enzymes that fungi produce, *cellulases*, *hemicellulases*, and *ligninases*, as well as their functions in biomass appraisal, are discussed in this section.

2.1.1 Cellulases

Enzymes known as *cellulases* convert cellulose, the most prevalent organic polymer on Earth, into glucose units. *Endoglucanases*, *β-glucosidases*, and *exoglucanases* (sometimes called *cellobiohydrolases*) are the three main forms of *cellulases*. *Endoglucanases* cleave internal bonds in cellulose chains to produce free chain ends. *Exoglucanases* on these free ends catalyze the hydrolysis of *cellobiose*, a disaccharide, into glucose by *β-glucosidases*. Fungi such as *Trichoderma reesei* are useful model organisms for many industrial applications because of their large cellulase production. Research has shown that cellulose or its derivatives, which function as inducers, control the expression of the cellulase genes in *T. reesei*. Large-scale biomass conversion operations have been made easier by the considerable increases in cellulase yields brought about by the genetic engineering of *T. reesei* and the modification of fermentation conditions (Berka et al., 2011).

2.1.2 Ligninases

Ligninases, which include *laccases, manganese peroxidases* (MnP), and *lignin peroxidases* (LiPs), are essential for breaking down lignin, a complex aromatic polymer that gives plant cell walls their structural integrity. To get to the cellulose and hemicellulose embedded in the lignocellulosic matrix, lignin must be broken down. White-rot fungi, such as *Pleurotus ostreatus* and *Phanerochaete chrysosporium* (Rytioja et al., 2014), are well known for their capacity to break lignin (Janusz et al., 2017). These fungi generate *ligninases*, which can break down lignin into smaller, easier-to-manage pieces by oxidizing its intricate structure. When nitrogen is scarce, lignin-degrading enzymes are expressed and their production increases (Hatakka, 2001; Janusz et al., 2017).

2.1.3 Hemicellulases

The class of enzymes that are known as Hemicellulases (like xylanohydrolases and glucuronoxylanases). Breaks down the heterogeneous polymer hemicellulose, which is made up of different sugars like galactose, mannose, and xylose. Similar to *mannanases*, *β-xylosidases*, and *xylanases*, *hemicelluloseases* are specialized enzymes that target different hemicellulose components. Two examples of fungi that produce *hemicellulases*, which are required for the full hydrolysis of lignocellulosic biomass, are *Penicillium funiculosum* and *Aspergillus niger*. Together, these enzymes degrade hemicellulose into simple sugars that can be fermented to create biofuels and other valuable substances. Because these fungi can produce a wide range of *hemicellulases* under different growing circumstances, they are commonly used in industrial operations (de Vries & Visser, 2001).

2.2 *Environmental Adaptability*

2.2.1 Versatility in Diverse Environmental Conditions

Fungi are highly adaptable and can thrive in diverse environmental conditions, making them extremely valuable in numerous biomass utilization processes. Their utility in industrial applications is due to their ability to thrive in harsh environments, including those with extreme pH levels, temperatures, and moisture content (Hölker et al., 2004).

pH Tolerance

Fungi may grow and survive in a variety of pH range environments, from very acidic to neutral and alkaline. This adaptability is necessary for the breakdown of lignocellulosic biomass, which typically occurs in a range of pH levels. For instance, because they can efficiently break down lignin at both acidic and neutral pH values, white-rot fungi like *Trametes versicolor* are helpful for a range of industrial operations (Friedrich et al., 2007). The fungal pH adaptation is caused by their ability to produce extracellular enzymes that are stable and active across a wide pH range. This particular characteristic fungus allows for continuous enzymatic activity without the need for strict pH control, which reduces operating costs and complexity in industrial applications (Kirk et al., 1978; Verma et al., 2021).

Temperature Tolerance

Additionally, fungi can survive in a variety of temperature ranges, including thermophilic (loving heat) and psychrophilic (loving cold) conditions. Specifically, *thermophilic fungi* are useful for biomass valorization because they can make enzymes that can withstand high temperatures. This temperature stability lowers the possibility of microbial contamination, a typical problem in large-scale biomass-processing, and speeds up enzymatic reactions. For instance, *Thermomyces lanuginosus* is very useful for breaking down hemicellulose in industrial settings because it generates *thermostable xylanases*, which work best at temperatures above 60 °C (Maheshwari et al., 2000).

Moisture Tolerance

Because they can adjust to different moisture conditions, fungi are especially useful in biomass valorization processes where substrate moisture content is subject to change. *Aspergillus oryzae* is one of the fungi that can grow and produce enzymes in both liquid-submerged and solid-state (low moisture) fermentation environments (Pandey et al., 2000a, b). This adaptability makes it possible to handle a variety of biomass feedstocks—including ones with low moisture content, like agricultural residues—efficiently. Compared to submerged fermentation, solid-state fermentation (SSF) produces better enzyme yields and requires less energy since it resembles the native habitat of many fungi, which makes it particularly beneficial for the valorization of lignocellulosic biomass (Pandey et al., 2000a, b).

3 Metabolic Pathways and Substrate Utilization

3.1 Pathways for High-Value Chemical Synthesis

Because of their exceptional metabolic adaptability, fungi can synthesize a wide range of valuable compounds from renewable biomass sources, including medicines, biofuels, and bioplastics. The complex metabolic pathways that underpin these actions are examined in this section, emphasizing the critical roles that fungus enzymes and metabolic intermediates play.

3.1.1 Biofuels Production

Biofuels such as bioethanol and biodiesel provide a sustainable substitute for fossil fuels since they are made from renewable biomass. Through the enzymatic breakdown of complex carbohydrates into fermentable sugars, which are then converted into ethanol or other biofuels, fungi play a crucial role in the generation of biofuels. *Alcohol dehydrogenases* (ADHs), hemicellulases, and cellulases are among the important enzymatic activities involved. The process of glycolysis, which turns glucose into pyruvate, is the first step in the metabolic pathway for producing bioethanol.

ADH enzymes next catalyze alcoholic fermentation, which produces carbon dioxide and ethanol. Fungi such as *Aspergillus niger* and *Trichoderma reesei* produce cellulases, which hydrolyze cellulose into glucose units. Plant cell walls include a stiff cellulose structure that must be broken down by an enzyme process. Hemicellulases break down hemicellulose components further into *xylose* and *mannose*, which are then fermented to make bioethanol (Himmel et al., 2007). Examples of these enzymes are *xylanases* and *mannanases.*

3.1.2 Bioplastics Synthesis

Bioplastics are sustainable alternatives to traditional plastics since they are biodegradable polymers made from renewable biomass sources. Through their metabolic processes, fungi produce *polyhydroxyalkanoates* (PHAs) and *polylactic acid* (PLA), which are essential for the synthesis of bioplastics. To produce PHA, acetyl-CoA—which is obtained from the metabolism of glucose through the citric acid cycle and glycolysis—must condense into different PHA monomers, including *polyhydroxybutyrate* (PHB) and *polyhydroxyvalerate* (PHV) (Anderson & Dawes, 1990). Under the right cultivation conditions, fungi such as *Aspergillus nidulans* and *Mucor circinelloides* can accumulate large amounts of PHA (Steinbüchel & Lütke-Eversloh, 2003). Contrarily, PLA is produced by fungal fermentation, which converts carbohydrates into lactic acid. Enzymatic and chemical reactions then polymerize the lactic acid into PLA. *Aspergillus awamori* and *Rhizopus oryzae* are two examples of fungal strains that have been modified to produce PLA effectively (De Albuquerque et al., 2021).

3.1.3 Pharmaceutical Compounds

Secondary metabolites produced by fungi are widely used in medicine as immunosuppressants, antibiotics, and anticancer medicines. These substances are produced via specific metabolic pathways that are different from those of primary metabolism, frequently in reaction to stressors or environmental cues. Penicillin, a widely used antibiotic, is produced by fungi belonging to the genus *Penicillium* through a complex biochemical route that involves multiple enzymatic steps (Demain, 2009). The fact that fungi like *Streptomyces* spp. and *Acremonium chrysogenum* can also synthesize additional antibiotics like *cephalosporins* and *macrolides* shows the breadth of fungal metabolic capabilities in pharmaceutical manufacturing. To produce immunosuppressants like *cyclosporine A*, *Tolypocladium inflatum* employs a non-ribosomal peptide synthesis pathway including specific enzymes and precursor components. These compounds have transformed medical treatments, showcasing the impact of fungal metabolic diversity on the development of new pharmaceuticals by comparing the diversity of metabolisms in fungi and bacteria (Bushley et al., 2013).

3.2 Comparison of Fungal and Bacterial Metabolic Diversity

Fungi and bacteria are two distinct groups of microorganisms with diverse metabolic capabilities, enabling them to thrive in various ecological niches and significantly contribute to industrial processes.

3.2.1 Metabolic Diversity

The ability of fungi to produce a wide variety of enzymes that can break down complicated chemical molecules like lignocellulose is well known. Species such as *Trichoderma reesei*, which yields a range of cellulolytic enzymes essential for the breakdown of biomass, are prime examples of this enzymatic diversity. These enzymes, which include lignin-modifying enzymes (LMEs), cellulases, and hemicellulases, allow fungi to effectively break down plant cell walls into sugars that may be fermented. On the other hand, bacteria frequently display highly specialized metabolic pathways designed to make use of certain substrates found in their environments. For example, *Pseudomonas* species are useful in environmental bioremediation because of their well-known capacity to break down hydrocarbons. These bacteria have genes that encode enzymes that help break down complex organic contaminants, such as *dioxygenases* and *alkane hydroxylases* (Dunlop, 2011).

3.2.2 Substrate Utilization

When it comes to using their substrates, fungi are remarkably adaptable. They may use a wide variety of carbon sources, from cellulose and lignin to more complex biopolymers like chitin. Its flexibility is essential to fungi's functions in the cycle of nutrients and the resilience of ecosystems. Fungal metabolic pathways are also important for industry and ecology because they produce secondary metabolites like *mycotoxins* and *antibiotics*. Although they tend to have more specialized substrate preferences, bacteria are excellent at efficiently utilizing simpler carbon molecules. *Escherichia coli*, for instance, has undergone considerable engineering to generate a range of biofuels and biochemicals from renewable feedstocks, including *glucose* and *glycerol. Bacteria* are desirable candidates for industrial bioprocesses that need high substrate conversion rates because of their metabolic flexibility and quick proliferation (Steinbüchel & Lütke-Eversloh, 2003).

3.2.3 Efficiency and Yield of Product Synthesis

Because of differences in their metabolic methods and regulatory mechanisms, fungi and bacteria differ greatly in their efficiency in product synthesis. Fungi are used in biocatalysis and medicines because they frequently produce larger amounts of secondary metabolites and enzymes. For example, *Aspergillus niger*'s effective metabolic pathways and high citrate synthase activity are used in the industrial synthesis of citric acid (de Vries & Visser, 2001). On the other hand, bacteria are preferred because of their quick rates of development and streamlined metabolic pathways that are tailored for particular industrial uses. For example, *Corynebacterium glutamicum* is used in the manufacture of amino acids because, in regulated environments, it can effectively convert sugars into amino acids (Wendisch et al., 2006). In a similar vein, *polyhydroxyalkanoates* (PHAs), biodegradable polymers derived from renewable resources, are produced using *Ralstonia eutropha* (Steinbüchel & Lütke-Eversloh, 2003).

4 Biochemical Processes in Fungal Biomass Breakdown

4.1 Mechanisms of Lignocellulose Degradation

The main components of lignocellulose-containing raw materials are cellulose, hemicellulose, and lignin; however, these feedstocks also contain trace amounts of pectin, nitrogen compounds, and mineral leftovers. The levels of the listed elements vary depending on the origin. Because of its compact structure and the many bonds between cellulose, hemicellulose, and lignin, lignocellulose is a very difficult substrate for enzymes. Hydrogen bonds are the primary means of coupling those constituents. Chemical linkages, which are mainly those between galactose and arabinose residues and carbohydrates, are found between hemicellulose and lignin (Andlar et al., 2018).

Fungi, primarily white-rot and brown-rot basidiomycetes, are used in the biological degradation of raw materials containing lignocellulose. It uses a portion of the plant polysaccharides and has a lengthy application period with a fungal breakdown rate that is too slow for industrial use. Table 2 shows the key enzymes involved in fungal biomass breakdown (Cherry & Fidantsef, 2003; Lynd et al., 2002).

Enzymatic degradation is faster and more selective than fungal degradation, although it costs more when done on a big scale. Multiple carbohydrate-active enzymes, which typically work in tandem with complementary, synergistic activities and modes of action, are required for the enzymatic degradation of raw materials containing lignocellulose (Andlar et al., 2018). The primary enzymes and enzymatic breakdown mechanisms of the principal lignocellulose constituents are discussed in the sections that follow.

Table 2 Key enzymes involved in fungal biomass breakdown (Cherry & Fidantsef, 2003; Lynd et al., 2002)

Enzyme	Function	Fungal source
Cellulases	Hydrolyze cellulose into glucose	*Trichoderma reesei*
Ligninases	Degrade lignin	*Phanerochaete chrysosporium*
Hemicellulases	Break down hemicellulose into simple sugars	*Aspergillus niger*
Pectinases	Decompose pectin in plant cell walls	*Aspergillus* spp.
Amylases	Convert starch into maltose	*Aspergillus oryzae*

4.1.1 Cellulose Degradation

β-1,4-linked glucose molecules make up the linear polymer known as cellulose. Three different forms of cellulases work together to break down cellulose in the presence of fungi: endoglucanases, exoglucanases (also known as *cellobiohydrolases*), and *β-glucosidases. Endoglucanases* produce free chain ends by cleaving the internal bonds in the cellulose chains. After acting on these free ends, *exoglucanases* release cellobiose, which *β-glucosidases* hydrolyze into glucose (Lynd et al., 2002). This group of enzymes should be enough to break down cellulose entirely into glucose in theory, and commercial cellulase mixtures usually contain these enzymes in different amounts. However, hydrolytic enzymes only target a fraction of the crystalline portion of cellulose, and as a result, the destruction of this portion of cellulose happens slowly and incompletely. Early on in the study of cellulose degradation, it was hypothesized that full hydrolysis could not occur without extra ingredients. Reese first proposed the C1-Cx two-enzyme concept in the 1950s. He realized that the activity of a component ("C1") shortens native cellulose without producing soluble *cellulose-oligomers*, which facilitates the terminal hydrolyzing step ("Cx" activity) that produces small soluble molecules like glucose. Both of the enzyme's components were unknown at the time, but some of their hydrolyzing elements had been found and were grouped under the name "Cx" (Morgenstern et al., 2014). Later, the idea was broadened by mentioning how the structure of crystalline cellulose is disrupted by an unidentified oxidative enzyme (Morgenstern et al., 2014).

Efficient generation of cellulases is a well-known characteristic of fungi like *Trichoderma reesei*. The presence of cellulose or its derivatives, which function as inducers, affects the control of cellulase gene expression in *T. reesei*. Cellulase production in *T. reesei* has been greatly increased through genetic changes and fermentation environment optimization, making it easier to use in industrial biomass conversion processes.

4.1.2 Hemicellulose Degradation

A heterogeneous set of polysaccharides, comprising *hexoses* (glucose, mannose, and galactose), *pentoses* (xylose and arabinose), and a few other saccharides (*rhamnose* and *fructose*), make up hemicellulose, a biopolymer (Wang et al., 2007).

Hemicellulose also contains a small amount of acetyl groups and uronic acids. A vast range of hemicellulose polysaccharides with various compositions and architectures can be produced by assembling various functional groups. Hemicellulose can be divided into four main classes according to the contents of these sugar units: *mannan*, *xyloglucan*, *mixed-linkage β-glucans*, and *xylan* (Sun et al., 2022). In hardwoods and herbaceous plants, *xylan* is the most prevalent hemicellulose component, whereas *mannan* is the predominant one in softwoods.

Hemicellitate degradation is quite challenging and requires a solution. The enzyme responsible for breaking down hemicellulose is called *hemicellulase*. The glycosyl hydrolase (GH) families, which include GH2, GH10, GH11, GH16, GH26, GH30, GH31, GH39, GH42, and GH43, are groups of enzymes having hemicellulase activity (Sun et al., 2022).

Hemicellulases are produced in large quantities by fungi like *Penicillium funiculosum* and *Aspergillus niger*. Together, these enzymes break down hemicellulose and increase the overall effectiveness of the breakdown of lignocellulose. These fungi are useful in industrial applications because they may produce a wide range of hemicellulases under different growing circumstances (Polizeli et al., 2005).

4.1.3 Lignin Degradation

The most prevalent aromatic biopolymer on Earth, lignin, is incredibly resistant to breakdown. It inhibits the passage of lignocellulolytic enzymes into the interior of the lignocellulosic structure by binding to both hemicellulose and cellulose, forming a barrier to any solutions or enzymes. Lignin can be effectively broken down by some basidiomycetes white-rot fungi by combining extracellular ligninolytic enzymes with organic acids, mediators, and auxiliary enzymes.

Enzymes referred to as "ligninases" are secreted by fungi, which break down lignin. Ligninases fall into one of two categories: heme peroxidases (which include versatile peroxidase (VP), manganese peroxidase (MnP), and phenol oxidases (laccase)). Generally speaking, *peroxidases* employ hydrogen peroxide as a co-substrate, whereas laccases use molecular oxygen as an electron acceptor (Dashtban et al., 2010). *White-rot fungi* release additional substances required for efficient lignin breakdown together with one or more of the lignin-modifying enzymes (LMEs). It has been demonstrated that *P. chrysosporium* lacks laccase but produces several *UP and MnP isoenzymes*. In accordance, *P. chrysosporium*'s genome has five *MnP* and ten *UP genes*. *Glyoxal oxidase (GLOX)*, an enzyme that produces H_2O_2, has also been discovered in *P. chrysosporium cultures*. The most effective lignin-degrading microorganisms that have been examined are white-rot *basidiomycetes*, including *P. chrysosporium*, *T. versicolor*, and *Coriolus versicolor*. Several lignin-degrading fungi lack LiP, even though it may oxidize the non-phenolic portion of lignin, which makes up 80–90% of its composition (Dashtban et al., 2010).

4.1.4 Synergistic Action of Enzymes

Organisms break down lignocellulose by a variety of processes that involve several complimentary enzymes, especially *glycoside hydrolases* (GHs). Both the Fenton chemistry of *brown-rot fungus* and the *lytic polysaccharide monooxygenases* (LPMOs) generated by bacteria and fungi aid in deconstruction. *White-rot fungi* and certain bacteria use *laccases* and *peroxidases* to depolymerize lignin. The intricacy of bacterial degradative activity in lignocellulose-rich settings is currently being revealed by meta-omics. Various oomycetes and termite gut protozoa produce lignocellulolytic enzymes in diverse forms. While certain GHs are secreted by lignocellulosic-consuming animals, the majority of them are home to a varied gut microbiota that secretes enzymes, a mutualism that is especially intricate in termites. On the other hand, microorganisms that secrete GH and LPMO are found inside shipworms, apart from the isopod and the digestive site. Only endogenous enzymes are necessary for *limnoria* (Cragg et al., 2015).

4.2 Specific Enzymes Involved and Their Roles

Multiple enzymes that are active in carbohydrates are used to break down lignocellulose-containing raw materials enzymatically. These enzymes typically work in tandem with each other, exhibiting complimentary and synergistic activities and modes of action. The primary enzymes and enzymatic breakdown mechanisms of the principal lignocellulose constituents are discussed in the sections that follow.

4.2.1 Cellulases

Three hydrolytic enzymes are thought to make up the cellulase system in fungi: (i) *endo-(1,4)-b-D-glucanase* (also known as *endoglucanase, endo-cellulase*, and *carboxymethyl cellulase*), which cleaves b-linkages randomly, usually in the amorphous parts of cellulose; (ii) *exo-(1,4)-b-D-glucanase* (also known as *cellobiohydrolase, exocellulase, microcrystalline cellulase*, and *Avicelase*), which releases cellobiose from either the non-reducing or the reducing end, usually from the crystalline parts of cellulose; and (iii) *b-glucosidase* (also known as *cellobiase*), which releases glucose from *cellobiose* and *short-chain cello-oligosaccharides*. Despite not directly interacting with cellulose, b-glucosidase is recognized as a part of the *cellulase* system because of its ability to promote cellulose hydrolysis (Bhat & Bhat, 1997).

Use of Cellulolytic Enzymes in Baking

When making dough, cellulases can be added to give it a nice texture and quality. Because *endoglucanases* can help with pentosan hydrolysis, they are employed. *Pentoses* are tiny components of wheat, but because of their great water-holding capacity, they play a major role in determining the rheology of the dough and the quality of the bread. The mass's elasticity increases with the quantity of soluble *pentosans* (De Souza & Kawaguti, 2021).

Use of Cellulases in the Beverage Industry
In the procedures involved in producing beverages, cellulases and hemicellulases are employed singly or in combination with other enzymes. Fruit and vegetable juice extraction and clarification involve the use of several *cellulolytic* and *hemicellulolytic enzymes. Endoglucanases* can be added to the fermentation process to facilitate the hydrolysis of glucan, which lowers the viscosity of the wort and increases its filterability in fermented beverages like beer (De Souza & Kawaguti, 2021).

4.2.2 Hemicellulases

Hemicellulases are highly branched polymers that contain various monomers of *pentoses*, such as *beta-D-xylose* and *alpha-L-arabinose*, and *hexoses*, such as *beta-D-glucose, alpha-D-galactose,* and *beta-D-mannose. Hemicelluloses* strengthen the primary cell walls of plants through their interactions with cellulose.

The solubility of hemicelluloses in water or alkalis is attributed to their side chains, which are made up of acetic acid, pentoses, hexuronic acids, and deoxyhexoses like *alpha-L-rhamnose* and *alpha-L-fucose* (Bonechi et al., 2017; Toushik et al., 2017).

Xylanases The most prevalent structural polysaccharide in nature is xylan. A range of hydrolytic enzymes must work together to completely break it down: the *en-do xylanases*, which cleave *b-1,4-linked xylose* (the xylan backbone) at random; the *β-xylosidases*, which hydrolyse *xylooligomers*; and the various side-branch-splitting enzymes, such as *α-glucuronidase* and *α-arabinosidase*, *acetyl xylan esterase*, and *acetyl esterase*, which release other sugars (glucuronic acid arabinose) that are attached as branches to the backbone (Biely, 1985).

Mannanases *Mannans* are widely distributed in plant tissues and are the primary constituents of the hemicellulosic fraction of softwoods. The main enzymes that break down mannan are called *β-mannanases*, and they are produced by a variety of organisms, including bacteria, fungi, actinomycetes, and plants. These enzymes are capable of operating in a broad pH and temperature range.

Fruits, plant seeds, and softwood hemicelluloses have higher concentrations of *mannans.* The polymers known as *mannans* are divided into four groups:

galactomannans, which are made up of a main chain formed by *mannoses* and branches with galactoses linked in alpha-1,6; linear mannans, which are made up of *beta-1,4-linked mannose units*; *glucomannans*, which are made up of the main chain containing beta-1,4-linked mannose and glucose residues; and *galactoglucomannans*, which have the main chain containing *beta-1,4-linked mannoses* and glucose as well as alpha-1,6-linked galactose branches. The most thermally and chemically sensitive structure, hemicellulose, joins lignin to cellulose fibers, giving the cellulose–hemicellulose–lignin complex more stability, flexibility, and elasticity (De Souza & Kawaguti, 2021).

Use of *Hemicellulases* in the Beverage Industry

Xylanases can be employed in combination with cellulases to decrease viscosity, clarify fruit juice, and hydrolyze hemicellulose in fruit peels. Hydrolytic enzymes can gradually break down the middle lamella between cells by weakening the cell wall. This allows intracellular materials, such as water, to be released, which improves juice recovery. *Xylanases* are used to treat fruit juices because they release units that reduce carbohydrates. This improves pulp processing and increases fruit yield contents. Accordingly, the amount of reducing sugars released serves as a marker for the enzyme's breakdown of hemicellulosic materials (De Souza & Kawaguti, 2021).

4.2.3 Ligninases

By secreting enzymes known as *ligninases*, or lignin-modifying enzymes, fungi break down lignin. These enzymes are divided into two groups: laccases, which are phenol oxidases, and *heme-peroxidases*, which include lignin peroxidases, manganese peroxidases, versatile peroxidases, and dye *decolourizing peroxidases*. *Laccases* are multi-copper *oxidoreductases*, an equally important enzyme required for completely breaking down lignin, while *peroxidases* are *heme-containing enzymes* with catalytic cycles that are activated by peroxide. *Accessory oxidases* also produce hydrogen peroxide required for peroxidase activity. *Heme-containing and* H_2O_2*-dependent lignin peroxidase (LiP), manganese peroxidase (MnP), multifunctional peroxidases*, and *Cu-containing laccase* are the main enzymes that break down lignin (Gunjal et al., 2020).

Lignin Peroxidases (LiPs) *Lignin peroxidases* (*LiPs*) are glycosylated ligninolytic biocatalysts that contain heme and catalyze the oxidative breakdown of lignin and lignin-related recalcitrant aromatic compounds into a variety of end products. Numerous isozymes of *LiP* have been identified. For example, in the culture broth of *P. chrysosporium BKM-F-1767*, Farrell and colleagues found the existence of six *LiP isozymes*, namely H1, H2, H6, H7, H8, and H10. Similarly, the same strain's five *LiP isozymes* were characterized, varying in terms of substrate selectivity, stability, isoelectric point, and sugar content. This monomeric hemoprotein has a molecular weight of 38 to 43 kDa and an isoelectric point range of 3.3 to 4.7. It functions best at an acidic pH (around 3.0) when veratryl alcohol is present as the substrate. LiPs with a high redox potential can oxidize a variety of non-phenolic lignin structures, such as *acylglycerols-aryl* ethers, which account for over 90% of lignin degradation. Their oxidative characteristics include the creation of a radical cation based on one-electron oxidation, which can cause side-chain breakage, intramolecular addition, demethylation, and rearrangements. Notably, the stability, substrate selectivity, and catalytic efficiency of the enzyme are altered when different tryptophan residues are present on the surface of the enzyme molecule (Bilal & Iqbal, 2020).

Manganese Peroxidases (MnPs) MnPs use Mn2+ as a common electron donor and have a catalytic cycle resembling that of other heme-containing *peroxidases* like LiP or HRP (Horseradish Peroxidase). By concurrently oxidizing Mn2+ to a highly reactive Mn3+, MnP catalyzes the preferential oxidation of phenolic structures to phenoxy radicals in an H_2O_2-assisted enzyme reaction. This highly reactive Mn3+ can then degrade a variety of phenolic molecules, such as phenols, cleaves, and dyes. However, with the introduction of low-molecular-weight mediators, such as ethyl or lipid radicals, the degrading potential of MnPs can be extended to non-phenolic structures. Moreover, the roles of these biocatalysts in fungal ligninolysis have been expanded by numerous additional proteins that work in concert with MnPs. Many studies on the MnP-assisted oxidative depolymerization of synthetic and natural lignins as well as various other refractory compounds were found in the literature review (Bilal & Iqbal, 2020).

Laccases Enzymes called *laccases*, which contain copper, catalyze the oxidation of phenolic compounds, which is followed by the reduction of oxygen to water. Every laccase oxidizes substrates in a different range. One electron is removed from the phenolic nucleus to oxidize it, producing a phenoxy-free radical product that then goes through polymerization. Compost has been shown to have laccase-like activity. The strong oxidation capacity of laccases from microbial sources, particularly those from WRF (White-Rot Fungi), and their broader range of substrate specificity have drawn a lot of attention. These enzymes are classified into three classes, type 1 (blue), type 2 (normal), and type 3 (coupled binuclear), based on the presence of copper centers. Types 1 and 2 each have one Cu atom, while type 3 has two Cu atoms. The type 1 copper ion's oxidation-reduction potential, which is where substrate oxidation occurs, is essential to laccases' biocatalytic activity. *Microbial laccases* have a higher redox potential than plant-based enzymes, which suggests that they have more catalytic activity and efficiency than enzymes derived from plants. In contrast, types 2 and 3 transmit electrons from T1 to the tri-nuclear site to form a tri-nuclear cluster (T2/T3) for molecular oxygen reduction to water. Typically, the oxidation of four substrates and the reduction of oxygen to two molecules of water are involved in the biocatalytic reaction of *laccases*. *Laccase-mediated* catalysis benefits more from the use of ambient oxygen as an electron acceptor than from *peroxidases*' usage of H_2O_2. However, to effectively remove a variety of resistant and non-phenolic environmental pollutants, these enzymes require the utilization of redox mediators, such as *acetosyringone*, *2,2′-casino-bis (3-ethylbenzothiazoline-6-sulphonic acid) (ABTS)*, *syringaldehyde (SA)*, *vanillin*, and *1-hydroxy benzotriazole*. Notwithstanding this drawback, *laccases* have garnered significant attention because of their enormous biotechnological potential (Bilal & Iqbal, 2020).

5 Industrial Applications and Technologies

5.1 Current Applications in Various Industries

Fungal technology is utilized extensively in the manufacture of chemicals and biofuels due to its advantages in lignin breakdown, saccharification, fermentation, and lipid accumulation. It offers a gentle, low-cost, ecologically benign option for biomass conversion. To assist us in comprehending the proper application of fungal technologies in biomass conversion, it is necessary to identify the reasons and effects of doing so. Table 3 shows the industrial applications of fungal biomass valorization (Pandey et al., 2000a, b; Cherubini, 2010).

5.1.1 Biotechnology Industry

Fungi are widely used in the biotechnology sector to produce enzymes, antibiotics, and other useful substances. The synthesis of cellulases and hemicellulases by *Trichoderma reesei* for the bioconversion of lignocellulosic biomass into fermentable sugars, which are then utilized in the synthesis of bioethanol, is one of the most noteworthy examples (Singhania et al., 2010).

Furthermore, the generation of citric acid—an essential organic acid utilized in the food and pharmaceutical industries—is facilitated by fungi such *as Aspergillus*

Table 3 Industrial applications of fungal biomass valorization (Pandey et al., 2000a, b; Cherubini, 2010)

Application	Fungal strain	Product	Benefits
Biofuel production	*Saccharomyces cerevisiae*	Ethanol	Renewable energy source
Bioplastics	*Aspergillus niger*	Polyhydroxyalkanoates (PHAs)	Biodegradable plastics
Pharmaceuticals	*Penicillium chrysogenum*	Antibiotics	Medical advancements
Enzyme production	*Trichoderma reesei*	Industrial enzymes	Cost-effective production
Bioremediation	*Phanerochaete chrysosporium*	Degradation of pollutants	Environmental cleanup

niger. Because *A. niger* fermentation is so effective, it is the method of choice for producing citric acid on a large scale (Papagianni, 2007).

5.1.2 Agriculture Industry

Fungi are used as biopesticides and biofertilizers in agriculture to increase crop productivity and guard against pests and diseases. By developing symbiotic relationships with plant roots, *mycorrhizal fungi* increase phosphorus uptake and promote plant development. To promote eco-friendly farming, *mycorrhizal fungi* are used in sustainable agriculture practices to assist decrease reliance on chemical fertilizers (Smith & Read, 2008).

Additionally, fungi that are employed as biocontrol agents against insect pests include *Beauveria bassiana* and *Metarhizium anisopliae*. By infecting and killing insects, these entomopathogenic fungi provide a sustainable substitute for chemical pesticides. These fungi are useful in integrated pest control techniques because of their efficiency and selectivity in addressing pests (Faria & Wraight, 2007).

5.1.3 Pharmaceutical Industry

Fungal metabolites, which function as antibiotics, immunosuppressants, and cholesterol-lowering agents, are very beneficial to the pharmaceutical business. Since the discovery of penicillin from *Penicillium notatum*, fungi have been the source of many antibiotics, including cephalosporins and griseofulvin (Demain & Sanchez, 2009). This discovery transformed modern medicine.

In organ transplantation, immunosuppressants such as cyclosporine—which is generated by *Tolypocladium inflatum*—are essential because they inhibit the immune system and avert organ rejection. Furthermore, lovastatin, which is generated by *Aspergillus terreus*, is one of the statins that is commonly used to decrease cholesterol and prevent cardiovascular disorders (Manzoni & Rollini, 2002).

5.1.4 Bioenergy Industry

The bioenergy sector relies heavily on fungi, especially when it comes to the lignocellulosic biomass that is used to produce biofuels. Fungal cellulases and hemicellulases catalyze the enzymatic hydrolysis of biomass to produce fermentable sugars, which microbial fermentation subsequently transforms into bioethanol. By lowering greenhouse gas emissions and reliance on nonrenewable energy sources, this method provides a sustainable substitute for fossil fuels (Lynd et al., 2002).

Fungi are involved in the anaerobic digestion process, which produces biogas in addition to bioethanol. Anaerobic digestion methods can produce more biogas and be more efficient when certain fungal species are used to break down complex organic materials (Arora et al., 2016).

5.1.5 Food and Beverage Industry

Fungi are widely used in fermentation processes in the food and beverage sector to create a wide range of goods. Three types of yeast are used in bread, beer, and wine production: *Saccharomyces cerevisiae*. Yeasts' fermentation skills aid in the formation of food products' flavor and texture as well as their preservation (Fleet, 2007).

Additionally, traditional fermented foods like miso, tempeh, and soy sauce are made using fungi like *Aspergillus oryzae* and *Rhizopus oligosporus*. These fungi improve the nutritional content and digestibility of fermented foods by aiding in the breakdown of complex dietary components (Steinkraus, 2002).

5.1.6 Environmental Applications

In environmental applications, fungi are also essential, especially in bioremediation. Some fungi are capable of breaking down and purifying various environmental contaminants, such as pesticides, heavy metals, and hydrocarbons. A useful tool in the remediation of contaminated settings is the white-rot fungus *Phanerochaete chrysosporium*, which is well known for its capacity to break down lignin and a variety of xenobiotic chemicals (Harms et al., 2011).

5.2 *Technological Advancements and Innovations*

In recent years, there have been notable technological developments and innovations in the use of fungi for biomass valorization. The efficiency and economic viability of turning biomass into useful products like biofuels, bioplastics, and other biochemicals have improved as a result of these advancements. Some significant technological developments in the field of fungal biomass valorization are examined in this section.

5.2.1 Genetic Engineering of Fungal Strains

The development of fungi's potential for industrial use has been greatly aided by genetic engineering. Researchers have been able to boost the production of particular enzymes involved in the breakdown of biomass by altering the genomes of fungal species. For instance, *Trichoderma reesei* has been genetically modified to overproduce cellulases, which greatly increases the fungus's capacity to convert cellulose into sugars that can be fermented (Peterson & Nevalainen, 2012).

Progress in CRISPR/Cas9 technology has enabled more accurate genetic engineering of fungi. According to Liu et al. (2015), this method enables the targeted interruption or augmentation of particular genes, which results in the generation of fungal strains with optimized metabolic pathways for the synthesis of biofuels and other bioproducts.

5.2.2 Enzyme Optimization and Production

Another important area of technical improvement has been the optimization of enzyme manufacturing. Scientists have devised strategies to improve fungal enzyme stability and activity in industrial settings. Enzyme variations with increased temperature stability and catalytic performance have been produced through methods like directed evolution and protein engineering (Cherry & Fidantsef, 2003).

In microbial hosts like *Escherichia coli* and *Pichia pastoris*, the large-scale manufacture of fungal enzymes has also been made possible by the application of recombinant DNA technology. According to Nevalainen et al. (2014), this strategy has greatly decreased the cost of producing enzymes and boosted the supply of high-quality enzymes for processes involving the conversion of biomass.

5.2.3 Solid-State Fermentation (SSF) Technologies

Fungal biomass valorization appears to be a viable use of solid-state fermentation (SSF). To replicate their natural environment, SSF entails the development of fungi on solid substrates with minimal moisture content. Because it frequently generates better yields than submerged fermentation, this approach is especially beneficial for the generation of enzymes and other secondary metabolites (Pandey et al., 2000a, b).

The creation of bioreactors that offer ideal conditions for fungal growth and enzyme production is one example of an innovation in SSF technology. According to Couto and Sanromán (2006), these bioreactors are made to maintain the proper humidity, temperature, and aeration levels to guarantee reliable and effective fermentation operations.

5.2.4 Bioinformatics and Omics Technologies

The study and use of fungus in biomass valorization have been transformed by the fusion of bioinformatics and omics technology. Proteomics, metabolomics, transcriptomics, and genomics offer a wealth of information about the metabolic networks and regulatory processes of fungi. According to Bischof et al. (2016), this information is essential for identifying the important genes and enzymes involved in the breakdown of biomass and for developing plans to increase their expression and activity.

The identification of new enzymes and metabolic pathways in fungi has been made possible by high-throughput sequencing and analysis technologies. New biocatalysts have been developed as a result of these discoveries, and they may find use in a variety of industrial processes, such as the generation of biofuel and bioremediation (Lynd et al., 2002).

5.2.5 Biorefinery Integration

The integration of biorefineries has become popular as a comprehensive strategy for biomass valorization. Similar to a petroleum refinery, a biorefinery processes biomass to produce a variety of products from a single feedstock. By converting lignocellulosic biomass into fermentable sugars, which can subsequently be transformed into biofuels, bioplastics, and other compounds, fungi play a critical part in this process (Kamm & Kamm, 2004).

The creation of integrated processes combining microbial fermentation, enzymatic hydrolysis, and fungal pretreatment is one of the most recent developments in biorefinery technologies. These integrated methods lessen the manufacturing of bioproducts' negative environmental effects while increasing the overall efficiency of biomass conversion (Cherubini, 2010).

6 Challenges and Future Directions

6.1 Technical and Economic Challenges

Although fungi have a bright future in biomass valorization, several technological and financial obstacles need to be overcome before their full commercial application can be realized. This section addresses some of the major issues and looks at possible ways to get over them in the future.

6.1.1 Technical Challenges

Enzyme Production and Activity

The effective synthesis and activity of enzymes is one of the main technological obstacles in the fungal biomass valorization process. The production of numerous enzymes required for the breakdown of lignocellulose by fungi is possible, but several variables can influence the quantity and activity of these enzymes. For example, the growth medium's composition, pH, temperature, and the presence of inducers or repressors can all affect the development of cellulases and hemicellulases (Juturu & Wu, 2014). To maximize the production and activity of enzymes, it is imperative to optimize these circumstances.

Furthermore, one major challenge is the stability of fungal enzymes in industrial settings. Enzymes are frequently needed in industrial operations where they must operate in harsh pH ranges, high temperatures, or the presence of several inhibitors.

Research on engineering enzymes to improve their stability and activity in these circumstances is essential (Cherry & Fidantsef, 2003).

Pretreatment of Biomass

For the enzymatic hydrolysis of cellulose and hemicellulose to occur efficiently, lignocellulosic biomass must be effectively pretreated. Nevertheless, some disadvantages of the pretreatment techniques currently in use, such as steam explosion, acid hydrolysis, and alkaline pretreatment, include high energy requirements, the production of inhibitory compounds, and the requirement for costly chemicals (Kumar et al., 2009). The development of pretreatment techniques that are both economical and ecologically sound is still a major technical problem.

Genetic Engineering of Fungal Strains

Even though genetic engineering has enhanced fungi's potential for biomass valorization, some obstacles still need to be overcome. It can be difficult and time-consuming to genetically modify fungi, and adding foreign genes can occasionally cause unexpected metabolic changes or slower growth rates (Liu et al., 2015). Furthermore, great thought must be given to the safety and regulatory implications of utilizing genetically modified organisms (GMOs) in industrial processes.

6.1.2 Economic Challenges

Cost of Enzyme Production

The primary economic obstacle to the widespread implementation of fungal biomass valorization is the elevated expense associated with enzyme manufacturing. The cost of producing fungal enzymes is influenced by several processes, such as formulation, fermentation, and downstream processing (Singhania et al., 2013a, b). The creation of more effective production processes, the use of less expensive substrates for the synthesis of enzymes, and the recycling of enzymes are some methods to lower these expenses.

Market Competition

Fungal biomass valorization processes have to contend with items and technology that are already well established in the market. For example, fungal biomass conversion produces biofuels that need to compete with fossil fuels and other renewable energy sources. In a similar vein, petroleum-based goods must face competition from bioplastics and other biochemicals produced from fungi. A major economic hurdle is achieving cost parity with these well-established items (Cherubini, 2010).

Scale-Up and Commercialization

There are many obstacles to overcome when transferring laboratory-scale procedures to commercial production scale. Maintaining process consistency, optimizing bioreactor design, and guaranteeing product quality are just a few of the technical challenges that come with scaling up from bench-scale to industrial-scale operations (Kamm & Kamm, 2004). These are also major capital investments. Furthermore, obtaining the governmental licenses necessary for activities on a commercial scale can be expensive and time-consuming.

6.1.3 Future Directions

Advanced Genetic Engineering Techniques

Utilizing cutting-edge genetic engineering methods like CRISPR/Cas9 can help generate fungal strains with improved biomass valorization capabilities. With the use of these techniques, the fungal genome can be precisely modified to include advantageous characteristics including higher enzyme production, enhanced inhibitor tolerance, and improved metabolic pathways that lead to the generation of biofuels and biochemicals (Liu et al., 2015).

Integrated Biorefinery Approaches

Including fungal biomass valorization processes in biorefineries is a promising way to increase these technologies' economic viability. An integrated biorefinery maximizes resource use and minimizes waste by producing a variety of products from a single feedstock. The overall efficiency of biomass conversion can be increased by developing integrated methods that incorporate enzymatic hydrolysis, microbial fermentation, and fungal pretreatment (Cherubini, 2010).

Sustainable and Cost-Effective Pretreatment Methods

Research on economic and environmental-friendly pretreatment techniques is crucial for the development of fungal biomass valorization. The overall cost of converting biomass can be considerably decreased by creating pretreatment techniques that employ renewable chemicals, produce fewer inhibitors, and need less energy. Process engineering and green chemistry developments are probably going to be major factors in this field (Kumar et al., 2009).

Public–Private Partnerships and Policy Support

Innovations in fungal biomass valorization technology can be commercialized with the help of public–private partnerships and policy backing. Government subsidies, R&D funding, and encouraging regulatory regimes can all aid in bridging the knowledge gap between lab-based research and large-scale industrial production. To overcome the technological and financial obstacles related to fungal biomass valorization, cooperation between government, business, and academia is essential (Cherubini, 2010).

6.2 *Research Gaps and Future Research Directions*

Fungi have shown a great deal of promise for biomass valorization, but to fully realize their potential, several issues still need to be resolved. With an emphasis on overcoming current obstacles and breaking new ground, this section addresses the research gaps and future directions in fungal biomass valorization.

6.2.1 Genetic Diversity and Engineering

Advancements in genetic engineering techniques, such as CRISPR/Cas9, offer promising tools for developing fungal strains with enhanced capabilities. However, the complexity of fungal genomes and the regulatory mechanisms governing gene expression pose significant challenges (Liu et al., 2015). Future research should focus on optimizing these genetic tools and developing robust methods for the genetic modification of diverse fungal species.

Novel strategies for creating fungal strains with improved properties include genetic engineering techniques like CRISPR/Cas9. However, there are a lot of obstacles to overcome, including the intricacy of fungal genomes and the regulatory systems controlling gene expression (Liu et al., 2015). Subsequent investigations ought to concentrate on refining these genetic instruments and creating resilient techniques for the genetic alteration of varied fungus species.

6.2.2 Enzyme Efficiency and Stability

Fungal enzyme stability and efficiency are essential for the economic conversion of biomass. Even though they work well, current fungal enzymes frequently perform poorly in industrial settings with high temperatures and extreme pH values (Cherry & Fidantsef, 2003). The goal of future research should be to use directed evolution and protein engineering to create more stable and active enzymes that are more resilient.

Additionally, the efficiency of biomass breakdown may be greatly increased by the creation of enzyme cocktails that replicate the natural synergistic activity of fungal enzymes. To increase conversion rates and lower the need for enzyme loading, research should concentrate on improving these enzyme mixes.

6.2.3 Process Optimization and Scale-Up

One major problem that still exists is translating laboratory-scale procedures to industrial-scale applications. Many methods for utilizing fungal biomass are not yet optimized for large-scale production, leading to inefficiencies and high costs. (Pandey et al., 2000a, b). Process optimization, including the creation of scalable fermentation technologies and effective biomass pretreatment techniques, should be the main emphasis of future research.

For the manufacture of fungal enzymes, solid-state fermentation (SSF) is a promising approach; nevertheless, additional tuning is needed for practical application. The efficiency and scalability of SSF systems could be improved by innovations in bioreactor design and process management (Couto & Sanromán, 2006).

6.2.4 Integration with Biorefinery Systems

There are benefits and drawbacks to integrating fungus biomass valorization techniques into biorefinery systems. Although fungi can play a major role in the degradation of lignocellulosic biomass, it is important to optimize their integration with other microbial processes and chemical transformations (Cherubini, 2010). Future studies ought to focus on creating integrated biorefinery systems that take advantage of fungal processes' advantages while maintaining their compatibility with other chemical and bioconversion processes.

6.2.5 Environmental and Economic Sustainability

One major challenge is the fungal biomass valorization processes' viability from an economic and environmental standpoint. Although using fungi instead of chemicals is a sustainable and environmentally beneficial option, these procedures still need to be proven to be scalable and economical. To evaluate the overall sustainability of fungal biomass valorization, life cycle assessments and techno-economic analyses are crucial (Kamm & Kamm, 2004).

The use of waste biomass and inexpensive substrates for fungal growing should also be investigated in future studies as this could greatly lower production costs and improve the economic feasibility of these procedures (Soccol et al., 2017).

7 Environmental and Sustainability Considerations

7.1 Environmental Impact of Fungal Processes

When it comes to the environment, using fungi instead of conventional chemical and bacterial processes for biomass valorization has various advantages. These advantages result from the intrinsic qualities of fungi, which include their capacity to break down complex organic compounds, their low energy needs, and their ability to grow on waste products. This section explains how fungal processes affect the environment and outlines the benefits and drawbacks of using them.

7.1.1 Reduction in Greenhouse Gas Emissions

A major advantage of fungal biomass valorization for the environment is that it may lead to a decrease in greenhouse gas emissions. Conventional biomass-processing techniques frequently entail chemical treatments at high temperatures that produce a substantial quantity of CO_2 and other greenhouse gases. On the other hand, fungal activities can efficiently break down biomass without the use of harsh chemicals because they usually work at lower temperatures (Mohanram et al., 2013). For instance, it has been demonstrated that using white-rot fungi, like *Phanerochaete chrysosporium*, to break down lignin considerably lowers the carbon footprint of processing biomass (Lynd et al., 2002).

7.1.2 Utilization of Waste Biomass

Among the many waste biomass types that fungi are excellent at decomposing include forestry waste, industrial by-products, and agricultural residues. This capability aids in efficient waste management in addition to offering a sustainable source of raw materials. To lessen the environmental impact of trash disposal, fungus, for example, can transform lignocellulosic waste into useful products like biofuels and bioplastics (Chandel et al., 2018). Reducing the dependence on food crops for biofuel generation through the use of waste biomass as a substrate for fungal activities can also help to resolve the dispute between fuel and food.

7.1.3 Lower Energy Consumption

In general, fungal processes use less energy than traditional chemical ones. Fungi's enzymatic activities take place in moderate environments, lowering the requirement for significant energy inputs. Reduced operating expenses and a lessening of the environmental effect are the results of this energy efficiency. For instance, considerable energy savings are achieved when fungi like *Trichoderma reesei* produce cellulases and other hydrolytic enzymes at room temperature (Peterson & Nevalainen, 2012). Fungal processes are energy-efficient in various researches as well. One example of how fungal biomass valorization might save energy is the production of enzymes by *Aspergillus niger* at low temperatures (Singh et al., 2011). Furthermore, the application of fungus in the mild bioconversion of agricultural residues into bioethanol is an example of the lower energy needs and advantages for the environment.

7.1.4 Biodegradability and Nontoxic By-products

When compared to the hazardous waste produced by chemical procedures, the by-products of fungal biomass valorization processes are nontoxic and biodegradable, making them environmentally beneficial. One of the best examples of this benefit is the application of fungal enzymes in wastewater treatment and bioremediation. Toxic contaminants and dyes in industrial effluents have been broken down by fungi like *Pleurotus ostreatus* and *Aspergillus niger*, resulting in cleaner water discharge.

7.1.5 Challenges and Future Directions

Despite the advantages for the environment, implementing fungal processes on a wide scale comes with certain problems. Optimizing fungal growth conditions and enzyme synthesis to achieve high output and efficiency is one of the major problems. On an industrial scale, it can be challenging to maintain the ideal temperature, pH, and aeration levels that fungi require.

Furthermore, a detailed assessment of the economic viability of fungal processes is required. The advantages for the environment are obvious, but producing enzymes and optimizing processes can be expensive. According to Lynd et al. (2002), future studies should concentrate on creating affordable techniques for producing enzymes and investigating the application of mixed fungal cultures to improve the effectiveness of biomass breakdown.

7.2 *Sustainability of Fungal Biomass Valorization*

Fungal biomass valorization offers a sustainable substitute for conventional techniques of converting biomass. By using fungi's innate ability to break down complex organic materials, they create useful goods with less of an impact on the environment. The benefits of fungal biomass valorization for the environment, its economic feasibility, and its potential to support a circular bio-economy are all examined in this section, which addresses the sustainability of this process.

7.2.1 Environmental Benefits

Comparing fungal biomass valorization to chemical and microbiological processes, there are a number of environmental benefits. A notable advantage is the decrease in greenhouse gas emissions. Conventional techniques for converting biomass, like chemical hydrolysis, frequently call for large energy inputs and result in noticeable CO_2 emissions. By comparison, fungal activities usually need less energy and produce less carbon emissions since they work with naturally occurring enzymes and operate in milder environments (Bayer et al., 2007).

Fungi can also grow on a variety of substrates, such as industrial by-products, forestry debris, and agricultural wastes. Reducing the environmental impact of waste disposal and keeping waste out of landfills are two benefits of using these low-value or waste materials as feedstocks for fungal biomass valorization (Pandey et al., 2000a, b). Additionally, by encouraging the recycling of organic matter and nutrients, this strategy improves soil health and lessens the demand for synthetic fertilizers.

7.2.2 Economic Viability

For fungal biomass valorization to be widely used, it must be economically sustainable. The affordability of producing enzymes is one of the primary benefits of fungal methods. Extracellular enzymes that are secreted by fungi in significant amounts can be captured and used for biomass conversion without requiring a lot of purification (Cherry & Fidantsef, 2003). When compared to chemical or microbiological approaches, this lowers the total cost of producing enzymes.

The economic viability of fungal biomass valorization is further enhanced by the utilization of abundant and low-cost feedstocks. The low or free cost of agricultural and forestry leftovers makes them appealing substrates for large-scale production (Lynd et al., 2002). Because fungus may grow on a variety of substrates, methods can be adjusted to make use of local biomass supply, which lowers transportation costs and supports regional bio economies.

7.2.3 Contribution to a Circular Bio-economy

Fungal biomass valorization is in line with the circular economy's tenets, which stress material recycling and the sustainable use of biological resources. Fungal processes help to reduce waste and make optimal use of resources by transforming waste biomass into useful products, including biofuels, bioplastics, and biochemicals (Cherubini, 2010).

Furthermore, the total sustainability of bio-based output can be improved by including fungal biomass valorization in biorefinery systems. A biorefinery combines several processes to enhance biomass value extraction while reducing waste and enhancing resource efficiency (Kamm & Kamm, 2004). In this integrated strategy, fungi can be quite helpful because they offer sustainable and effective ways to pretreat and convert biomass.

7.2.4 Challenges and Future Directions

Even if the sustainability of fungal biomass valorization seems promising, several issues need to be resolved before its full potential can be reached. The ability of fungal processes to scale up is one significant obstacle. Fungal biomass conversion is effective in laboratory settings, but transferring these processes to industrial settings will require a large amount of investment and optimization.

The creation of resilient and effective fungal strains that can function consistently in a range of environmental circumstances is another challenge. By making it possible to create fungus with improved properties and stress tolerance, developments in genetic engineering and synthetic biology offer hope for overcoming this obstacle (Meyer et al., 2020).

Future studies should concentrate on techno-economic analyses and life cycle assessments (LCAs) to measure the financial and environmental advantages of fungal biomass valorization. These evaluations can assist in pinpointing areas in need of development and offer insightful information about how sustainable various procedures are (Cherubini, 2010).

8 Conclusion

The substantial benefits of using fungus for biomass valorization over conventional bacterial approaches have been emphasized in this chapter. The production of a diverse range of extracellular enzymes, such as cellulases, ligninases, and hemicelluloses, which are vital for the breakdown of complex organic molecules like lignocellulose, is one of the distinctive physiological and biochemical capacities of fungi. These enzymes enable fungi to efficiently transform biomass into useful goods, including medicines, biofuels, and bioplastics. Fungi also have a remarkable ability to adapt to their surroundings, growing in a variety of often harsh environments, which increases their usefulness in a wide range of industrial applications. Their superiority over bacterial systems in biomass valorization is highlighted by their metabolic plasticity and efficiency in substrate conversion.

Significant environmental advantages come from using fungal processes for biomass valorization, such as lower energy consumption, effective use of waste biomass, fewer greenhouse gas emissions, and the creation of nontoxic and biodegradable by-products. A sustainable and profitable method of turning waste biomass into useful goods is fungus biomass valorization. Its potential for integration into biorefinery systems, cost-effectiveness, and environmental advantages make it a promising technology for promoting a circular bio-economy. To fully realize the promise of fungal biomass valorization, research and innovation in the areas of scalability and strain development are essential.

9 The Potential Future of Fungi in Biomass Valorization

Fungi in biomass valorization appear to have a bright future ahead of them, with several interesting directions for future study and advancement. Developments in synthetic biology and genetic engineering could greatly expand the potential of fungal strains, increasing their resilience and productivity in a range of environmental settings. The creation of genetically altered fungi with enhanced stress tolerance, metabolic pathways, and enzyme production has the potential to transform the field and enable more effective biomass conversion procedures.

Furthermore, a sustainable method of resource exploitation is provided by the incorporation of fungal biomass valorization processes into biorefinery systems. Biorefineries can optimize resource efficiency and reduce waste by taking advantage of fungi's capacity to break down and transform an extensive variety of waste materials. This integrated strategy supports the sustainable use of biological resources and aids in environmental protection. It is consistent with the tenets of a circular bio-economy.

The broad use of fungal biomass valorization will also depend on ongoing research into improving industrial-scale fungal processes, including developments in fermentation technology and bioreactor design. Techno-economic evaluations and life cycle assessments will help assess the financial and environmental advantages of these procedures, offering important information for their widespread adoption.

In conclusion, fungi provide a flexible and long-lasting approach to biomass valorization, and with continued study and technical development, they have the potential to revolutionize the industry. The use of fungi can result in more economical, eco-friendly, and efficient biomass conversion processes by tackling present issues and breaking new ground. This will have a major impact on the growth of a sustainable bio-based economy.

References

Anderson, A. J., & Dawes, E. A. (1990). Occurrence, metabolism, metabolic role, and industrial uses of bacterial polyhydroxyalkanoates. *Microbiological Reviews, 54*(4), 450–472. https://doi.org/10.1128/mmbr.54.4.450-472.1990

Andlar, M., Rezić, T., Marđetko, N., Kracher, D., Ludwig, R., & Šantek, B. (2018). Lignocellulose degradation: An overview of fungi and fungal enzymes involved in lignocellulose degradation. *Engineering in Life Sciences, 18*(11), 768–778. https://doi.org/10.1002/elsc.201800039

Arora, N., Patel, A., Pruthi, P. A., & Pruthi, V. (2016). Synergistic dynamics of nitrogen and phosphorous influences lipid productivity in Chlorella minutissima for biodiesel production. *Bioresource Technology, 213*, 79–87. https://doi.org/10.1016/j.biortech.2016.02.112

Bayer, E. A., Lamed, R., Himmel, M. E., & Ding, S. Y. (2007). The potential of cellulases and cellulosomes for cellulosic waste management. *Current Opinion in Biotechnology, 18*(3), 237–245. https://doi.org/10.1016/j.copbio.2007.04.004

Berka, R. M., Grigoriev, I. V., Otillar, R., Salamov, A., Grimwood, J., Reid, I., Ishmael, N., John, T., Darmond, C., Moisan, M.-C., Henrissat, B., Coutinho, P. M., Lombard, V., Natvig, D. O., Lindquist, E., Schmutz, J., Lucas, S., Harris, P., Powlowski, J., Bellemare, A., Taylor, D., Butler, G., De Vries, R. P., Allijn, I. E., Van Den Brink, J., Ushinsky, S., Storms, R., Powell, A. J., Paulsen, I. T., Elbourne, L. D. H., Baker, S. E., Magnuson, J., LaBoissiere, S., Clutterbuck, A. J., Martinez, D., Wogulis, M., De Leon, A. L., Rey, M. W., & Tsang, A. (2011). Comparative genomic analysis of the thermophilic biomass-degrading fungi Myceliophthora thermophila and Thielavia terrestris. *Nature Biotechnology, 29*(10), 922–927.

Bhat, M. K., & Bhat, S. (1997). Cellulose degrading enzymes and their potential industrial applications. *Biotechnology Advances, 15*(3–4), 583–620. https://doi.org/10.1016/s0734-9750(97)00006-2

Biely, P. (1985). Microbial xylanolytic systems. *Trends in Biotechnology, 3*(11), 286–290. https://doi.org/10.1016/0167-7799(85)90004-6

Bilal, M., & Iqbal, H. M. N. (2020). Ligninolytic enzymes mediated ligninolysis: An untapped biocatalytic potential to deconstruct lignocellulosic molecules in a sustainable manner. *Catalysis Letters, 150*(2), 524–543. https://doi.org/10.1007/s10562-019-03096-9

Bischof, R. H., Ramoni, J., & Seiboth, B. (2016). Cellulases and beyond: The first 70 years of the enzyme producer Trichoderma reesei. *Microbial Cell Factories, 15*(1), 106. https://doi.org/10.1186/s12934-016-0507-6

Bonechi, C., Consumi, M., Donati, A., Leone, G., Magnani, A., Tamasi, G., & Rossi, C. (2017). Biomass: An overview. In F. Dalena, A. Basile, & C. Rossi (Eds.), *Bioenergy Systems for the Future* (pp. 3–42). Elsevier. https://doi.org/10.1016/B978-0-08-101031-0.00001-6

Bozell, J. J., & Petersen, G. R. (2010). Technology development for the production of biobased products from biorefinery carbohydrates—The US Department of Energy's "Top 10" revisited. *Green Chemistry, 12*(4), 539. https://doi.org/10.1039/b922014c

Bushley, K. E., Raja, R., Jaiswal, P., Cumbie, J. S., Nonogaki, M., Boyd, A. E., Owensby, C. A., Knaus, B. J., Elser, J., Miller, D., & Di, Y. (2013). The genome of Tolypocladium inflatum: Evolution, organization, and expression of the cyclosporin biosynthetic gene cluster. *PLoS Genetics, 9*(6), e1003496.

Chandel, A. K., Garlapati, V. K., Singh, A. K., Antunes, F. A. F., & Da Silva, S. S. (2018). The path forward for lignocellulose biorefineries: Bottlenecks, solutions, and perspective on commercialisation. *Bioresource Technology, 264*, 370–381. https://doi.org/10.1016/j.biortech.2018.06.004

Cherry, J. R., & Fidantsef, A. L. (2003). Directed evolution of industrial enzymes: An update. *Current Opinion in Biotechnology, 14*(4), 438–443. https://doi.org/10.1016/S0958-1669(03)00099-5

Cherubini, F. (2010). The biorefinery concept: Using biomass instead of oil for producing energy and chemicals. *Energy Conversion and Management, 51*(7), 1412–1421. https://doi.org/10.1016/j.enconman.2010.01.015

Couto, S. R., & Sanromán, M. A. (2006). Application of solid-state fermentation to food industry—a review. *Journal of Food Engineering, 76*(3), 291–302. https://doi.org/10.1016/j.jfoodeng.2005.05.022

Cragg, S. M., Beckham, G. T., Bruce, N. C., Bugg, T. D., Distel, D. L., Dupree, P., Etxabe, A. G., Goodell, B. S., Jellison, J., McGeehan, J. E., McQueen-Mason, S. J., Schnorr, K., Walton, P. H., Watts, J. E., & Zimmer, M. (2015). Lignocellulose degradation mechanisms across the Tree of Life. *Current Opinion in Chemical Biology, 29*, 108–119. https://doi.org/10.1016/j.cbpa.2015.10.018

Dashtban, M., Schraft, H., Syed, T. A., & Qin, W. (2010). Fungal biodegradation and enzymatic modification of lignin. *International Journal of Biochemistry and Molecular Biology, 1*(1), 36–50.

De Albuquerque, T. L., Júnior, J. E. M., De Queiroz, L. P., Ricardo, A. D. S., & Rocha, M. V. P. (2021). Polylactic acid production from biotechnological routes: A review. *International Journal of Biological Macromolecules, 186*, 933–951. https://doi.org/10.1016/j.ijbiomac.2021.07.074

De Bhowmick, G., Sarmah, A. K., & Sen, R. (2018). Lignocellulosic biorefinery as a model for sustainable development of biofuels and value-added products. *Bioresource Technology, 247*, 1144–1154. https://doi.org/10.1016/j.biortech.2017.09.163

De Souza, T. S. P., & Kawaguti, H. Y. (2021). Cellulases, hemicellulases, and pectinases: Applications in the food and beverage industry. *Food and Bioprocess Technology, 14*(8), 1446–1477. https://doi.org/10.1007/s11947-021-02678-z

de Vries, R. P., & Visser, J. (2001). Aspergillus enzymes involved in the degradation of plant cell wall polysaccharides. *Microbiology and Molecular Biology Reviews, 65*(4), 497–522. https://doi.org/10.1128/MMBR.65.4.497-522.2001

Demain, A. L. (2000). Microbial biotechnology. *Trends in Biotechnology, 18*(1), 26–31.

Demain, A. L. (2009). Antibiotics: Natural products essential to human health. *Medicinal Research Reviews, 29*(6), 821–842. https://doi.org/10.1002/med.20154

Demain, A. L., & Sanchez, S. (2009). Microbial drug discovery: 80 years of progress. *The Journal of Antibiotics, 62*(1), 5–16. https://doi.org/10.1038/ja.2008.16

Dunlop, M. J. (2011). Engineering microbes for tolerance to next-generation biofuels. *Biotechnology for Biofuels, 4*(1), 32. https://doi.org/10.1186/1754-6834-4-32

Faria, M. R., & Wraight, S. P. (2007). Mycoinsecticides and Mycoacaricides: A comprehensive list with worldwide coverage and international classification of formulation types. *Biological Control, 43*(3), 237–256. https://doi.org/10.1016/j.biocontrol.2007.08.001

Fleet, G. H. (2007). Yeasts in foods and beverages: Impact on product quality and safety. *Current Opinion in Biotechnology, 18*(2), 170–175. https://doi.org/10.1016/j.copbio.2007.01.010

Friedrich, J., Zalar, P., Mohorčič, M., Klun, U., & Kržan, A. (2007). Ability of fungi to degrade synthetic polymer nylon-6. *Chemosphere, 67*(10), 2089–2095. https://doi.org/10.1016/j.chemosphere.2006.09.038

Ghorai, S., Banik, S. P., Verma, D., Chowdhury, S., Mukherjee, S., & Khowala, S. (2009). Fungal biotechnology in food and feed processing. *Food Research International, 42*(5–6), 577–587.

Gunjal, A. B., Patil, N. N., & Shinde, S. S. (2020). Ligninase in degradation of lignocellulosic wastes. In *Springer eBooks* (pp. 55–70). https://doi.org/10.1007/978-3-030-44671-0_4

Harms, H., Schlosser, D., & Wick, L. Y. (2011). Untapped potential: Exploiting fungi in bioremediation of hazardous chemicals. *Nature Reviews Microbiology, 9*(3), 177–192. https://doi.org/10.1038/nrmicro2519

Hassan, S. S., Williams, G. A., & Jaiswal, A. K. (2019). Moving towards the second generation of lignocellulosic biorefineries in the EU: Drivers, challenges, and opportunities. *Renewable and Sustainable Energy Reviews, 101*, 590–599. https://doi.org/10.1016/j.rser.2018.11.041

Hatakka, A. (2001). In M. Hofrichter & A. Steinbuchel (Eds.), *Biopolymers Online, Biodegradation of lignin*. Wiley-VCH. https://doi.org/10.1002/3527600035.bpol1005

Himmel, M. E., Ding, S. Y., Johnson, D. K., Adney, W. S., Nimlos, M. R., Brady, J. W., & Foust, T. D. (2007). Biomass recalcitrance: Engineering plants and enzymes for biofuels production. *Science, 315*(5813), 804–807. https://doi.org/10.1126/science.1137016

Hölker, U., Höfer, M., & Lenz, J. (2004). Biotechnological advantages of laboratory-scale solid-state fermentation with fungi. *Applied Microbiology and Biotechnology, 64*(2), 175–186. https://doi.org/10.1007/s00253-003-1504-3

Howard, R. L., Abotsi, E., Jansen, V. R. E. L., & Howard, S. (2003). Lignocellulose biotechnology: Issues of bioconversion and enzyme production. *African Journal of Biotechnology, 2*(12), 602–619.

Janusz, G., Pawlik, A., Sulej, J., Świderska-Burek, U., Jarosz-Wilkołazka, A., & Paszczyński, A. (2017). Lignin degradation: Microorganisms, enzymes involved, genomes analysis and evolution. *FEMS Microbiology Reviews, 41*(6), 941–962. https://doi.org/10.1093/femsre/fux049

Jones, D. T., & Woods, D. R. (1986). Acetone-butanol fermentation revisited. *Microbiological Reviews, 50*(4), 484–524.

Juturu, V., & Wu, J. C. (2014). Microbial cellulases: Engineering, production and applications. *Renewable and Sustainable Energy Reviews, 33*, 188–203. https://doi.org/10.1016/j.rser.2014.01.077

Kamm, B., & Kamm, M. (2004). Principles of biorefineries. *Applied Microbiology and Biotechnology, 64*(2), 137–145. https://doi.org/10.1007/s00253-003-1537-7, https://doi.org/10.1007/s00253-003-1537-9

Kirk, T. K., Schultz, E., Connors, W. J., Lorenz, L. F., & Zeikus, J. G. (1978). Influence of culture parameters on lignin metabolism byPhanerochaete chrysosporium. *Archives of Microbiology, 117*(3), 277–285. https://doi.org/10.1007/bf00738547

Kumar, P., Barrett, D. M., Delwiche, M. J., & Stroeve, P. (2009). Methods for pretreatment of lignocellulosic biomass for efficient hydrolysis and biofuel production. *Industrial & Engineering Chemistry Research, 48*(8), 3713–3729. https://doi.org/10.1021/ie801542g

Lee, K. H., Park, J. H., Kim, T. Y., Kim, H. U., & Lee, S. Y. (2007). Systems metabolic engineering of Escherichia coli for L-threonine production. *Molecular Systems Biology, 3*(1), 149.

Liu, R., Chen, L., Jiang, Y., Zhou, Z., Xu, Z., Wei, X., Zhang, S., Chen, S., & Liu, G. (2015). CRISPR/Cas9-mediated genome editing and targeted gene regulation in Aspergillus niger. *Biotechnology Journal, 10*(8), 1155–1163. https://doi.org/10.1002/biot.201400749

Lynd, L. R., Weimer, P. J., van Zyl, W. H., & Pretorius, I. S. (2002). Microbial cellulose utilization: Fundamentals and biotechnology. *Microbiology and Molecular Biology Reviews, 66*(3), 506–577. https://doi.org/10.1128/MMBR.66.3.506-577.2002

Maheshwari, R., Bharadwaj, G., & Bhat, M. K. (2000). Thermophilic fungi: Their physiology and enzymes. *Microbiology and Molecular Biology Reviews, 64*(3), 461–488. https://doi.org/10.1128/MMBR.64.3.461-488.2000

Mankar, A. R., Pandey, A., Modak, A., & Pant, K. K. (2021). Pretreatment of lignocellulosic biomass: A review on recent advances. *Bioresource Technology, 334*, 125235.

Manzoni, M., & Rollini, M. (2002). Biosynthesis and biotechnological production of statins by filamentous fungi and application of these cholesterol-lowering drugs. *Applied Microbiology and Biotechnology, 58*(5), 555–564. https://doi.org/10.1007/s00253-002-0932-9

Meyer, V., Basenko, E. Y., Benz, J. P., Braus, G. H., Caddick, M. X., Csukai, M., De Vries, R. P., Endy, D., Frisvad, J. C., Gunde-Cimerman, N., Haarmann, T., Hadar, Y., Hansen, K., Johnson, R. I., Keller, N. P., Kraševec, N., Mortensen, U. H., Perez, R., Ram, A. F. J., et al. (2020). Growing a circular economy with fungal biotechnology: A white paper. *Fungal Biology and Biotechnology, 7*(1), 5. https://doi.org/10.1186/s40694-020-00095-z

Mohanram, S., Amat, D., Choudhary, J., Arora, A., & Nain, L. (2013). Novel perspectives for evolving enzyme cocktails for lignocellulose hydrolysis in biorefineries. *Sustainable Chemical Processes, 1*(1), 1–10. https://doi.org/10.1186/2043-7129-1-15

Morgenstern, I., Powlowski, J., & Tsang, A. (2014). Fungal cellulose degradation by oxidative enzymes: From dysfunctional GH61 family to powerful lytic polysaccharide monooxygenase family. *Briefings in Functional Genomics, 13*(6), 471–481. https://doi.org/10.1093/bfgp/elu032

Nevalainen, O., Hakala, T., Suomalainen, J., Mäkipää, R., Peltoniemi, M., Krooks, A., & Kaasalainen, S. (2014). Fast and nondestructive method for leaf level chlorophyll estimation using hyperspectral LiDAR. *Agricultural and Forest Meteorology, 198*, 250–258.

Okolie, J. A., Epelle, E. I., Tabat, M. E., Orivri, U., Amenaghawon, A. N., Okoye, P. U., & Gunes, B. (2022). Waste biomass valorization for the production of biofuels and value-added products: A comprehensive review of thermochemical, biological and integrated processes. *Process Safety and Environmental Protection/Transactions of the Institution of Chemical Engineers. Part B, Process Safety and Environmental Protection/Chemical Engineering Research and Design/Chemical Engineering Research & Design, 159*, 323–344.

Pandey, A., Soccol, C. R., & Mitchell, D. (2000a). New developments in solid-state fermentation: I-bioprocesses and products. *Process Biochemistry, 35*(10), 1153–1169. https://doi.org/10.1016/S0032-9592(00)00152-7

Pandey, A., Soccol, C. R., Nigam, P., Soccol, V. T., Vandenberghe, L. P. S., & Mohan, R. (2000b). Biotechnological potential of agro-industrial residues. I: Sugarcane bagasse. *Bioresource Technology, 74*(1), 69–80. https://doi.org/10.1016/S0960-8524(99)00142-X

Papagianni, M. (2007). Advances in citric acid fermentation by Aspergillus niger: Biochemical aspects, membrane transport and modelling. *Biotechnology Advances, 25*(3), 244–263. https://doi.org/10.1016/j.biotechadv.2007.01.002

Peterson, R., & Nevalainen, H. (2012). Trichoderma reesei RUT-C30 – thirty years of strain improvement. *Microbiology, 158*(1), 58–68. https://doi.org/10.1099/mic.0.054031-0

Polizeli, M. L. T. M., Rizzatti, A. C. S., Monti, R., Terenzi, H. F., Jorge, J. A., & Amorim, D. S. (2005). Xylanases from fungi: Properties and industrial applications. *Applied Microbiology and Biotechnology, 67*(5), 577–591. https://doi.org/10.1007/s00253-005-1904-7

Rytioja, J., Hildén, K., Yuzon, J., Hatakka, A., De Vries, R. P., & Mäkelä, M. R. (2014). Plant-polysaccharide-degrading enzymes from basidiomycetes. *Microbiology and Molecular Biology Reviews, 78*(4), 614–649. https://doi.org/10.1128/mmbr.00035-14

Saravanan, A., Yaashikaa, P. R., Kumar, P. S., Thamarai, P., Deivayanai, V. C., & Rangasamy, G. (2023). A comprehensive review on techno-economic analysis of biomass valorization and conversional technologies of lignocellulosic residues. *Industrial Crops and Products, 200*, 116822.

Sarkar, N., Ghosh, S. K., Bannerjee, S., & Aikat, K. (2012). Bioethanol production from agricultural wastes: An overview. *Renewable Energy, 37*(1), 19–27. https://doi.org/10.1016/j.renene.2011.06.045

Schindler, D. (2020). Genetic engineering and synthetic genomics in yeast to understand life and boost biotechnology. *Bioengineering, 7*(4), 137.

Singh, A., Kuhad, R. C., Ward, O. P., & Kumar, A. (2011). Eco-friendly pretreatment of lignocellulosic residues for enhanced delignification and enzymatic digestibility. *Indian Journal of Microbiology, 51*(1), 12–18. https://doi.org/10.1007/s12088-011-0149-1

Singhania, R. R., Patel, A. K., Soccol, C. R., & Pandey, A. (2010). Recent advances in solid-state fermentation. *Biochemical Engineering Journal, 44*(1), 13–18. https://doi.org/10.1016/j.bej.2008.10.019

Singhania, R. R., Patel, A. K., Soccol, C. R., & Pandey, A. (2013a). Recent advances in solid-state fermentation. *Biochemical Engineering Journal, 81*, 146–161. https://doi.org/10.1016/j.bej.2013.10.013

Singhania, R. R., Sukumaran, R. K., Patel, A. K., Larroche, C., & Pandey, A. (2013b). Advancement and comparative profiles in the production technologies using solid-state and submerged fermentation for microbial cellulases. *Enzyme and Microbial Technology, 53*(6–7), 428–440. https://doi.org/10.1016/j.enzmictec.2013.04.004

Smith, S. E., & Read, D. J. (2008). *Mycorrhizal Symbiosis* (3rd ed.). Academic Press. https://doi.org/10.1023/A:1000902213906

Soccol, C. R., da Costa, E. S. F., Letti, L. A. J., Karp, S. G., Wojciechowski, A. L., & de Souza Vandenberghe, L. P. (2017). Recent developments and innovations in solid state fermentation. *Biotechnology Research and Innovation, 1*(1), 52–71. https://doi.org/10.1016/j.biori.2017.01.002

Steinbüchel, A., & Lütke-Eversloh, T. (2003). Metabolic engineering and pathway construction for biotechnological production of relevant polyhydroxyalkanoates in microorganisms. *Biochemical Engineering Journal, 16*(2), 81–96. https://doi.org/10.1016/S1369-703X(03)00036-6

Steinkraus, K. H. (2002). Fermentations in world food processing. *Comprehensive Reviews in Food Science and Food Safety, 1*(1), 23–32. https://doi.org/10.1111/j.1541-4337.2002.tb00004.x

Sun, D., Lv, Z.-W., Rao, J., Tian, R., Sun, S.-N., & Peng, F. (2022). Effects of hydrothermal pretreatment on the dissolution and structural evolution of hemicelluloses and lignin: a review. *Carbohydrate Polymers, 281*, 119050. https://doi.org/10.1016/j.carbpol.2021.119050

Toushik, S. H., Lee, K.-T., Lee, J.-S., & Kim, K.-S. (2017). Functional applications of lignocellulolytic enzymes in the fruit and vegetable processing industries: Applications of lignocellulolytic enzymes. *Journal of Food Science, 82*(3), 585–593. https://doi.org/10.1111/1750-3841.13636

Verma, N., Kumar, V., & Bansal, M. C. (2021). Valorization of waste biomass in fermentative production of cellulases: A review. *Waste and Biomass Valorization, 12*, 613–640. https://doi.org/10.1007/s12649-020-01048-8

Wang, M., Wu, M., & Huo, H. (2007). Life-cycle energy and greenhouse gas emission impacts of different corn ethanol plant types. *Environmental Research Letters, 2*(2), 024001. https://doi.org/10.1088/1748-9326/2/2/024001

Wendisch, V. F., Bott, M., & Eikmanns, B. J. (2006). Metabolic engineering of Escherichia coli and Corynebacterium glutamicum for biotechnological production of organic acids and amino acids. *Current Opinion in Microbiology, 9*(3), 268–274. https://doi.org/10.1016/j.mib.2006.03.001

Extraction and Purification of Essential Oils from Food Waste Biomass

Vineeta Gautam and Umesh Kumar

Abstract Essential oils are aromatic, oily liquids derived from various parts of plants, including leaves, peels, bark, flowers, buds, and seeds. Volatile odoriferous oils are extracted through methods such as steam distillation and expression. The flavor and odor of essential oils vary based on the types and quantities of their components, especially in fragrant plants. The differing quantities that are extracted from various plants have an impact on the price of essential oils. In addition to aromatic ingredients, natural pigments contribute to their different colors. Essential oils, primarily sourced from plants, have long been recognized for their aromatic and medicinal benefits. The extraction and purification of essential oils from biomass food waste significantly enhance value addition and sustainable resource use. Various extraction techniques yield essential oils from different plants, with the botanical material determining the process used. The state and presentation of the material are also key factors to consider. The quality of an essential oil is largely determined by the extraction process, as improper extraction can alter or damage its chemical signature. The rapidly growing field of essential oils presents unique chemical compositions influenced by extraction methods and specific fungal species. Due to their abundance in various volatile molecules, these oils have potential uses in cuisine, cosmetics, and health. Ongoing research into their extraction techniques, chemical makeup, and therapeutic properties further underscores their potential. This chapter examines various approaches and procedures for obtaining essential oils from food waste, along with strategies for refining these oils to improve quality and marketability. It covers the importance of essential oils in manufacturing processes, the problems food waste poses to the environment, and the possibility of utilizing waste biomass to produce goods with added value. Additionally, the chap-

V. Gautam (✉)
School of Chemical Technology, Harcourt Butler Technological University, Kanpur, India
e-mail: vgautamch@hbtu.ac.in

U. Kumar
Department of Civil Engineering, Indian Institute of Technology Banaras Hindu University, Varanasi, India

D. B. Pal, S. L. Bansal (eds.), *Fungal Waste Biomass Management for Energy, Environment and Value-Added Products*, Fungal Biology,
https://doi.org/10.1007/978-3-031-82599-6_6

ter explores potential research opportunities for enhancing these processes and evaluates the sustainability and economic aspects of extracting essential oils from food waste.

Keywords Essential oils · Extraction · Purification · Fungal · Sustainability · Food waste

1 Introduction

The extraction and purification of essential oils (EOs) from biomass food waste offer innovative and sustainable solutions to global challenges such as waste management, resource efficiency, and environmental sustainability. Because of their unique fragrances and therapeutic qualities, essential oils are volatile, aromatic chemicals that are widely employed in a variety of industries, including food processing, cosmetics, and pharmaceuticals. Essential oils are often derived from fresh plant material, a process that can be resource-intensive and harmful to the environment. Using food waste biomass as a source for essential oils is a powerful alternative that aligns with the principles of waste valorization and the circular economy. A wide range of organic materials that are wasted during the manufacture, processing, distribution, and consumption of food are included in food waste biomass (Girotto et al., 2015). This biomass consists of materials typically considered waste, such as fruit peels, vegetable scraps, seeds, and other by-products. With an estimated 1.3 billion tons of food lost yearly, food waste is a major problem on a global scale (Sahoo et al., 2023). This trash can be converted into useful goods like essential oils, which not only lessens the environmental impact of garbage disposal but also offers a sustainable supply of high-value components. Natural molecules with a variety of biological functions, primarily terpenes and their derivatives, are combined to create complex combinations known as essential oils. Essential oils, with their long-standing use in traditional medicine, are crucial for contemporary applications such as aromatherapy, flavorings, and preservatives. Their unique properties, including antibacterial, antifungal, and antioxidant effects, make them highly sought after across various industries. However, it is essential to produce these oils sustainably to meet increasing demand without depleting natural resources (Chouhan et al., 2017; Butnariu & Sarac, 2018). Regarding the economic, environmental, and systemic implications of oil extraction from medicinal plants, not a single review study has been published in the Indian region. Researchers will be able to use this to determine the most effective way to extract essential oils from a specific aromatic and medicinal plant (Lahiri et al., 2023). The optimum process for making rosemary oil is water distillation with full energy integration when looking at it from a techno-economic and environmental standpoint. Supercritical fluid extraction (SFE) is the ideal technique for oregano oil (Kant & Kumar, 2022).

Today, extraction operations are integral to production methods across industries such as materials, fine chemicals, food, biofuels, pharmaceuticals, cosmetics, and perfumes. While the extraction of natural products was initially considered "clean" compared to heavy chemical industries, experts now recognize that its environmental impact is significantly larger than previously thought. Products made from unprocessed plant matter that can only be separated physically are known as essential oils (Chemat et al., 2017). The physical techniques are dry distillation of natural materials, expression (sometimes called cold-pressing for citrus peel oils), and alembic distillation (steam, steam/water, and water). These procedures require a lot of water, energy, solvent, and time. Using petroleum solvent, lipids, fragrances, colors, and antioxidants are extracted. The Soxhlet extraction method, which commonly uses solvents like hexane, dichloromethane, acetone, or methanol, involves repeatedly percolating the solvent through the sample (Zhang et al., 2023). This technique has long been the standard for solid–liquid extraction and remains widely used in both laboratories and industry. However, Soxhlet extraction has several significant drawbacks. It requires large volumes of solvent, necessitates evaporation and concentration at the end of the extraction, and is not suitable for thermolabile analytes. Additionally, the process is time-consuming, often taking many hours to complete (Abubakar et al., 2022).

1.1 Importance of Essential Oils in Various Industries

Essential oils (EOs) are volatile, low molecular weight compounds with diverse biological functions, produced in various plant organs such as flowers, buds, leaves, branches, stems, seeds, fruits, bark, and roots. They contain a variety of active components, including alkaloids, tannins, steroids, glycosides, resins, phenols, volatile oils, and flavonoids (Bolouri et al., 2022). These oils have become more well-liked as natural substitutes for artificial antioxidants in recent years. Consumer worries about the harmful effects of synthetic antioxidants—which have been connected to toxicity and several cancers—are driving this change. As a result, there is a rising need for novel, reasonably priced natural antioxidant sources to improve consumer health and guarantee food security. The high concentration of volatile, aromatic, and bioactive components found in EOs and extracts makes them valuable in a variety of businesses. These useful substances are essential to the food, pharmaceutical, cosmetic, and agricultural industries because they naturally have antibacterial and antioxidant qualities (Raut & Karuppayil, 2014). EOs are flammable substances that dissolve in specific organic solvents and alcohol. Their primary components are terpenes, which can be either cyclic or linear in structure. The type of terpene produced depends on the number and arrangement of carbon atoms; for instance, monoterpenes contain 10 carbon atoms, sesquiterpenes have 15, and diterpenes possess 20. Key compounds found in essential oils include thymol, cinnamaldehyde, citral, geraniol, eugenol, carvacrol, and cinnamic acid (Ali et al., 2015).

The method of "plant extraction" is used to separate parts of plants. The intended extract is produced by solubilizing these specific plant components in a solvent. Plant oils and extracts have long been used for a variety of purposes. These oils have recently gained significant attention as a natural source of antimicrobial agents. The discovery of new medicinal compounds, including those derived from natural products, is increasingly important. Plants offer a wide range of applications across various industries, including food, cosmetics, healthcare, and agriculture. Since ancient times, EOs have been used in traditional medical systems. The wide range of biological characteristics of oils, including as their antibacterial, antiviral, antimutagenic, anticancer, antioxidant, anti-inflammatory, immunomodulatory, and antiprotozoal effects, are being studied by scientists all over the world (Raut & Karuppayil, 2014). "Aroma," which means fragrance or smell, and "therapy," which means treatment, are the words from which aromatherapy gets its name. The goal of this natural healing technique is to improve a person's health, mind, and soul. Aromatherapy has been utilized for at least 6,000 years as a popular supplemental and alternative therapy in ancient civilizations like Egypt, China, and India. Aromatherapy is now widely used to treat a wide range of issues and ailments. According to published research, this therapy became well known in the latter half of the twentieth century and is still widely used today. Owing to its significance, ubiquity, and extensive use, aroma science therapy is currently acknowledged (Pezantes-Orellana et al., 2024).

EOs are gaining increasing importance in fields such as medicine, cosmetics, aromatherapy, fragrance, and spiritual practices. Their well-documented antibacterial, antibiotic, and antiviral properties contribute to their broad range of applications. Numerous studies and traditional healing practices suggest that essential oils may offer therapeutic benefits for conditions such as Alzheimer's, cardiovascular disease, cancer, and labor pain during pregnancy. Their use in treating sleep disorders and cancer is also growing in popularity. The organic nature of essential oils and their gentle interaction with the body promote a sense of well-being (Sarkic & Stappen, 2018). Studies have demonstrated that the inhalation of essential oils of rosemary, which are utilized in phytotherapy as restorative treatments for exhaustion, dramatically enhances the locomotor activity of mice. Consequently, those have emerged as a promising resource for discovering novel, safe, and effective bioactive compounds with a range of medical applications, particularly those related to antioxidant and antibacterial properties. Due primarily to their antibacterial and anti-inflammatory properties, essential oils provide numerous health benefits. For example, cinnamon oil changes lipid profiles, suppresses cell division, and changes bacterial membranes—all of which may provide protection against colitis. Patients with metabolic syndrome experienced a significant improvement in diastolic pressure when supplementing with cumin essential oil. Especially in fish like tilapia, coriander essential oil has shown antibacterial activity, improving health and resistance to bacterial infections (Durczynska & Żukowska, 2024). Parsley essential oil acts as an antioxidant, cumin essential oil helps reduce inflammation, and orange peel oil possesses pro-oxidative, antibacterial, antifungal, and antiparasitic properties. Thyme essential oil protects the colon from damage and DNA methylation, while lemon essential oil has been shown to influence intestinal flora in mice. Additionally, carnosic acid in rosemary oil can reduce the viability of prostate cancer cells by altering the function of the endoplasmic reticulum. Dairy products are particularly susceptible to contamination, with raw milk, the dairy industry, and sometimes vegetable coagulants being common sources. Additives such as preservatives and antioxidants are used to enhance the overall quality and safety of products, as well as to extend their shelf life. Even though artificial chemicals have many benefits, they can cause allergic responses, asthma attacks, nausea, diarrhea, and in certain situations, even cancer in consumers (El-Aziz et al., 2023). Essential oils are among the best natural alternatives to pasteurization or chemical additives for preventing bacterial spoilage in various food products. For instance, incorporating extracts from the *Thymus capitatus* plant has been shown to inhibit bacterial growth in pasteurized milk, thereby extending its shelf life. This suggests that *T. capitatus* enhances the effectiveness of pasteurization in preserving raw milk quality (Shaaban, 2023). Encapsulated essential oils, with their potent antibacterial and antioxidant properties, offer a promising solution for improving quality control and maintaining high standards in milk preservation within the dairy industry. Numerous studies have also highlighted the therapeutic benefits of plant-derived oils, which are advantageous in fields such as medicine, aromatherapy, microbiology, agriculture, livestock, and the food industry (Misra et al., 2022).

1.2 Environmental Impact of Food Waste Disposal

Food waste is a serious problem owing to the loss of food as well as the waste of resources such as energy, water, and land utilized in production. This exacerbates food insecurity and contributes to environmental degradation, especially in countries already vulnerable during the global COVID-19 pandemic. As to the United Nations (UN), the global food industry presently uses 30% of the world's energy and generates 25% of all greenhouse gas (GHG) emissions (Roy et al., 2023). Regrettably, a third of the food that is produced is wasted. There is no universal definition of "food waste" that applies across all countries; the term can have different meanings, depending on the context. Some definitions focus on food items that are safe and nutritious for human consumption but are discarded at various stages of the food supply chain, including both edible and inedible parts (Xue et al., 2024). Climate change is mostly caused by the food business, which emits 26% of the world's greenhouse gases (GHGs). Conventional food waste management techniques, including burning or landfilling, generate harmful chemicals that can contaminate air and water sources, including dioxins, methane (CH_4), carbon monoxide (CO), and ammonia (NH_3) (Kohli et al., 2024). Although various methods exist to manage food waste, the primary objective of these solutions should be to minimize environmental impact rather than create additional issues. Cutting down on food waste is crucial for environmental sustainability, emphasizing the importance of applying the principles of reduce, reuse, recycle, and recover. These guidelines ought to be followed to reduce food waste production and the environmental damage (Rakhasiya et al., 2023).

The demand for high-quality food is expanding along with the global population growth and gradual increase in purchasing power. Farmers and agricultural companies are growing and processing more food than is required to address the issue of food shortage, to maintain their competitiveness, and to provide a range of options. Uncontrolled greenhouse gas emissions from food production and processing are increasingly challenging sustainable food supply efforts. The food production industry is responsible for 50% of pollution on habitable land and 78% of water pollution worldwide, primarily due to nutrient-rich pollutants. Therefore, food production needs to become sustainable to reduce the use of fossil fuels, to protect animals, and to solve water pollution (Sarker et al., 2024). However, a major barrier to minimizing environmental damage is the growing demand for highly processed, fast-produced food. The absence of cutting-edge technologies to minimize pollution is a significant problem (Batool et al., 2024). One strategy being used to reduce carbon dioxide (CO_2) emissions is the introduction of renewable energy sources, which can help reduce reliance on fossil fuels. However, the industries that produce and supply food have not yet adopted comparable alternatives (Al-Rumaihi et al., 2020).

1.3 Significance of Utilizing Food Waste Biomass for Essential Oil Extraction

Food waste, both in its dry and wet forms, is a significant global issue, largely driven by the food industry. Due to its severe economic and environmental consequences, coupled with rising concerns about food security, this problem has garnered considerable attention. With the global population increasing, shifting demographics, and the adverse effects of climate change on food supply, addressing food waste has become an urgent priority. Recovering and distributing food from waste streams is complex and costly. Current food waste management systems often lack efficiency and financial sustainability, making them economically unviable. Perishable items like fruits and vegetables are more prone to spoilage, increasing waste compared to durable foods. This highlights the need for innovative and sustainable solutions to reduce food waste while addressing environmental and food security concerns (Roy et al., 2023). The term "food loss" describes the inadvertent decrease in the amount or quality of food, which frequently happens because of spills, bruises, spoiling, or other damages sustained during several phases of the supply chain, such as harvesting, processing, distribution, and storage (Lahiri et al., 2023). This type of loss typically happens when food is burned or thrown away after harvest, significantly decreasing the likelihood that the lost food can be upgraded into valuable things. There are many different places where food waste is generated, including residences, workplaces, and hospitals. It can include leftovers, expired goods, or precooked food. At every point in the food supply chain, an astounding 1.3 billion tons of food are lost annually, and the quantity of food waste keeps rising, according to the Food and Agriculture Organization (FAO). According to the FAO, a third of the food produced globally is lost or wasted (Kumar et al., 2022). In response to the growing issue of food waste, the concept of a circular bio-economy has emerged. This approach integrates green technology to create commercially viable products by repurposing waste as a resource. Major contributors to food loss include dairy products, fruits, vegetables, meat, and baked goods (Mohanty et al., 2022).

2 Extraction Techniques

2.1 Steam Distillation

To commercialize essential oils, it is crucial to maximize yield while ensuring high quality. One of the main challenges is the extraction process, which is often labor- and energy-intensive. Solvent extraction and azeotropic distillation are commonly used methods for obtaining essential oils. Azeotropic distillation encompasses techniques such as steam distillation, hydrodistillation (HD), and hydrodiffusion, with steam distillation being the most frequently employed due to its relatively low investment and operating costs, despite some risks related to yield and process. Steam distillation remains popular because it uses water steam instead of organic solvents, which can require explosion-proof installations, produce waste, and necessitate air treatment and exhaust systems (Kant & Kumar, 2022). This makes steam distillation more environmental-friendly. However, it is often associated with high resource consumption, long extraction times, and lower yields. Laboratory-scale tests typically show higher oil content in the plant matrix compared to commercial processes, suggesting that industrial equipment could be optimized to improve quality, reduce extraction times, and increase yields. Despite this, few advancements have been made beyond mechanical improvements aimed at reducing energy use. There is a need for further research and innovation to enhance the steam distillation process. Traditionally, steam distillation can extract approximately 93% of the volatile compounds from plants, with the remaining 7% typically obtained through other methods (Machado et al., 2024). The ease of scalability, operational safety, and environmental friendliness of this technology make it popular. In comparison to other extraction techniques, steam distillation is also efficient at yielding pure chemicals with less contaminants. Steam distillation does have some disadvantages, though (Kabuba & Huberts, 2009). To completely recover the EOs, the extraction method is laborious and slow. Residual oil that dissolves in the water during distillation can also produce unpleasant odors. To maximize essential oil recovery, the water used in steam distillation must be redistilled using a process called cohabitation, which recaptures the dissolved oil components. However, this redistillation increases utility costs, particularly in terms of energy and heating (Mejri et al., 2018). Figure 1 shows the steam distillation technique.

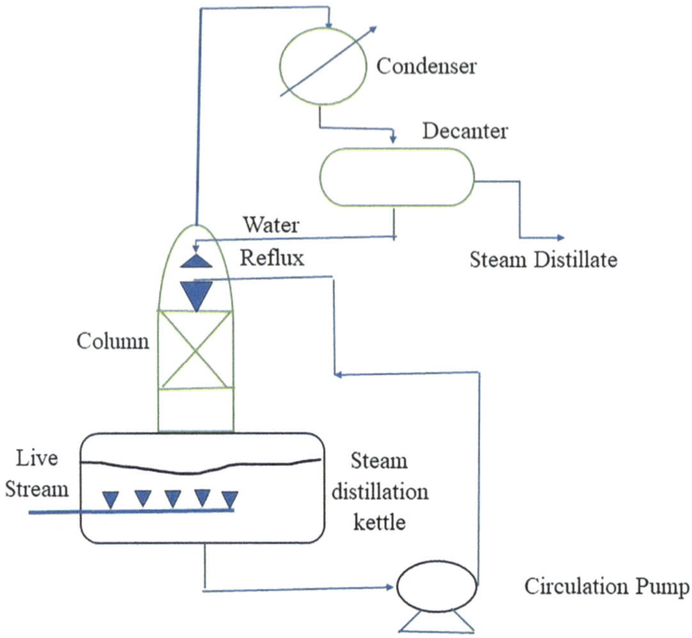

Fig. 1 Steam distillation technique

2.2 *Hydrodistillation*

Hydrodistillation is one of the oldest and most established methods for extracting essential oils, with a history extending over 5,000 years. It remains a widely used technique today and is well documented in the literature. One of its main advantages is its simplicity and ease of maintenance. In this method, plant materials are boiled in water, producing steam that carries the volatile oils into a cooling system where they condense and are collected (Fagbemi et al., 2021). Compared to more sophisticated extraction techniques, this simple process is accessible and less expensive because it requires less specialized equipment. Moreover, hydrodistillation is preferred because it produces high-quality results while maintaining the integrity of essential oils. Additionally, this technique is preferred because it produces high-quality products while maintaining the integrity of essential oils (Mahawer et al., 2022). Because it is dependable, simple to use, and less expensive to run, it remains a popular extraction method even in the face of technological developments like supercritical CO_2 extraction and solvent extraction. Its significant history and ongoing applicability in the essential oil sector highlight its value. Despite advancements in extraction technologies like solvent extraction and supercritical CO_2 extraction, hydrodistillation remains a popular method due to its low operating costs, ease of use, and reliability. Its long history and continued relevance in the essential oil industry underscore its enduring value (Aziz et al., 2018). The conventional and popular process of obtaining essential oils is hydrodistillation. The plant material is mixed with water or other solvents and heated in this procedure, which causes the essential oils to evaporate. The resulting vapor—which is made up of water and essential oils—is then transported into a condenser to cool and solidify. The liquid combination is gathered in a different chamber, where the water and essential oil separate organically because of differences in density (Katekar et al., 2022). One of the main advantages of hydrodistillation is its ease of use and adaptability. This method is suitable for both small- and large-scale operations due to its ability to scale efficiently. It also consumes less energy than traditional extraction techniques and minimizes the risk of losing chemical compounds during prolonged processes (Immaroh et al., 2021). In a study to extract eucalyptus essential oil, 300 milliliters of water were combined with 15 grams of eucalyptus leaves and distilled for five hours (Mohamed & Alotaibi, 2022). This approach primarily captured volatile compounds, while methods like Soxhlet and supercritical fluid extraction (SFE) yielded both volatile and heavier molecules (Joco et al., 2023). The yield of eucalyptus oil increased from 3.1% after one hour to 3.8% after five hours, demonstrating the effectiveness of this technique, particularly in extracting volatile components from eucalyptus oil (Rahman et al., 2023).

2.3 *Solvent Extraction*

Solvent extraction is a widely used technique across various industries, including hydrometallurgy, food processing, waste management, and pharmaceutical manufacturing. It involves separating a specific liquid component from a mixture of solid and liquid using a chemical solvent. The choice of solvent—whether polar or nonpolar—depends on the nature of the substance being extracted (Aziz et al., 2018). A variety of solvents is frequently utilized in industrial settings, such as acetone, diethyl ether, ethanol, methanol, hexane, chloroform, and petroleum ether. These solvents have different properties that determine which kinds of extractions they are appropriate for, based on variables such as solvent affinity, boiling point, and polarity (Bitwell et al., 2023). Solvent extraction can be considered as a mass transfer procedure in which the target chemical, like oil, moves from the solid phase into the solvent phase. A dynamic interaction between the phases may result from the solid substance simultaneously absorbing the solvent. This keeps going until the system reaches equilibrium, after which there is no more transfer (Costa et al., 2012). Solvent extraction can be performed either continuously or in batches. Continuous extraction, often using countercurrent flow, is ideal for large-scale operations as it improves efficiency by maximizing contact between phases. In batch extraction, methods like Soxhlet or fixed column setups are common. In a fixed column, the solvent slowly passes through a stationary solid sample, while the Soxhlet extractor continuously cycles solvent through the sample for thorough extraction (Patel et al., 2019). Solvent extraction is a fundamental technique in both laboratory and industrial settings because it offers a flexible and efficient way to separate and purify chemicals (Geow et al., 2021).

2.4 Supercritical Fluid Extraction

One of the most advanced and widely used methods for extracting and isolating essential oils from aromatic plants is supercritical fluid extraction (SFE) (Yousefi et al., 2019). This technology has gained increasing popularity due to its numerous advantages over traditional extraction methods, making it the preferred choice for both industrial and research applications (Sojic et al., 2023). SFE's capacity to offer quick and effective extraction is one of its main advantages. The method makes use of the special qualities of supercritical fluids, which combine the features of liquid and gas, to enhance the penetration and solvation of plant materials. This leads to an extraction method that is incredibly efficient and produces high-quality essential oils fast (Uwineza & Waskiewicz, 2020). Supercritical fluid extraction (SFE) is particularly notable for its ability to operate at moderate temperatures, which is advantageous when dealing with heat-sensitive substances such as those found in essential oils. Unlike traditional methods that often require high temperatures and extended extraction times, SFE minimizes thermal degradation and helps preserve the delicate aromatic and medicinal properties of essential oils (Ahangari et al., 2021). Another significant advantage of supercritical fluid extraction (SFE) is its ability to eliminate the need for cumbersome post-extraction cleanup. The high selectivity of supercritical fluids allows for precise targeting of molecules, reducing the extraction of unwanted materials and streamlining downstream processing. This not only simplifies the overall extraction process but also enhances the purity of the final product (Kamali et al., 2015). The use of hazardous organic solvents is notably avoided by SFE, making it a safer and more ecologically responsible choice. Conventional solvent-based extraction techniques frequently use volatile and possibly dangerous compounds, which can be harmful to the environment and human health. SFE, on the other hand, significantly improves both sustainability and safety by using carbon dioxide (CO_2) as the main solvent (Siddiqui et al., 2024).

CO_2 is an ideal solvent for extracting and separating essential oils from plants due to its effectiveness. It is a cost-effective and safe choice for large-scale operations, being nontoxic, nonexplosive, and readily available. Additionally, simply reducing the pressure allows for the easy removal of the solvent from the extracted compounds, resulting in a pure, solvent-free essential oil ready for use. This simple removal process is a key benefit of the technique (Machado et al., 2013). As CO_2 is a gas at room temperature, it can be easily and completely removed, ensuring that essential oils are free from residual solvents. This results in a pure, high-quality product that meets stringent safety and purity standards, making it ideal for the cosmetic, food, and pharmaceutical industries (Capuzzo et al., 2013). The extraction of essential oils from aromatic plants has been transformed by supercritical fluid extraction, to sum up. Since carbon dioxide is a perfect solvent and SFE is an efficient, safe, and environmentally friendly technique of extracting high-purity essential oils, it is the method of choice for both commercial and research contexts (Fornari et al., 2012).

2.5 *Microwave-Assisted Extraction (MAE)*

The rapidly shifting electric and magnetic fields at a frequency of 2450 MHz, which is frequently employed in commercial microwave systems, cause microwave heating to interact with dipoles in polar and polarizable materials. The dipoles in polar molecules are constantly trying to align themselves with the changing field due to the rapid direction changes of the electric and magnetic components of the microwave field. Because of internal friction and molecular collisions, this endeavor to continuously reorient in response to the oscillating electromagnetic field results in the formation of heat (Drinic et al., 2020). Polar solvents heat efficiently with microwave radiation due to their dipole moments interacting with the electric field. In contrast, nonpolar solvents, lacking polarizable groups, experience reduced heating efficiency. In these cases, only dielectric absorption occurs, leading to less effective and lower-quality heating (Haffizi et al., 2020). Even though microwave radiation's thermal effects happen extremely instantly at the molecular level, they typically only affect a little region close to the material's surface. Large particles or clusters of smaller particles do not heat evenly due to this limited depth of penetration (Mazzara et al., 2022). Microwave energy is primarily absorbed by the surface of the material, requiring conduction to transfer heat to the core. A significant drawback of microwave heating technology is the potential for uneven heating. Despite the use of high-power microwave sources to improve penetration depth, the efficiency of this method is limited because microwave radiation exponentially decreases as it penetrates deeper into the material (Akhbari et al., 2018).

There are two main ways that materials receive energy from microwave heating. Dipole rotation: at 2450 MHz, an alternating electromagnetic field causes polar molecules to reorient. Internal friction is created by the fast movement (around 4.93 billion times per second), which results in heat. Ionic conduction: electrophoretic migration is the movement of charged ions in a substance in response to an electromagnetic field. Frictional heat is generated by the resistance that ions encounter in the surrounding medium. In practical applications, dipole rotation and ionic conduction can occur simultaneously. When compared to conventional methods, microwave heating offers several advantages: it shortens processing times and reduces temperatures, leading to enhanced product quality (Hu et al., 2021); better yields of target components due to increased marker recovery; increased purity of extracts, precise temperature control prevents further deterioration; decreased processing expenses, less time and energy needed; faster extraction, much faster than with traditional techniques; reduced energy and solvent consumption, more effective, less harmful to the environment and waste; and possibility of new products, product development might be innovative, thanks to special heating properties.

2.5.1 Microwave-Assisted Extraction (MAE) Parameters

Multiple factors influence the effectiveness of MAE, including solvent-to-feed ratio (S/F) and solvent selection, the ratio and choice of solvents influence the extraction efficiency; solvents need to absorb microwaves; extraction time and cycle, the length of time and the number of cycles maximize purity and yield (Tran et al., 2020); microwave power and extraction temperature, quality and efficiency can be improved by varying the power and temperature parameters; water content and contact surface area effects on microwave; efficiency of heating and energy absorption; and stirring reduces hotspots and ensures uniform energy dispersion.

2.5.2 Microwave-Assisted Extraction's Drawbacks

Microwave-Assisted Extraction (MAE) has certain disadvantages despite its benefits. For instance, the method requires solvents that absorb microwaves, which restricts the selection of suitable solvents. Additionally, a cleanup step is often necessary after extraction, which can increase both the time and cost of the process. Lastly, the need for a cooling period extends processing times, as the containers and their contents must cool before handling (Son et al., 2023).

2.6 Comparison of Methods: Principles, Advantages, and Limitations

This chapter provides a comprehensive analysis of modern and traditional techniques for obtaining essential oils from plants. Conventional techniques such as solvent extraction, steam distillation, and cold-pressing have been used for decades. These methods are widely commended for their simplicity of use and ability to produce oils that faithfully convey the plant's scent. These techniques are frequently praised for their ease of use and their capacity to yield oils that accurately capture the aroma of the plant (Bitwell et al., 2023). However, because of the heat or chemicals required, they can also be time-consuming, less effective, and occasionally cause sensitive materials to degrade. On the other hand, contemporary methods like as ultrasonic, microwave, and CO_2 extraction offer greater yields, enhanced efficiency, and better preservation of the aromatic qualities of the oil. For instance, supercritical CO_2 extraction is commended for its capacity to yield pure, superior oils without the need of heat or solvents while maintaining the entire range of aromatic compounds found in the plant (Oubannin et al., 2024). These advanced techniques are more expensive and less accessible to small-scale producers due to the required specialized equipment and technical expertise (Souiy, 2023). The chapter examines the fundamentals of each extraction method, focusing on how these processes impact the yield, quality, and environmental effects of essential oil production. Additionally, it provides a comprehensive comparison to assist both industrial and artisanal producers in selecting the most appropriate extraction method.

3 Food Waste Biomass as a Source of Essential Oils

To fully understand the concepts of "food waste" and "food loss," it is essential to grasp the broader meanings of "food" within moral, ethical, socioeconomic, and nutritional contexts. "Food" refers to any material intended for human or animal consumption, whether raw, partially processed, or fully processed. Meals often come with "inedible parts" that are unsuitable for consumption or use. As a vital necessity for life, concerns about waste and loss require careful consideration (Lahiri et al., 2023). Reducing these issues is crucial to mitigating their impact on people, the environment, and ecosystems. Waste can be liquid, semi-solid, or solid, with solid waste typically comprising the food products themselves, their components, or related parts. Such waste often contains high levels of organic substances like monosaccharides, cellulose, starch, and lignin, found in various food sources (Pour & Makkawi, 2021).

3.1 Types of Food Waste Biomass Suitable for EO Extraction

Food waste biomass ideal for essential oil extraction includes a range of by-products from food processing and consumption, such as fruit peels, vegetable scraps, seeds, and other plant residues. Fruit peels, particularly those from oranges, lemons, and grapefruits, are notably rich in essential oils due to their high content of volatile and aromatic compounds. These peels are frequently discarded in large amounts, making them a valuable resource for extraction. Likewise, vegetable scraps, including leftover herbs and leafy greens, also contain essential oils that can be utilized for their flavor and medicinal benefits (Sahoo et al., 2023). The diverse range of aromatic compounds present in these materials makes them suitable for various applications, from flavoring to therapeutic uses. Additionally, there is potential for essential oil extraction from seeds and other plant wastes, such as those from spices and herbs. Rich oil content found in the seeds of plants like coriander and cumin, as well as avocados, is recognized to have uses in both cosmetic and culinary arts (Karne et al., 2023). Moreover, leftover grains from brewing processes or oilseeds from extraction procedures are examples of food processing by-products that can be repurposed to produce essential oils. The residual oils and chemicals in these materials can be recovered and processed, thereby reducing waste and enhancing resource efficiency. By identifying and utilizing these diverse types of food waste biomass, the potential for sustainable essential oil production can be significantly increased (Munir et al., 2024).

3.2 Chemical Composition of Essential Oils in Different Food Waste Sources

The type of biomass used and the extraction methods employed significantly influence the chemical composition of essential oils derived from various food waste sources. Essential oils are complex mixtures containing a variety of chemicals, including phenols, alcohols, aldehydes, ketones, esters, terpenes, and other aromatic compounds (Gonzalez-Mas et al., 2019). This provides a comprehensive overview of the chemical makeup of essential oils extracted from different kinds of food waste (Bora et al., 2020). Citrus essential oils are widely used as flavoring agents in culinary, medicinal, and cosmetic products, as well as in the creation of natural fruity perfumes. These oils are primarily extracted from the flavedo, or the outer rind of the fruit, though the flowers and foliage can also be utilized. The oil composition consists of 85–99% volatile and semi-volatile compounds, comprising over 200 different substances. Hydrocarbon and derivative monoterpenes and sesquiterpenes are the most reported components (Zhang et al., 2022). These are followed by a range of aromatic compounds and aliphatic and olefinic C_6–C_{12} non-terpene chemicals, including alcohols, acids, ketones, esters, and aldehydes. Sterols, fatty acids, diterpenoids, coumarins, and flavonoids make up most of the nonvolatile residue (Agarwal et al., 2022).

3.3 Factors Influencing Extraction Efficiency and Oil Quality

The efficiency and quality of essential oil extraction are affected by several factors, which can be broadly grouped into plant-related factors, process-related factors, and environmental conditions. Understanding these factors is essential for optimizing the extraction process and producing high-quality essential oils with desirable properties. Key variables include the variety and maturity of the raw materials, sample preparation and post-harvest processing techniques, the type of drying method used, and storage conditions such as temperature and packaging. These factors play a critical role in maintaining the quality of essential oils during storage (Samadi et al., 2020). Over time, chemical reactions such as oxidation, evaporation, and isomerization can change the makeup of the oil. For example, spearmint essential oil keeps its freshness longer under cold storage, especially in the freezer (Kant & Kumar, 2022). On the other hand, plastic packaging in a controlled environment for six to nine months was shown to be the ideal storage conditions for *Ocimum gratissimum* leaves. Overall, the quality of essential oils throughout storage is dependent on maintaining appropriate storage conditions and limiting exposure to factors that degrade them (Ngo et al., 2020).

4 Purification Strategies

4.1 *Fractional Distillation*

Fractional distillation is a separation method used for separating two or more components based on their volatility. The process is influenced by factors such as temperature, pressure, and the unique physical and chemical properties of the components. To enhance separation efficiency, fractional distillation can be performed alongside steam or hydrodistillation, or independently using a vacuum distillation apparatus and a fractionating column. By lowering the boiling points of the components, this technique reduces the degradation of sensitive chemical compounds and allows for separation at lower temperatures (Zhou et al., 2023). The success of the fractionation process largely depends on the boiling points of each component. By carefully regulating the pressure and temperature within the distillation apparatus, different fractions are obtained at varying temperatures. The process is further enhanced under vacuum conditions, which reduce atmospheric pressure and lower the boiling points of the components. This approach is particularly beneficial for temperature-sensitive compounds, as it prevents thermal degradation and preserves the integrity of the separated substances. Additionally, the development of batch distillation techniques has made fractional distillation more precise and efficient. By processing specific quantities of material in distinct batches, this method allows for greater control over the distillation process and the quality of the separated fractions. Overall, fractional distillation remains a versatile and effective method for separating volatile substances based on their boiling points. By regulating temperature, pressure, and other operational parameters, this method ensures the precise and efficient separation of components while minimizing the risk of degradation and preserving the quality of the final product (Silvestre et al., 2019).

4.2 *Microwave Distillation Technique*

Microwave is a type of electromagnetic wave with frequencies ranging from 0.3 to 300 GHz and wavelengths between 0.01 and 1 meter. These waves are known for their strong penetrating ability and significant thermal effects. What makes microwaves unique is their ability to penetrate materials efficiently due to their perpendicular electric and magnetic fields. This leads to a rapid increase in both internal and external temperatures of the material, promoting cell rupture. The breakdown of the cell structure facilitates the diffusion of internal components into the surrounding extraction solution (Boudraa et al., 2021). Microwave heating is particularly unique and effective for the extraction of EOs. Unlike conventional heating methods, which primarily transfer heat from the outside to the inside of the material, microwave heating transfers heat in both directions—some flows from the outside in, while some flows from the inside out. This two-way heat transfer enhances the extraction efficiency of EOs by aligning with the mass transfer direction of volatile oil components. During microwave-assisted extraction, the interaction between heat and mass transfer significantly boosts both the yield and quality of the extracted oils (Farhat et al., 2011). The potential of microwave-assisted extraction (MAE) to efficiently extract bioactive chemicals from a variety of plant materials has drawn a lot of attention lately. Examples include the extraction of essential oils from *Thymus vulgaris* L., *Humulus lupulus*, and *Citrus limon* (Lisbon variety) peels, among other plants. According to studies, an extraction yield that is eight times higher than that of conventional hydrodistillation (HD) techniques can be obtained by utilizing microwave-assisted hydrodistillation (MAHD) for just 30 minutes. Microwave time is the most important factor influencing the yield of essential oils in MAE, followed by microwave power and the solid-to-liquid ratio (Mukherjee et al., 2024). Moreover, using higher microwave power settings in MAHD can lower CO_2 emissions and quickly and dramatically enhance the output of essential oils. But it is important to remember that using too much microwave power can change the extracted oils' chemical makeup, which could have an impact on their effectiveness and purity. The solvent-free microwave extraction (SFME) technology was created to meet the need for quicker energy transfer, more effective heating, eco-friendly operations, and less time and money spent (Amaresh et al., 2017). SFME is considered a promising method for essential oil extraction. It utilizes dry distillation combined with microwave heating, eliminating the need for water or chemical solvents, making it a more environmental-friendly and sustainable approach. This technique minimizes the environmental impact of the extraction process while preserving the natural qualities of the oils. Overall, the advancement of microwave technology and its application in essential oil extraction represents a significant step forward in both sustainability and efficiency, offering a valuable alternative to traditional extraction methods (Yingngam et al., 2021).

4.3 Solvent-Assisted Technique

In solvent-assisted procedures for essential oil purification, solvents are used to selectively remove contaminants and enhance the quality of the oils. Solvents such as ethanol, methanol, hexane, or supercritical CO_2 are commonly employed because of their ability to dissolve specific components with similar chemical properties. The process typically begins by mixing the essential oil with the chosen solvent and allowing them to interact. Depending on the method used—such as liquid–liquid extraction, solid-phase extraction, or adsorption—the mixture is then separated into layers or phases, with the contaminants either dissolved in the solvent or adsorbed onto a solid material (Fan & Li, 2022). Next, the refined essential oil is extracted from the leftover layer or phase. By focusing on removing components or contaminants, these processes are very useful in enhancing the purity and sensory qualities of essential oils. To guarantee that the procedure is efficient and environmental-friendly while preserving the integrity of the intended aromatic components in the essential oil, they do, however, need careful solvent selection (Turek & Stintzing, 2013).

4.4 *Chromatography (Column Chromatography, Thin-Layer Chromatography [TLC])*

Chromatography is a fundamental and widely used technology in separation science, valued for its broad applications in the chemical process industry and laboratories. It is supported by a robust theoretical foundation and is known for its simplicity and versatility. The core principle of chromatography involves separating components of a mixture based on their differing affinities for a stationary phase, which selectively retains the components, and a mobile phase, which moves the mixture through the system (Coskun, 2016). This makes it possible to efficiently separate, analyze, isolate, and purify a variety of substances from big macromolecules to tiny chemical compounds. The various forms of chromatography, such as gas chromatography (GC), liquid chromatography (LC), thin-layer chromatography (TLC), and high-performance liquid chromatography (HPLC), each designed for uses and analyte kinds, demonstrate the method's adaptability. For example, HPLC is better suited for nonvolatile or thermally unstable molecules, whereas GC works best for volatile substances (Attimarad et al., 2011). Chromatography is a vital tool in research, quality control, and production settings because of its simple operational approach, ability to produce exact and repeatable results, and advancements in automated systems and advanced detectors. Furthermore, advancements in chromatographic materials and technologies continue to expand its range of applications, offering enhanced resolution, faster analysis times, and the ability to handle complex mixtures. These improvements bolster the instrument's reputation as a crucial and widely utilized tool in both preparatory and analytical environments (D'Atri et al., 2018).

4.5 *Membrane Filtration*

Membrane filtering is a crucial step in the purification of essential oils, preserving their integrity and potency while removing undesirable chemicals, water residues, and contaminants. This technique employs semipermeable membranes with varying pore sizes to filter out impurities based on their chemical composition or molecular size. The essential oil passes through a series of membranes, including those used for reverse osmosis (RO), microfiltration (MF), ultrafiltration (UF), and nanofiltration (NF) (Olkiewicz et al., 2022). Each type of membrane serves a specific purpose: reverse osmosis and nanofiltration are effective at removing tiny contaminants and solvents, while microfiltration targets larger particles and bacteria. By avoiding excessive heat or chemicals that could compromise the oil's quality, this method ensures that only the pure essential oil components are retained. Membrane filtering thus offers an effective and environmental-friendly means of refining essential oils, making it suitable for various applications in aromatherapy, natural products, and perfumery (Divakar et al., 2022).

4.6 Importance of Purification in Enhancing Essential Oil Quality and Purity

The global essential oils market has experienced steady and robust growth across various major end-use industries, including food and beverage, personal care, cosmetics, and aromatherapy. The anticipated health benefits of essential oils are expected to drive increased demand in pharmaceutical and medical applications. Unlike many prescription medications, essential oils generally do not have significant adverse effects, which is likely to contribute to the market's expansion. Additionally, the rising prevalence of age-related and stress-related health issues, such as anxiety, diabetes, Alzheimer's disease, and cardiovascular conditions, has heightened the demand for therapeutic essential oils in aromatherapy. However, due to their notable medicinal effectiveness and market value, essential oils are sometimes subject to counterfeiting or contamination with inferior, less expensive alternatives (Jackson-Davis et al., 2023). The variation in essential oils from the same botanical species is mostly due to the extraction and purification process, but it is also influenced by the chemical composition, which is influenced by climate and geography. As a result, verifying the identity of essential oils can be challenging. A practical approach involves using mass spectrometric detection alongside other chromatographic techniques, such as gas chromatography and high-performance liquid chromatography, to accurately quantify the chemical components. With exceptional accuracy and precision, analytical techniques such as gas chromatography (GC), high-performance liquid chromatography (HPLC), nuclear magnetic resonance (NMR) spectroscopy, and electroanalytical methods are frequently employed for the analysis and assessment of consumable product authenticity. Unfortunately, some of these techniques have limitations that hinder their routine use for control analysis. These drawbacks include lengthy analysis times, high costs, and a lack of portable equipment (Marriot et al., 2001). Moreover, these methods often require detailed tuning of chromatographic separation parameters for each type of essential oil. Interpreting a GC/mass spectrometry (MS) chromatogram also demands significant expertise and experience. The quantification process further requires standards and predefined ranges for each component to determine if their concentrations fall within acceptable limits (Rubiolo et al., 2008).

To overcome these obstacles, chemometric analysis, in conjunction with nondestructive, user-friendly, analytical spectroscopic techniques like infrared (IR) spectroscopy—a green tool—has been utilized for food product quality assurance and identification. Fourier-transform infrared (FTIR) spectroscopy has become a popular nontargeted fingerprinting method as a substitute for traditional GC-based techniques since it provides a quick, environmental-friendly, nondestructive, and economical way to evaluate the quality of essential oils (Agatonovic-Kustrin et al., 2020). The primary advantage of infrared spectroscopy is its capacity for rapid and straightforward data analysis with minimal to no sample preparation (Alighiri et al., 2018). Most IR spectrometers are lightweight, portable, and cost-effective, enabling online sample analysis (Mahawer et al., 2022).

5 Economic and Sustainability Considerations

Essential oils present a complex interplay of sustainability and economic considerations, balancing substantial market potential with social and environmental responsibilities. The essential oil industry significantly contributes to the global economy, particularly in sectors such as medicine, aromatherapy, and cosmetics. However, the production of essential oils is resource-intensive, often requiring large quantities of plant material (Ferraz et al., 2022). This raises concerns about overharvesting wild species and allocating agricultural land for essential oil crops instead of food production, which can lead to soil degradation and loss of biodiversity. From a sustainability perspective, it is crucial to balance economic gains with environmental stewardship by ensuring fair labor practices and ethical sourcing. Thus, the essential oils sector embodies a nuanced mix of sustainability and economic factors, requiring careful consideration of both social and environmental impacts (Maurya et al., 2021). The world economy benefits economically from the essential oil industry, which is highly demanded in industries like medicines, aromatherapy, and cosmetics. Nevertheless, the extraction of essential oils requires a lot of resources and frequently calls for substantial amounts of plant material, which increases (Carrubba & Catalano, 2009). Certain impacts can be mitigated by adopting sustainable agricultural practices, including organic farming, water-efficient irrigation, and minimization of waste during the distillation process. Furthermore, the production of essential oils can have a smaller environmental impact by using renewable energy sources for transportation and distillation, which will help the sector meet global sustainability targets (Wells, 2024).

5.1 Cost Analysis of Extraction and Purification Processes

The cost analysis of essential oil extraction and purification methods involves a detailed assessment of capital, operating, and labor expenses across techniques such as solvent extraction, CO_2 extraction, steam distillation, and cold-pressing. Steam distillation, though the most widely used method, incurs significant capital costs for equipment like boilers and distillation units, along with ongoing expenses for fuel, water, and maintenance. While cold-pressing offers reduced operational costs due to lower energy and equipment demands, its lower extraction efficiency makes it less viable for certain oils, particularly those requiring higher yields (Wainer et al., 2022). Solvent extraction and CO_2 extraction, on the other hand, entail higher capital and operational costs due to the need for specialized equipment and solvents, especially in supercritical CO_2 systems that involve precise temperature and pressure control. Purification processes, such as fractional distillation and membrane filtration, significantly influence overall expenditures based on the required purity levels and process complexity. These stages are crucial for targeting high-value markets but introduce additional costs that must be weighed against yield and quality outcomes. Additionally, the extraction and purification process affect labor expenses, as more labor-intensive techniques require skilled professionals, increasing total costs. The choice of extraction and purification methods ultimately depends on the desired quality and purity of the essential oil, considering both the potential market value of the final product and the associated cost factors (Ahmadi et al., 2024).

5.2 Energy Consumption and Waste Management

The extraction and purification of essential oils from food waste biomass require stringent waste management practices and energy-intensive procedures. Techniques such as solvent extraction, steam distillation, and cold-pressing are major sources of energy consumption, demanding significant amounts of heat, electricity, and occasionally chemicals. While these methods are often designed to be energy-efficient, they can still be costly, particularly on a large scale. Additionally, managing residual biomass and by-products from the extraction process is crucial. Effective waste management can transform these by-products into valuable resources, such as compost or biofuels, enhancing the overall sustainability of the process. Conversely, improper disposal can have detrimental environmental impacts (Paritosh et al., 2017). The environmental impact of producing essential oil from food waste biomass can be greatly decreased by putting strong waste management procedures and energy-efficient technologies into place. This makes the process a more sustainable method of using resources. To properly comprehend the ideas of "food loss" and "food waste," one must first comprehend the nutritional value, socioeconomic significance, morality, and ethics of "food." Any material that is mostly meant for human or animal consumption, whether it be raw, semi-processed, or processed, is referred to as food. Food often includes "inedible parts" that are typically discarded. Given the essential role of food for survival, addressing issues like "food loss" and "food waste" is crucial. Reducing these instances is important not only for human well-being but also for maintaining ecological balance and protecting the environment (Mohanty et al., 2022).

5.3 *Marketability of Purified Essential Oils Derived from Food Waste*

As consumers and businesses grow more environmentally conscious, the marketability of refined essential oils made from food waste is growing. As the market for ethically and environmentally generated goods grows, these oils—which are made from biomass that would otherwise be discarded—offer a sustainable substitute for traditional sources (Sniegonova et al., 2023). The unique selling proposition of these essential oils lies in their ability to convert waste into valuable products, appealing to both environmentally conscious businesses and sustainability-minded consumers. Furthermore, these oils can be competitively positioned in industries such as cosmetics, aromatherapy, and natural health products due to their quality and purity meeting or exceeding industry standards (Kumar et al., 2017). However, their marketability also depends on factors like certification, consumer awareness, and the ability to scale production profitably. With effective marketing strategies and a focus on sustainability, these purified essential oils have the potential to capture a significant share of the global market, addressing food waste issues and advancing a circular economy (Torres-Leon et al., 2018).

5.4 *Sustainability Metrics and Environmental Impact Assessment*

The extraction of essential oils from food waste biomass is evaluated using sustainability metrics and environmental impact assessments to ensure both economic and environmental viability. These metrics provide measurable insights into various production aspects, including waste reduction, water and energy consumption, and carbon emissions. By monitoring these parameters, producers can optimize their processes, reduce resource usage, and improve the efficiency of converting food waste into valuable essential oils. This approach not only minimizes the environmental footprint by reusing waste that would otherwise end up in landfills but also supports a circular economy (Campion & Essel, 2013). Sustainability metrics also offer a way to measure financial gains, such as savings from reduced waste. Evaluating the economic benefits of sustainability initiatives, including potential revenue from high-value essential oils and cost savings from waste reduction, reinforces their viability (Meshalkin et al., 2023). A comprehensive examination of the potential environmental impacts associated with extracting essential oils from food waste biomass helps identify and address concerns related to ecosystem effects, soil and water contamination, and the impact of increased production. Incorporating these findings into mitigation efforts allows producers to implement environmental-friendly extraction techniques, effective waste management plans, and eco-friendly disposal or utilization of by-products. Together, sustainability metrics and impact assessments promote responsible resource use and environmental stewardship while minimizing waste (Gulis et al., 2022).

6 Future Perspectives and Research Directions

Future research on essential oil production from food waste biomass is promising, focusing on improving extraction methods, using eco-friendly technologies like green solvents and supercritical fluid extraction, and exploring diverse waste streams. Integrated biorefinery approaches could produce additional by-products like biofuels and bioplastics (Wang et al., 2022). Investigating the bioactivity and health benefits of these oils may open new markets in wellness and natural medicine. Collaboration among academia, industry, and government is crucial for scaling processes and developing supportive regulatory frameworks (Quevedo-Amador et al., 2024).

6.1 Innovations in Extraction and Purification Technologies

Advancements in essential oil extraction and purification are enhancing productivity, sustainability, and quality. Innovations like supercritical fluid extraction (SFE) with CO_2 provide eco-friendly alternatives to traditional methods, while microwave-assisted extraction (MAE) and ultrasound-assisted extraction (UAE) cut extraction times and energy use (Rodriguez-Marquez et al., 2022). Improved purification techniques, such as molecular distillation and membrane filtration, ensure precise refinement and maintain therapeutic properties (Bhadange et al., 2024). These developments support sustainability and enable the production of high-quality oils to meet growing demand for natural and organic products (Suwal & Marciniak, 2019).

6.2 Optimization of Processes for Enhanced Efficiency and Sustainability

Optimizing essential oil production with advanced technologies and streamlined processes enhances efficiency and sustainability. Techniques like supercritical CO_2 extraction provide higher purity with fewer hazardous chemicals (Kamali et al., 2015). Improvements in distillation, raw material handling, automation, and closed-loop systems reduce waste, energy use, and carbon footprint (Utomo et al., 2018). Utilizing renewable energy and sustainable sourcing further boosts efficiency. These advancements support eco-friendly product demands and align with global sustainability goals (Ara et al., 2015).

6.3 Exploration of Novel Food Waste Sources and Their Potential for Essential Oil Extraction

Exploring food waste for essential oil extraction offers an innovative solution to both waste management and the demand for natural products (Crescente et al., 2023). Utilizing peels, seeds, and other by-products can yield valuable oils for culinary, cosmetic, and pharmaceutical uses, reducing waste disposal impact and adding value to discarded materials (Skendi et al., 2022). The diverse chemical compositions of these oils may uncover unique compounds with special antimicrobial, antioxidant, or therapeutic properties, while also lowering production costs and enhancing the economic and environmental sustainability of essential oil extraction (Lankanayaka et al., 2021).

6.4 Integration of Circular Economy Principles in Valorizing Food Waste Biomass

Applying circular economy principles to food waste biomass transforms waste into valuable resources, such as biogas, fertilizers, essential oils, and natural colors (Ouro-Salim & Guarnieri, 2021). This approach reduces environmental impact, lowers waste management costs, and creates new revenue streams while fostering innovation in resource management and product development, like biodegradable packaging and bio-based chemicals (Sharma et al., 2023). It also promotes industry collaboration by repurposing one sector's waste as another's resource, enhancing resource efficiency and supporting a more sustainable and resilient economy (Roy et al., 2023). It promotes resource efficiency while supporting more eco-friendly practices within the industry (Lucarini et al., 2021). By utilizing waste as a resource, businesses can reduce disposal costs, enhance their sustainability credentials, and contribute to the development of a more sustainable food system. Additionally, producers can reduce their reliance on chemical additives by employing essential oils as natural taste enhancers and preservatives. This will appeal to health-conscious consumers and expand the market for sustainable products (Nirmal et al., 2023). The ramifications hold similar weight in the fragrance sector. The process of extracting essential oils from food waste offers a chance to create novel, environmental-friendly scent items that meet the increasing demands of consumers for ethical and sustainable goods. A more sustainable supply chain that derives raw materials from trash can help the business by lowering the environmental effect of conventional farming methods and lowering the demand for water and land resources (Sharmeen et al., 2021). Furthermore, because food waste has distinct chemical compositions, essential oils made from it may have novel scents with distinctive characteristics, providing perfumers with new creative possibilities. The fragrance industry has the potential to set the standard for environmental stewardship by putting these sustainable strategies into reality and producing products that are not only aesthetically pleasing but also environmental-friendly (Mahmud et al., 2024).

6.5 Recommendations for Future Research and Industry Initiatives

To enhance yield and preserve the chemical integrity of essential oils, future research on extracting these oils from food waste biomass should focus on optimizing extraction techniques. Advanced methods such as enzymatic, microwave-assisted, and supercritical fluid extraction should be explored and compared with traditional techniques like steam distillation, assessing both their efficiency and environmental impact. Additionally, studies should evaluate the qualities of essential oils derived from various types of food waste to identify the biomass sources that produce the most potent or unique oils. This may involve analyzing how pretreatment processes, such as grinding or drying, influence the yield and quality of the extracted oils. Further research is also needed to assess the stability and shelf life of these essential oils to ensure their commercial viability and explore their potential applications in industries like food preservation, cosmetics, and pharmaceuticals. Collaboration between industry, academia, and government is crucial for advancing sustainable practices in the extraction and purification of essential oils from food waste. Investment in research and development by industry stakeholders will be the key to scaling these extraction methods in a way that is both environmentally and economically sustainable. Enterprises with food processors and farmers can help secure a reliable supply of food waste biomass. Additionally, creating certification standards for essential oils derived from food waste, along with ensuring supply chain transparency and traceability, can attract eco-conscious consumers. Advocating for incentives and supportive regulations will further integrate these sustainable practices into the broader market, helping to maximize the value of food waste biomass and support a circular economy.

7 Conclusion

Extracting essential oils from food waste biomass offers a sustainable way to repurpose waste while producing valuable bioactive compounds. This process reduces environmental impact and supports the development of natural products in industries like food, cosmetics, and pharmaceuticals. Improving extraction methods enhances both the yield and quality of these oils, highlighting the potential for industrial innovation and environmental conservation. The growing interest in essential oils, particularly from biomass sources, underscores the importance of sustainable practices in maximizing the value of natural resources. As research advances, the extraction and purification of essential oils from food waste biomass offer promising solutions for resource efficiency, waste management, and environmental sustainability. This process reduces the environmental impact of food waste while providing a valuable source of essential oils with unique scents and medicinal properties. Food waste biomass, such as fruit peels and vegetable scraps, is rich in compounds like terpenes, sought-after in the food, cosmetics, and pharmaceutical industries for their antibacterial, antifungal, and antioxidant qualities. Supporting the circular economy and waste valorization, this method meets the growing demand for essential oils without depleting natural resources. These chapters explore the extraction processes from food waste, focusing on different techniques and their impact on the chemical composition and quality of essential oils. The reuse of food waste reduces its ecological footprint and creates high-quality, marketable products, offering both environmental and economic benefits. The chapter also evaluates the sustainability and industrial feasibility of essential oil extraction from biomass, highlighting research opportunities to further enhance these methods. Using food waste to extract essential oils for the food and fragrance industries carries significant implications for sustainability. By repurposing biomass like fruit peels, vegetable scraps, and other by-products, this approach aligns with the principles of the circular economy and waste valorization. This is especially relevant in the food industry, where waste is a major issue. This method not only reduces the environmental impact of food waste but also provides an affordable source of high-quality ingredients for flavoring, packaging, and food preservation.

References

Abubakar, I. R., Maniruzzaman, K. M., Dano, U. L., AlShihri, F. S., AlShammari, M. S., Ahmed, S. M. S., Al-Gehlani, W. A. G., & Alrawaf, T. I. (2022). Environmental sustainability impacts of solid waste management practices in the global South. *International Journal of Environmental Research and Public Health, 19*(19), 12717. https://doi.org/10.3390/ijerph191912717

Agarwal, P., Sebghatollahi, Z., Kamal, M., Dhyani, A., Shrivastava, A., Singh, K. K., Sinha, M., Mahato, N., Mishra, A. K., & Baek, K. (2022). Citrus essential oils in aromatherapy: Therapeutic effects and mechanisms. *Antioxidants, 11*(12), 2374. https://doi.org/10.3390/antiox11122374

Agatonovic-Kustrin, S., Ristivojevic, P., Gegechkori, V., Litvinova, T. M., & Morton, D. W. (2020). Essential oil quality and purity evaluation via FT-IR spectroscopy and pattern recognition techniques. *Applied Sciences, 10*(20), 7294. https://doi.org/10.3390/app10207294

Ahangari, H., King, J. W., Ehsani, A., & Yousefi, M. (2021). Supercritical fluid extraction of seed oils – A short review of current trends. *Trends in Food Science & Technology, 111*, 249–260. https://doi.org/10.1016/j.tifs.2021.02.066

Ahmadi, K. E., Haboubi, K., Abdouni, A. E., Alaoui, H. E., Dira, I., Bastrioui, M. E., Dimane, F., & Hammoudani, Y. E. (2024). Evaluation of a new process for extracting essential oil from aromatic, medicinal, and pharmaceutical plants. *E3S Web of Conferences, 527*, 01008. https://doi.org/10.1051/e3sconf/202452701008

Akhbari, M., Masoum, S., Aghababaei, F., & Hamedi, S. (2018). Optimization of microwave assisted extraction of essential oils from Iranian Rosmarinus officinalis L. using RSM. *Journal of Food Science and Technology, 55*(6), 2197–2207. https://doi.org/10.1007/s13197-018-3137-7

Ali, B., et al. (2015). Essential oils used in aromatherapy: A systemic review. *Asian Pacific Journal of Tropical Biomedicine, 5*(8), 601–611. https://doi.org/10.1016/j.apjtb.2015.05.007

Alighiri, D., Cahyono, E., Eden, W. T., Kusuma, E., & Supardi, K. I. (2018). Study on the improvement of essential oil quality and its repellent activity of betel leaves oil (Piper betle l.) from Indonesia. *Oriental Journal of Chemistry, 34*(6), 2913–2926. https://doi.org/10.13005/ojc/340631

Al-Rumaihi, A., et al. (2020). Environmental impact assessment of food waste management using two composting techniques. *Sustainability, 12*(4), 1595. https://doi.org/10.3390/su12041595

Amaresh, A., Guha, P., Khan, S., & Zari, S. R. (2017). Comparative study of microwave assisted hydro-distillation with conventional hydro-distillation for extraction of essential oil from Piper beetle L. *Biosciences Biotechnology Research Asia, 14*(1), 401–407. https://doi.org/10.13005/bbra/2458

Ara, K. M., Jowkarderis, M., & Raofie, F. (2015). Optimization of supercritical fluid extraction of essential oils and fatty acids from flixweed (Descurainia Sophia L.) seed using response surface methodology and central composite design. *Journal of Food Science and Technology, 52*(7), 4450–4458. https://doi.org/10.1007/s13197-014-1353-3

Attimarad, M., Ahmed, K. K. M., Aldhubaib, B. E., & Harsha, S. (2011). High-performance thin layer chromatography: A powerful analytical technique in pharmaceutical drug discovery. *Pharmaceutical Methods, 2*(2), 71–75. https://doi.org/10.4103/2229-4708.84436

Aziz, Z. A. A., Ahmad, A., Setapar, S. H. M., Karakucuk, A., Azim, M. M., Lokhat, D., Rafatullah, M., Ganash, M., Kamal, M. A., & Ashraf, G. M. (2018). Essential oils: Extraction techniques, pharmaceutical and therapeutic potential – A review. *Current Drug Metabolism, 19*(13), 1100–1110. https://doi.org/10.2174/1389200219666180723144850

Batool, F., et al. (2024). Environmental impacts of food waste management technologies: A critical review of life cycle assessment (LCA) studies. *Trends in Food Science & Technology, 143*, 104287. https://doi.org/10.1016/j.tifs.2023.104287

Bhadange, Y. A., Carpenter, J., & Saharan, V. K. (2024). A comprehensive review on advanced extraction techniques for retrieving bioactive components from natural sources. *ACS Omega, 9*(29), 31274–31297. https://doi.org/10.1021/acsomega.4c02718

Bitwell, C., Indra, S. S., Luke, C., & Kakoma, M. K. (2023). A review of modern and conventional extraction techniques and their applications for extracting phytochemicals from plants. *Scientific African, 19*, e01585. https://doi.org/10.1016/j.sciaf.2023.e01585

Bolouri, P., et al. (2022). Applications of essential oils and plant extracts in different industries. *Molecules, 27*(24), 8999. https://doi.org/10.3390/molecules27248999

Bora, H., Kamle, M., Mahato, D. K., Tiwari, P., & Kumar, P. (2020). Citrus essential oils (CEOs) and their applications in food: An overview. *Plants, 9*(3), 357. https://doi.org/10.3390/plants9030357

Boudraa, H., Kadri, N., Mouni, L., & Madani, K. (2021). Microwave-assisted hydrodistillation of essential oil from fennel seeds: Optimization using Plackett–Burman design and response surface methodology. *Journal of Applied Research on Medicinal and Aromatic Plants, 23*, 100307. https://doi.org/10.1016/j.jarmap.2021.100307

Butnariu, M., & Sarac, I. (2018). Essential oils from plants. *Journal of Biotechnology and Biomedical Science, 1*(4), 35–43. https://doi.org/10.14302/issn.2576-6694.jbbs-18-2489

Campion, B. B., & Essel, G. (2013). Environmental impact assessment and sustainable development: A critical review. *Environment and Natural Resources Research, 3*(2). https://doi.org/10.5539/enrr.v3n2p37

Capuzzo, A., Maffei, M., & Occhipinti, A. (2013). Supercritical fluid extraction of plant flavors and fragrances. *Molecules, 18*(6), 7194–7238. https://doi.org/10.3390/molecules18067194

Carrubba, A., & Catalano, C. (2009). *Essential oil crops for sustainable agriculture – A review* (pp. 137–187). Springer eBooks. https://doi.org/10.1007/978-90-481-2716-0_8

Chemat, F., Rombaut, N., Sicaire, A., Meullemiestre, A., Fabiano-Tixier, A., & Abert-Vian, M. (2017). Ultrasound assisted extraction of food and natural products. Mechanisms, techniques, combinations, protocols and applications. A review. *Ultrasonics Sonochemistry, 34*, 540–560. https://doi.org/10.1016/j.ultsonch.2016.06.035

Chouhan, S., Sharma, K., & Guleria, S. (2017). Antimicrobial activity of some essential oils—Present status and future perspectives. *Medicine, 4*(3), 58. https://doi.org/10.3390/medicines4030058

Coskun, O. (2016). *Separation tecniques: Chromatography*. Northern Clinics of Istanbul. https://doi.org/10.14744/nci.2016.32757

Costa, R., Dugo, P., & Mondello, L. (2012). *Sampling and sample preparation techniques for the determination of the volatile components of milk and dairy products* (pp. 43–59). Elsevier eBooks. https://doi.org/10.1016/b978-0-12-381373-2.00127-7

Crescente, G., Bouymajane, A., Cascone, G., Squillaci, G., Morana, A., & Moccia, S. (2023). Essential oils: An overview of extraction methods and food applications. *Current Nutraceuticals, 4*(1). https://doi.org/10.2174/2665978604666230601104713

D'Atri, V., Fekete, S., Clarke, A., Veuthey, J., & Guillarme, D. (2018). Recent advances in chromatography for pharmaceutical analysis. *Analytical Chemistry, 91*(1), 210–239. https://doi.org/10.1021/acs.analchem.8b05026

Divakar, S., Padaki, M., & Balakrishna, R. G. (2022). Review on Liquid–Liquid separation by Membrane Filtration. *ACS Omega, 7*(49), 44495–44506. https://doi.org/10.1021/acsomega.2c02885

Drinic, Z., Pljevljakusic, D., Zivkovic, J., Bigovic, D., & Savikin, K. (2020). Microwave-assisted extraction of O. vulgare L. spp. hirtum essential oil: Comparison with conventional hydrodistillation. *Food and Bioproducts Processing, 120*, 158–165. https://doi.org/10.1016/j.fbp.2020.01.011

Durczynska, Z., & Żukowska, G. (2024). Properties and applications of essential oils: A review. *Journal of Ecological Engineering, 25*(2), 333–340. https://doi.org/10.12911/22998993/177404

El-Aziz, M. A., Salama, H. H., & Sayed, R. S. (2023). Plant extracts and essential oils in the dairy industry: A review. *Foods and Raw Materials*, 321–337. https://doi.org/10.21603/2308-4057-2023-2-579

Fagbemi, K. O., Aina, D. A., & Olajuyigbe, O. O. (2021). Soxhlet extraction versus hydrodistillation using the Clevenger apparatus: A comparative study on the extraction of a volatile compound from Tamarindus indica seeds. *The Scientific World Journal, 2021*, 1–8. https://doi.org/10.1155/2021/5961586

Fan, Y., & Li, Q. (2022). An efficient extraction method for essential oil from angelica sinensis radix by natural deep eutectic solvents-assisted microwave hydrodistillation. *Sustainable Chemistry and Pharmacy, 29*, 100792. https://doi.org/10.1016/j.scp.2022.100792

Farhat, A., Fabiano-Tixier, A., Maataoui, M. E., Maingonnat, J., Romdhane, M., & Chemat, F. (2011). Microwave steam diffusion for extraction of essential oil from orange peel: Kinetic data, extract's global yield and mechanism. *Food Chemistry, 125*(1), 255–261. https://doi.org/10.1016/j.foodchem.2010.07.110

Ferraz, C. A., Pastorinho, M. R., Palmeira-De-Oliveira, A., & Sousa, A. C. (2022). Ecotoxicity of plant extracts and essential oils: A review. *Environmental Pollution, 292*, 118319. https://doi.org/10.1016/j.envpol.2021.118319

Fornari, T., Vicente, G., Vázquez, E., García-Risco, M. R., & Reglero, G. (2012). Isolation of essential oil from different plants and herbs by supercritical fluid extraction. *Journal of Chromatography A, 1250*, 34–48. https://doi.org/10.1016/j.chroma.2012.04.051

Geow, C. H., Tan, M. C., Yeap, S. P., & Chin, N. L. (2021). A review on extraction techniques and its future applications in industry. *European Journal of Lipid Science and Technology, 123*(4). https://doi.org/10.1002/ejlt.202000302

Girotto, F., Alibardi, L., & Cossu, R. (2015). Food waste generation and industrial uses: A review. *Waste Management, 45*, 32–41. https://doi.org/10.1016/j.wasman.2015.06.008

Gonzalez-Mas, M. C., Rambla, J. L., López-Gresa, M. P., Blázquez, M. A., & Granell, A. (2019). Volatile compounds in citrus essential oils: A comprehensive review. *Frontiers in Plant Science, 10*. https://doi.org/10.3389/fpls.2019.00012

Gulis, G., Krishnankutty, N., Boess, E. R., Lyhne, I., & Kørnøv, L. (2022). Environmental impact assessment, human health and the sustainable development goals. *International Journal of Public Health, 67*. https://doi.org/10.3389/ijph.2022.1604420

Haffizi, M., Sulaiman, S., Jimat, D. N., & Amid, A. (2020). Review: A comparison of conditions for the extraction of vegetable and essential oils via microwave-assisted extraction. *IOP Conference Series Materials Science and Engineering, 778*(1), 012172. https://doi.org/10.1088/1757-899x/778/1/012172

Hu, Q., He, Y., Wang, F., Wu, J., Ci, Z., Chen, L., Xu, R., Yang, M., Lin, J., Han, L., & Zhang, D. (2021). Microwave technology: A novel approach to the transformation of natural metabolites. *Chinese Medicine, 16*(1). https://doi.org/10.1186/s13020-021-00500-8

Immaroh, N. Z., Kuliahsari, D. E., & Nugraheni, S. D. (2021). Review: Eucalyptus globulus essential oil extraction method. *IOP Conference Series Earth and Environmental Science, 733*(1), 012103. https://doi.org/10.1088/1755-1315/733/1/012103

Jackson-Davis, A., White, S., Kassama, L. S., Coleman, S., Shaw, A., Mendonca, A., Cooper, B., Thomas-Popo, E., Gordon, K., & London, L. (2023). A review of regulatory standards and advances in essential oils as antimicrobials in foods. *Journal of Food Protection, 86*(2), 100025. https://doi.org/10.1016/j.jfp.2022.100025

Joco, R. A., Lavarias, J. A., Peneyra, R. G., & Somera, C. G. (2023). Recent development on the extraction process of plants essential oil and its effect on chemical composition: A review. *Advanced Journal of Graduate Research, 14*(1), 9–20. https://doi.org/10.21467/ajgr.14.1.9-20

Kabuba, J., & Huberts, R. (2009). Steam extraction of essential oils: Investigation of process parameters. *The Canadian Journal of Chemical Engineering, 87*(6), 915–920. https://doi.org/10.1002/cjce.20236

Kamali, H., Aminimoghadamfarouj, N., Golmakani, E., & Nematollahi, A. (2015). The optimization of essential oils supercritical CO_2 extraction from Lavandula hybrida through static-dynamic steps procedure and semi-continuous technique using response surface method. *Pharmacognosy Research, 7*(1), 57. https://doi.org/10.4103/0974-8490.147209

Kant, R., & Kumar, A. (2022). Review on essential oil extraction from aromatic and medicinal plants: Techniques, performance and economic analysis. *Sustainable Chemistry and Pharmacy, 30*, 100829. https://doi.org/10.1016/j.scp.2022.100829

Karne, H., Kelkar, V., Mundhe, A., Ikar, M., Betawar, S., & Chaudhari, N. (2023). Essential oil extraction from orange and lemon peel. *E3S Web of Conferences, 455*, 01005. https://doi.org/10.1051/e3sconf/202345501005

Katekar, V. P., Rao, A. B., & Sardeshpande, V. R. (2022). Review of the rose essential oil extraction by hydrodistillation: An investigation for the optimum operating condition for maximum yield. *Sustainable Chemistry and Pharmacy, 29*, 100783. https://doi.org/10.1016/j.scp.2022.100783

Kohli, K., et al. (2024). Food waste: Environmental impact and possible solutions. *Sustainable Food Technology, 2*(1), 70–80. https://doi.org/10.1039/d3fb00141e

Kumar, K., Yadav, A. N., Kumar, V., Vyas, P., & Dhaliwal, H. S. (2017). Food waste: A potential bioresource for extraction of nutraceuticals and bioactive compounds. *Bioresources and Bioprocessing, 4*(1). https://doi.org/10.1186/s40643-017-0148-6

Kumar, A., Patel, R., & Yadav, A. (2022). Experimental analysis of performance of diesel engine fuelled by biofuel made from waste vegetable oil. *Acta Periodica Technologica, 53*, 25–35. https://doi.org/10.2298/apt2253025k

Lahiri, A., Daniel, S., Kanthapazham, R., Vanaraj, R., Thambidurai, A., & Peter, L. S. (2023). A critical review on food waste management for the production of materials and biofuel. *Journal of Hazardous Materials Advances, 10*, 100266. https://doi.org/10.1016/j.hazadv.2023.100266

Lankanayaka, A. A. K., Matharage, D. S., Senanayake, C. M., Jayathunge, K. G. L. R., Bandara, P. C., & Manatunga, D. C. (2024). Extraction of bioactive compounds from food waste and their potential applications in the food industry: A review. *Brazilian Journal of Development, 10*(4), e69187. https://doi.org/10.34117/bjdv10n4-061

Lucarini, M., Durazzo, A., Bernini, R., Campo, M., Vita, C., Souto, E. B., Lombardi-Boccia, G., Ramadan, M. F., Santini, A., & Romani, A. (2021). Fruit wastes as a valuable source of value-added compounds: A collaborative perspective. *Molecules, 26*(21), 6338. https://doi.org/10.3390/molecules26216338

Machado, B. A. S., Pereira, C. G., Nunes, S. B., Padilha, F. F., & Umsza-Guez, M. A. (2013). Supercritical fluid extraction using CO_2: Main applications and future perspectives. *Separation Science and Technology, 48*(18), 2741–2760. https://doi.org/10.1080/01496395.2013.811422

Machado, C. A. T., Hodel, K. V. S., Lepikson, H. A., & Machado, B. A. S. (2024). Distillation of essential oils: An innovative technological approach focused on productivity, quality and sustainability. *PLoS One, 19*(2), e0299502. https://doi.org/10.1371/journal.pone.0299502

Mahawer, S. K., Himani, N., Arya, S., Kumar, R., & Prakash, O. (2022). Extractions methods and biological applications of essential oils. *Biochemistry*. https://doi.org/10.5772/intechopen.102955

Mahmud, M. Z. A., Mobarak, M. H., & Hossain, N. (2024). Emerging trends in biomaterials for sustainable food packaging: A comprehensive review. *Heliyon, 10*(1), e24122. https://doi.org/10.1016/j.heliyon.2024.e24122

Marriot, P. J., Shellie, R., & Cornwell, C. (2001). Gas chromatographic technologies for the analysis of essential oils. *Journal of Chromatography A, 936*(1–2), 1–22. https://doi.org/10.1016/s0021-9673(01)01314-0

Maurya, A., Prasad, J., Das, S., & Dwivedy, A. K. (2021). Essential oils and their application in food safety. *Frontiers in Sustainable Food Systems, 5*. https://doi.org/10.3389/fsufs.2021.653420

Mazzara, E., Carletti, R., Petrelli, R., Mustafa, A. M., Caprioli, G., Fiorini, D., Scortichini, S., Dall'Acqua, S., Sut, S., Nuñez, S., López, V., Zheljazkov, V. D., Bonacucina, G., Maggi, F., & Cespi, M. (2022). Green extraction of hemp (Cannabis sativa L.) using microwave method for recovery of three valuable fractions (essential oil, phenolic compounds and cannabinoids): A central composite design optimization study. *Journal of the Science of Food and Agriculture, 102*(14), 6220–6235. https://doi.org/10.1002/jsfa.11971

Mejri, J., Aydi, A., Abderrabba, M., & Mejri, M. (2018). Emerging extraction processes of essential oils: A review. *Asian Journal of Green Chemistry, 2*(3), 246–267. https://doi.org/10.22034/ajgc.2018.61443

Meshalkin, V., Zharov, V., Leontiev, L., Nzioka, A., & Belozersky, A. (2023). Sustainable environmental impact assessment using indicators for sustainable energy-intensive industrial production. *Energies, 16*(7), 3172. https://doi.org/10.3390/en16073172

Misra, S., Keer, P. K., & Roy, S. (2022). Essentials oils for a healthy environment. *International Journal of Health Sciences*, 9048–9055. https://doi.org/10.53730/ijhs.v6ns3.8220

Mohamed, A. A., & Alotaibi, B. M. (2022). Essential oils of some medicinal plants and their biological activities: A mini review. *Journal of Umm Al-Qura University for Applied Sciences, 9*(1), 40–49. https://doi.org/10.1007/s43994-022-00018-1

Mohanty, A., Mankoti, M., Rout, P. R., Meena, S. S., Dewan, S., Kalia, B., Varjani, S., Wong, J. W., & Banu, J. R. (2022). Sustainable utilization of food waste for bioenergy production: A step towards circular bioeconomy. *International Journal of Food Microbiology, 365*, 109538. https://doi.org/10.1016/j.ijfoodmicro.2022.109538

Mukherjee, S., Chouhan, K. B. S., & Mandal, V. (2024). Decrypting solvent-free microwave as a dual green extraction: Studying simultaneous extraction of essential oil and phenolics from the same biomass – Valorization and outperforming traditional approaches. *Journal of Chemical Technology & Biotechnology, 99*(4), 931–945. https://doi.org/10.1002/jctb.7597

Munir, H., Yaqoob, S., Awan, K. A., Imtiaz, A., Naveed, H., Ahmad, N., Naeem, M., Sultan, W., & Ma, Y. (2024). Unveiling the chemistry of citrus peel: Insights into nutraceutical potential and therapeutic applications. *Foods, 13*(11), 1681. https://doi.org/10.3390/foods13111681

Ngo, T. C. Q., Tran, T. H., & Le, X. T. (2020). The effects of influencing parameters on the Eucalyptus globulus leaves essential oil extraction by hydrodistillation method. *IOP Conference Series Materials Science and Engineering, 991*(1), 012126. https://doi.org/10.1088/1757-899x/991/1/012126

Nirmal, N., Khanashyam, A., Mundanat, A., Shah, K., Babu, K., Thorakkattu, P., Al-Asmari, F., & Pandiselvam, R. (2023). Valorization of fruit waste for bioactive compounds and their applications in the food industry. *Foods, 12*(3), 556. https://doi.org/10.3390/foods12030556

Olkiewicz, M., Cota, I., Wozniak-Budych, M., Litowczenko, J., Luczak, M., Monotrnes, J. M., & Tylkowski, B. (2022). Membrane-based processes in essential oils production. *Physical Sciences Reviews, 8*(12), 4915–4924. https://doi.org/10.1515/psr-2022-0057

Oubannin, S., Bijla, L., Ahmed, M. N., Ibourki, M., Kharrassi, Y. E., Devkota, K., Bouyahya, A., Maggi, F., Caprioli, G., Sakar, E. H., & Gharby, S. (2024). Recent advances in the extraction of bioactive compounds from plant matrices and their use as potential antioxidants for vegetable oils enrichment. *Journal of Food Composition and Analysis, 128*, 105995. https://doi.org/10.1016/j.jfca.2024.105995

Ouro-Salim, O., & Guarnieri, P. (2021). Circular economy of food waste: A literature review. *Environmental Quality Management, 32*(2), 225–242. https://doi.org/10.1002/tqem.21836

Paritosh, K., Kushwaha, S. K., Yadav, M., Pareek, N., Chawade, A., & Vivekanand, V. (2017). Food waste to energy: An overview of sustainable approaches for food waste management and nutrient recycling. *BioMed Research International, 2017*, 1–19. https://doi.org/10.1155/2017/2370927

Patel, K., Panchal, N., & Ingle, P. (2019). Review of extraction techniques extraction methods: Microwave, ultrasonic, pressurized fluid, soxhlet extraction, etc. *International Journal of Advanced Research in Chemical Sciences, 6*(3). https://doi.org/10.20431/2349-0403.0603002

Pezantes-Orellana, C., et al. (2024). Essential oils: A systematic review on revolutionizing health, nutrition, and omics for optimal well-being. *Frontiers in Medicine, 11*. https://doi.org/10.3389/fmed.2024.1337785

Pour, F. H., & Makkawi, Y. T. (2021). A review of post-consumption food waste management and its potentials for biofuel production. *Energy Reports, 7*, 7759–7784. https://doi.org/10.1016/j.egyr.2021.10.119

Quevedo-Amador, R. A., Escalera-Velasco, B. P., Arias, A. M. R., Reynel-Ávila, H. E., Moreno-Pirajón, J. C., Giraldo, L., & Bonilla-Petriciolet, A. (2024). Application of waste biomass for the production of biofuels and catalysts: A review. *Clean Technologies and Environmental Policy, 26*(4), 943–997. https://doi.org/10.1007/s10098-023-02728-4

Rahman, M. M., Wills, R. B. H., Bowyer, M. C., Vuong, V. Q., Golding, J. B., Kirkman, T., & Pristijono, P. (2023). Efficacy of Lemon Myrtle essential oil as a bio-fungicide in inhibiting citrus green mould. *Plants, 12*(21), 3742. https://doi.org/10.3390/plants12213742

Rakhasiya, B., Munisamy, S., Mathew, D. E., Tothadi, S., Yadav, A., & Mantri, V. A. (2023). Potential utility of industrially unwanted constituent under the framework of waste to wealth: Edible salt from commercial marine red seaweed Kappaphycus alvarezii (Doty) L.M. Liao. *Bioresource Technology Reports, 23*, 101529. https://doi.org/10.1016/j.biteb.2023.101529

Raut, J. S., & Karuppayil, S. M. (2014). A status review on the medicinal properties of essential oils. *Industrial Crops and Products, 62*, 250–264. https://doi.org/10.1016/j.indcrop.2014.05.055

Rodriguez-Marquez, C. D., Arteaga-Marin, S., Rivas-Sánchez, A., Autrique-Hernández, R., & Castro-Muñoz, R. (2022). A review on current strategies for extraction and purification of hyaluronic acid. *International Journal of Molecular Sciences, 23*(11), 6038. https://doi.org/10.3390/ijms23116038

Roy, P., Mohanty, A. K., Dick, P., & Misra, M. (2023). A review on the challenges and choices for food waste valorization: Environmental and economic impacts. *ACS Environmental Au, 3*(2), 58–75. https://doi.org/10.1021/acsenvironau.2c00050

Rubiolo, P., Liberto, E., Sgorbini, B., Russo, R., Veuthey, J., & Bicchi, C. (2008). Fast-GC–conventional quadrupole mass spectrometry in essential oil analysis. *Journal of Separation Science, 31*(6–7), 1074–1084. https://doi.org/10.1002/jssc.200700577

Sahoo, A., Dwivedi, A., Madheshiya, P., Kumar, U., Sharma, R. K., & Tiwari, S. (2023). Insights into the management of food waste in developing countries: With special reference to India. *Environmental Science and Pollution Research*. https://doi.org/10.1007/s11356-023-27901-6

Samadi, M., Abidin, Z. Z., Yoshida, H., Yunus, R., & Biak, D. R. A. (2020). Towards higher oil yield and quality of essential oil extracted from Aquilaria malaccensis wood via the subcritical technique. *Molecules, 25*(17), 3872. https://doi.org/10.3390/molecules25173872

Sarker, A., et al. (2024). A comprehensive review of food waste valorization for the sustainable management of global food waste. *Sustainable Food Technology, 2*(1), 48–69. https://doi.org/10.1039/d3fb00156c

Sarkic, A., & Stappen, I. (2018). Essential oils and their single compounds in cosmetics—A critical review. *Cosmetics, 5*(1), 11. https://doi.org/10.3390/cosmetics5010011

Shaaban, H. A. (2023). Potential use of essential oils and their individual components in cosmeceuticals: A review. *Annals of Biomedical Science and Engineering, 7*(1), 031–037. https://doi.org/10.29328/journal.abse.1001023

Sharma, P., Bano, A., Verma, K., Yadav, M., Varjani, S., Singh, S. P., & Tong, Y. W. (2023). Food waste digestate as biofertilizer and their direct applications in agriculture. *Bioresource Technology Reports, 23*, 101515. https://doi.org/10.1016/j.biteb.2023.101515

Sharmeen, J., Mahomoodally, F., Zengin, G., & Maggi, F. (2021). Essential oils as natural sources of fragrance compounds for cosmetics and cosmeceuticals. *Molecules, 26*(3), 666. https://doi.org/10.3390/molecules26030666

Siddiqui, T., Sharma, V., Khan, M. U., & Gupta, K. (2024). Terpenoids in essential oils: Chemistry, classification, and potential impact on human health and industry. *Phytomedicine Plus, 4*(2), 100549. https://doi.org/10.1016/j.phyplu.2024.100549

Silvestre, W. P., Medeiros, F. R., Agostini, F., Toss, D., & Pauletti, G. F. (2019). Fractionation of rosemary (Rosmarinus officinalis L.) essential oil using vacuum fractional distillation. *Journal of Food Science and Technology, 56*(12), 5422–5434. https://doi.org/10.1007/s13197-019-04013-z

Skendi, A., Irakli, M., Chatzopoulou, P., Bouloumpasi, E., & Biliaderis, C. G. (2022). Phenolic extracts from solid wastes of the aromatic plant essential oil industry: Potential uses in food applications. *Food Chemistry Advances, 1*, 100065. https://doi.org/10.1016/j.focha.2022.100065

Sniegonova, P., Szotkowski, M., Holub, J., Sikorová, P., & Márová, I. (2023). The effect of oil-rich food waste substrates, used as an alternative carbon source, on the cultivation of microalgae—A pilot study. *Microorganisms, 11*(7), 1621. https://doi.org/10.3390/microorganisms11071621

Sojic, B., Milosevic, S., Savanovic, D., Zekovic, Z., Tomović, V., & Pavlic, B. (2023). Isolation, bioactive potential, and application of essential oils and Terpenoid-rich extracts as effective antioxidant and antimicrobial agents in meat and meat products. *Molecules, 28*(5), 2293. https://doi.org/10.3390/molecules28052293

Son, H. L., Ngoc, L. T. H., & Van Minh, T. (2023). Comparative study of microwave-assisted extraction and hydro-distillation methods for extracting essential oil from Melaleuca cajuputi in Southern Vietnam: Yield, composition, and efficiency analysis. *Asian Journal of Biotechnology and Bioresource Technology, 9*(2), 14–21. https://doi.org/10.9734/ajb2t/2023/v9i2179

Souiy, Z. (2023). Essential oil extraction process. *Biochemistry*. https://doi.org/10.5772/intechopen.113311

Suwal, S., & Marciniak, A. (2019). Technologies for the extraction, separation and purification of polyphenols – A review. *Nepal Journal of Biotechnology, 6*(1), 74–91. https://doi.org/10.3126/njb.v6i1.22341

Torres-Leon, C., Ramírez-Guzman, N., Londoño-Hernandez, L., Martinez-Medina, G. A., Díaz-Herrera, R., Navarro-Macias, V., Alvarez-Pérez, O. B., Picazo, B., Villarreal-Vázquez, M., Ascacio-Valdes, J., & Aguilar, C. N. (2018). Food waste and byproducts: An opportunity to minimize malnutrition and hunger in developing countries. *Frontiers in Sustainable Food Systems, 2*. https://doi.org/10.3389/fsufs.2018.00052

Tran, T. H., Ngo, T. C. Q., Dao, T. P., Nguyen, P. T. N., Pham, T. N., Nguyen, T. D., Linh, H. T. K., Nguyen, N. H., & Cang, M. H. (2020). Optimizatoin of microwave-assisted extraction and compositional determination of essential oil from leaves of Eucalyptus globulus. *IOP Conference Series Materials Science and Engineering, 736*(2), 022040. https://doi.org/10.1088/1757-899x/736/2/022040

Turek, C., & Stintzing, F. C. (2013). Stability of essential oils: A review. *Comprehensive Reviews in Food Science and Food Safety, 12*(1), 40–53. https://doi.org/10.1111/1541-4337.12006

Utomo, E. P., Marina, M., Warsito, W., & Agustian, E. (2018). Optimization on supercritical CO_2 extraction process to improve the quality of patchouli oil by response surface methodology approach. *Indonesian Journal of Chemistry, 18*(2), 235. https://doi.org/10.22146/ijc.26605

Uwineza, P. A., & Waskiewicz, A. (2020). Recent advances in supercritical fluid extraction of natural bioactive compounds from natural plant materials. *Molecules, 25*(17), 3847. https://doi.org/10.3390/molecules25173847

Wainer, J., Thomas, A., Chimhau, T., & Harding, K. G. (2022). Extraction of essential oils from Lavandula × intermedia 'Margaret Roberts' using steam distillation, hydrodistillation, and cellulase-assisted hydrodistillation: Experimentation and cost analysis. *Plants, 11*(24), 3479. https://doi.org/10.3390/plants11243479

Wang, C., Wang, Z., Wang, X., Li, N., Tao, J., Zheng, W., Yan, B., Cui, X., Cheng, Z., & Chen, G. (2022). A review on the hydrothermal treatment of food waste: Processing and applications. *Processes, 10*(11), 2439. https://doi.org/10.3390/pr10112439

Wells, C. W. (2024). Effects of essential oils on economically important characteristics of ruminant species: A comprehensive review. *Animal Nutrition, 16*, 1–10. https://doi.org/10.1016/j.aninu.2023.05.017

Xue, L., Song, G., & Liu, G. (2024). Wasted food, wasted resources? A critical review of environmental impact analysis of food loss and waste generation and treatment. *Environmental Science & Technology, 58*(17), 7240–7255. https://doi.org/10.1021/acs.est.3c08426

Yingngam, B., Brantner, A., Treichler, M., Brugger, N., Navabhatra, A., & Nakonrat, P. (2021). Optimization of the eco-friendly solvent-free microwave extraction of Limnophila aromatica essential oil. *Industrial Crops and Products, 165*, 113443. https://doi.org/10.1016/j.indcrop.2021.113443

Yousefi, M., Rahimi-Nasrabadi, M., Pourmortazavi, S. M., Wysokowski, M., Jesionowski, T., Ehrlich, H., & Mirsadeghi, S. (2019). Supercritical fluid extraction of essential oils. *TrAC Trends in Analytical Chemistry, 118*, 182–193. https://doi.org/10.1016/j.trac.2019.05.038

Zhang, W., Liu, D., Fu, X., Xiong, C., & Nie, Q. (2022). Peel essential oil composition and antibacterial activities of Citrus x sinensis L. *Osbeck 'Tarocco' and Citrus reticulata Blanco. Horticulturae, 8*(9), 793. https://doi.org/10.3390/horticulturae8090793

Zhang, M., Zhao, J., Dai, X., & Li, X. (2023). Extraction and analysis of chemical compositions of natural products and plants. *Separations, 10*(12), 598. https://doi.org/10.3390/separations10120598

Zhou, W., Li, J., Wang, X., Liu, L., Li, Y., Song, R., Zhang, M., & Li, X. (2023). Research progress on extraction, separation, and purification methods of plant essential oils. *Separations, 10*(12), 596. https://doi.org/10.3390/separations10120596

Exploring the Potential of Fungal Biomass in the Evolution of Polylactic Acid (PLA) Biocomposites

Angbeen Ali, Wajeeha Noreen, Sahar Naveed Baig, Ika Rahmatul Layly, Marcelo Franco, and Muhammad Irfan

Abstract A biodegradable thermoplastic polymer made from renewable resources like sugarcane and maize starch, polylactic acid (PLA), is an environmentally beneficial substitute for plastics made from petroleum. PLA has drawbacks in mechanical strength, thermal stability, and moisture resistance despite its environmental benefits. Incorporating natural fibers and biomass has been the focus of recent research aimed at improving the characteristics of PLA, with fungal biomass emerging as a possible reinforcement. In addition to increasing PLA's mechanical strength, elasticity, and toughness, fungal biomass also decreases water absorption and improves thermal stability. Fungal biomass is rich in polysaccharides like chitin and glucans. Several fabrication methods are investigated, such as injection molding, compression molding, manual layup, filament winding, and additive manufacturing. The results show that PLA-based biocomposites have potential for use in sustainable applications; however, more study is required to solve production issues and enhance material performance. PLA is biodegradable through a combination of chemical and microbiological processes; industrial composting degrades PLA more quickly than home composting. This analysis looks at how different fillers, like metal oxides, ceramics, and natural fibers, affect PLA's functionality. Applications for PLA composites in 3D printing, construction, automotive, and medical fields are growing. To increase the applicability and effectiveness of PLA-based materials across a range of industries, future research should concentrate on strengthening mechanical qualities, enhancing filler compatibility, and evaluating long-term durability.

A. Ali · W. Noreen · S. N. Baig · M. Irfan (✉)
Department of Biotechnology, University of Sargodha, Sargodha, Pakistan
e-mail: irfan.ashraf@uos.edu.pk

I. R. Layly
Applied Microbiology Research Center, National Research and Innovation Agency (BRIN), Jakarta, Indonesia

M. Franco
Department of Exact Science, State University of Santa Cruz, Ilheus, Brazil

D. B. Pal, S. L. Bansal (eds.), *Fungal Waste Biomass Management for Energy, Environment and Value-Added Products*, Fungal Biology,
https://doi.org/10.1007/978-3-031-82599-6_7

Keywords Polylactic acid (PLA) · Fungal biomass · Nanoadditives · Mechanical properties · Biocomposites · Applications

1 Introduction

Polylactic acid (PLA), a biodegradable polymer made from renewable resources like corn starch and sugarcane, has become a vital material in the development of eco-friendly biocomposites. Its environmental advantages make it a compelling alternative to petroleum-based plastics, particularly in packaging, biomedical, and automotive applications. However, PLA's inherent weaknesses in mechanical strength, thermal stability, and moisture resistance limit its broader use. To overcome these challenges, recent studies have explored improving polylactic acid properties by incorporating natural fibers and biomass. Among these, fungal biomass stands out due to its renewable nature, availability, and composition, particularly in polysaccharides like chitin and glucans (Zhou et al., 2023).

Fungal biomass proves to be a highly effective reinforcement for polylactic acid biocomposites, enhancing not only the polymer's mechanical and thermal properties but also improving biodegradability for more sustainable disposal. It can be sourced economically from industrial and agricultural waste, providing an environmentally friendly solution for composite development. Research has shown that adding fungal biomass to PLA significantly increases its tensile strength, elasticity, and toughness, making it better suited for demanding applications such as packaging and construction (Nguyen et al., 2022). Moreover, the improved thermal stability and reduced water absorption of these composites allow them to perform well in conditions where PLA alone would be insufficient.

Using fungal biomass in PLA-based composites aligns with the principles of the circular economy by repurposing waste materials and enhancing biodegradability. This approach supports waste reduction and resource efficiency while being more cost-effective than synthetic fiber alternatives, making it suitable for large-scale industrial production (Singh et al., 2021). As research advances, further improvements in fungal biomass extraction and processing are expected to enhance the performance and scalability of these biocomposites, opening new opportunities for their industrial applications.

2 Biocomposites

Biocomposites are materials composed of natural fiber reinforcements, often sourced from plants or animals, combined with a synthetic or biopolymer matrix. The use of fibers as fillers or reinforcers in plastic composites has gained traction, with thermoplastics making up over 80% of these applications. The popularity of biocomposites is driven by their environmental benefits, including affordability, sustainability, and biodegradability (John & Thomas, 2008a).

These composites are increasingly utilized across various sectors such as packaging, construction, and automotive industries due to their high strength, lightweight nature, and eco-friendly properties (Faruk et al., 2012). Common natural fibers used in these composites include jute, flax, hemp, and sisal, each offering distinct mechanical advantages (Bledzki & Gassan, 1999). While biocomposites offer numerous benefits, they also face challenges like increased moisture absorption and reduced durability. Current research is focused on overcoming these limitations to expand their applications (Pickering et al., 2016).

2.1 *Significance of Biocomposites*

The significance of biocomposites lies in their positive environmental impact and cost efficiency. They play a crucial role in sustainability by minimizing adverse ecological effects, thanks to their renewable composition and natural degradation (Mohanty et al., 2002). The incorporation of natural fibers helps reduce carbon footprints as these fibers absorb carbon dioxide during growth, thus lowering greenhouse gas emissions (Faruk et al., 2012). Additionally, natural fibers are more cost-effective than synthetic alternatives, which helps lower both manufacturing and material costs (John & Thomas, 2008b).

The lightweight nature of biocomposites also enhances fuel efficiency in automotive applications (Bledzki & Gassan, 1999). Their biocompatibility and safety make them suitable for use in food and medical products, offering a balanced solution for various needs. Furthermore, by utilizing agricultural byproducts, biocomposites support a circular economy and reduce waste (Mohanty et al., 2002). Advances in their mechanical properties are making biocomposites increasingly competitive with traditional materials (Faruk et al., 2012). The advantages and disadvantages of biocomposites are listed in Table 1 (Asokan et al., 2012; Jawaid et al., 2017; Kargarzadeh et al., 2017).

3 Overview of Polylactic Acid as a Biopolymer

To produce polylactic acid, carbohydrates must first ferment into lactic acid. This is followed by polymerization using techniques including condensation and ring-opening polymerization (ROP). A sustainable substitute for plastics derived from petroleum is polylactic acid, a biodegradable thermoplastic polymer derived from renewable resources like sugarcane or maize starch (Auras et al., 2004). Polylactic acid is recognized for its biodegradability and environmental friendliness since, when composted in an industrial setting, it can break down into carbon dioxide and water (Nampoothiri et al., 2010). In addition to being environmentally friendly, it

Table 1 Advantages and disadvantages of biocomposites

Advantages	Disadvantages
Requires less energy for its synthesis.	Properties are not uniform.
Less toxic to the environment and fewer health hazards	Absorb high moisture
Resist the fungal attacks	Thermally less stable
Can be recycled and is biodegradable	Poor compatible
Produces less pollution	Modest strength
Abrasion-resistant to machinery	Poor wettability of hydrophilic reinforcements with the hydrophobic polymer matrix
Need low management	

has exceptional mechanical properties that are comparable to those of conventional polymers like polyethylene and polypropylene, such as a high modulus and tensile strength (Gupta & Kumar, 2007).

These qualities make it ideal for a range of applications, such as packaging, medical equipment, and 3D printing (Drumright et al., 2000). In addition, PLA is biodegradable and biocompatible. Furthermore, by reducing greenhouse gas emissions and dependence on fossil fuels, the use of it promotes a more sustainable and environmentally friendly economy (Vink et al., 2003a, b). It offers several advantages, but it also has disadvantages, including brittleness and lower heat resistance. To improve these characteristics and broaden its applications, research is still being conducted (Garlotta, 2001).

4 Integrating Fungal Biomass with Polylactic Acid

Fungal biomass is essential to the creation of biocomposites because of its distinct structural and functional characteristics. The vegetative portion of fungi, known as the mycelium, is made up of a web of hyphae, which are structures that resemble threads (Haneef et al., 2017). When added to polymer matrices like polylactic acid, this naturally occurring fiber network confers exceptional mechanical qualities like high tensile strength and elasticity, which can greatly improve the performance of biocomposites. Fungal biomass is biodegradable and lightweight, which is in line with the growing need for eco-friendly materials. Fungal biomass is added to PLA to increase the mechanical qualities of the resultant biocomposites as well as their impact resistance and thermal stability (Appels et al., 2019).

Additionally, fungal biomass can give the composites antibacterial qualities, which is very useful in the food packaging and healthcare sectors (Elsacker et al., 2019). A further factor in the sustainability of biocomposites is fungus biomass. Fungi are usually grown on agricultural waste substrates for the generation of fungal biomass, which is an economical and environmentally beneficial procedure. By doing this, waste is decreased and agricultural byproducts are given a useful purpose. Fungal biomass's ability to decompose guarantees that biocomposites can be disposed of responsibly at the end of their useful lives, minimizing the impact on landfills and promoting a circular economy (Attias et al., 2021).

5 Polylactic Acid (PLA)

History claims that Wallace et al.'s (1932) attempt to produce PLA involved the polymerization and depolymerization of oligomeric lactides. By changing the procedures, high molecular weight polylactic acid which shares properties with petroleum-based plastics was made in the 1960s. This production approach, however, was not economical. The manufacturing of polylactic acid entered the commercial market in 1997 following the technique's development by Cargill Dow Corporation (Dorgan et al., 2006). Polylactic acid was produced using the melting technique, as opposed to this method's creation of the solution. The first polylactic acid production started up consistently in 2002 and had a 300 lb. yearly capacity.

5.1 Sources of Polylactic Acid

One of the main sources of polylactic acid is corn starch, which may be fermented to produce lactic acid, the monomer needed for its polymerization. A biodegradable and bioactive thermoplastic made from renewable resources is polylactic acid (Lim et al., 2008). Sugarcane, a major source of polylactic acid, offers a renewable alternative for the synthesis of polymers and has less of an adverse environmental effect than plastics made from petroleum (Auras et al., 2004).

In addition to maize starch and sugarcane, other biomass feedstocks including cassava, wheat, and barley have also been explored for their production, expanding the range of sustainable raw materials available for polymer synthesis (Gupta & Kumar, 2007). Lactic acid, the precursor to PLA, can be produced by microbial fermentation of carbohydrates derived from biomass sources, providing a versatile pathway for its production (Nampoothiri et al., 2010).

5.2 Production of Polylactic Acid

The monomer used to create PLA is lactic acid (2-hydroxypropionic acid), which was discovered in 1780 by the Swedish chemist Scheele. The chemical structure of monomer lactic acid is shown in Fig. 1. Dextrorotatory form L (positive) isomer and levorotatory form D (negative) isomer are its two distinct stereoisomers. The direction of rotation of the plane-polarized light is indicated by the sign; the L (positive) isomer rotates the plane-polarized light in a clockwise direction, while the D

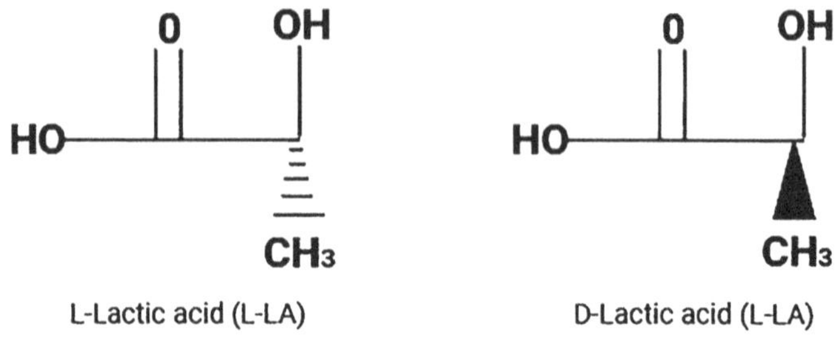

Fig. 1 Lactic acid stereoisomers

(negative) isomer rotates it in an anticlockwise direction (Rasal et al., 2010). The two isomers are mirror reflections of one other. These isomers are also known as L-lactic acid (L-LA) and D-lactic acid (D-LA). The CH_3 atom in the PLA's molecular structure is the reason behind its hydrophobic properties (John et al., 2008a, b; Chen et al., 2010).

The polylactic acid monomer can be transformed into polylactic acid, a thermoplastic polymer, via a variety of techniques including enzymatic polymerization, polycondensation, ring-opening polymerization (ROP), and azeotropic dehydration. The usual methods of manufacturing high molecular weight polylactic acid are labor-intensive and need high temperatures—180 °C to 200 °C and low pressures of 5 mm Hg. Technology advancements have shortened the time required to produce high molecular weight polylactic acid under benign reaction conditions. The two most popular methods for creating polylactic acid are the reaction of precipitation.

This procedure is not widely used because the lactic acid generated by the petrochemical method has an optically inactive mix of L and D lactic acid isomers (Rasal et al., 2010). The fermentation process that produced lactic acid monomers in an efficient and ecologically friendly way was developed in the 1990s (Akampumuza et al., 2016). The Lactobacillus bacteria convert maize starch to lactic acid through fermentation. The steps required to create PLA from agricultural waste are shown in Fig. 2. The fermentation of starch in a reactor containing bacteria and ammonia is the initial stage in the production of PLA.

This process produces ammonium lactate (Zhou et al., 2019). The next steps involve the separation of ammonium using the electrodialysis technique, and the formation of lactide from ammonium lactate using the water splitting chamber. Lactide is broken down into three subgroups: L-lactide, D-lactide, and meso-lactide. This process of prepolymerization involves the extraction of water from the lactide. The meso-lactide is removed by heat-assisted purification (Mladenovic et al., 2019). Fermentation produces two isomers of lactic acids: 99.5% is L-lactic acid and 0.5% is D-lactic acid. Lactic acid monomers can be polymerized to create polylactic acid by polycondensation and ring-opening polymerization. Figure 2 shows the polylactic acid synthesis.

This method transforms cyclic lactide monomers into polylactic acid (Robert & Aubrecht, 2008). Using this method, high molecular weight polylactic acid is

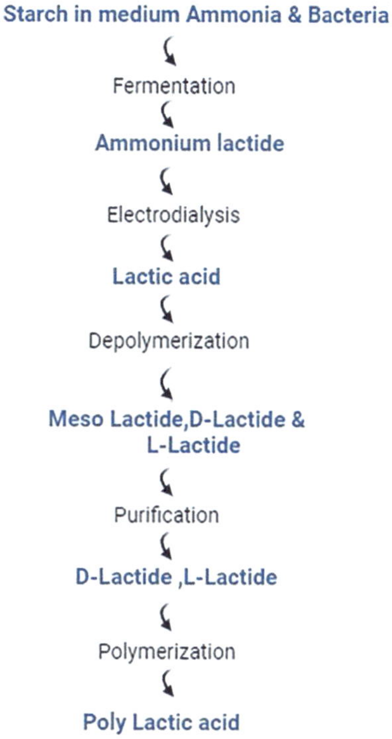

Fig. 2 Polylactic acid synthesis

created. Therefore, the ring-opening polymerization technique is used to generate polylactic acid on an industrial scale (Vink et al., 2003a, b).

5.3 *Mechanical Properties of Polylactic Acid*

Polylactic acid (PLA) is well-regarded for its excellent mechanical properties, which make it suitable for diverse applications, including biomedical devices and packaging. PLA demonstrates substantial tensile strength, generally between 50 and 70 MPa, which is comparable to that of traditional polymers like polyethylene and polypropylene, thereby positioning PLA as a strong alternative in various uses (Gupta et al., 2007).

Another critical mechanical property of PLA is its elastic modulus, which ranges from 2.7 to 16 GPa. This high modulus contributes to PLA's rigidity, making it appropriate for applications requiring structural stability (Garlotta, 2001).

In terms of thermal properties, PLA's performance is determined by its thermal processing parameters, specifically its melting temperature (Tm) and glass transition temperature (Tg). PLA's melting temperature usually falls between 150 and 160 °C, while its glass transition temperature is approximately 55 °C to 60 °C. These thermal properties are essential for determining PLA's compatibility with different thermal processing techniques and its stability across various temperature ranges.

5.4 Applications, Limitations, and Need for Enhancing Polylactic Acid

Polylactic acid finds extensive application across multiple industries. It is widely utilized for sutures, medication delivery systems, and scaffolds for tissue engineering in biomedical areas since it decomposes in the body into nontoxic metabolites. It is used frequently in packaging, in particular (Drumright et al., 2000; Farah et al., 2016).

Despite the advantages, it has certain disadvantages. Because of its relatively low impact resistance in comparison to typical polymers, its use in products requiring high toughness is restricted (Lim et al., 2008). PLA has a narrow processing temperature window due to its low glass transition temperature (Tg), which may limit its applicability for particular manufacturing processes (Auras et al., 2004). Moreover, because of its hydrophobic characteristics, it poses challenges in applications where moisture resistance is crucial, necessitating the employment of additive enhancement techniques (Drumright et al., 2000).

The mechanical properties of PLA are still being studied in copolymerization, mixing with other polymers, and the addition of reinforcing agents such as fibers or nanoparticles (NPs). By using these tactics, it should get stronger and more impact-resistant (Farah et al., 2016). Functionalization strategies such as chemical addition or surface chemistry modification can enhance PLA's hydrophilicity, adhesion, and barrier properties (Gupta & Kumar, 2007). As a result, it will be able to be utilized in more industrial processes and applications.

6 Development of Biocomposites Based on PLA

Natural fibers are being used as reinforcement in biocomposites amid environmental concerns and the expensive cost of products manufactured with synthetic fibers and petrochemical polymers. Biocomposites are divided into two categories: wood fibers and nonwood fibers. Natural fibers have high cellulose content and are comparatively lengthy, providing excellent tensile strength and crystallinity. However, these fibers have downsides because they contain hydroxyl groups (OH), which can draw water molecules. As a result, these fibers expand and create empty spaces at the composite's interface, with a reduction in its mechanical properties and dimensional stability. Biocomposites are used instead of synthetic materials that can be produced by the emission of pollutants that are not biodegradable or renewable (Christian et al., 2019). Natural fibers have drawn more attention over the last three decades. The pros and cons of using natural fibers are shown in Table 2.

Table 2 The pros and cons of natural fibers

Pros	Cons
Widely accessible with low cost	Matrix incompatibility
Minimal energy use during processing	Mild potency
It is reusable and is not noxious	Resins with inadequate wetting properties
Capable of sustainable use and environment-friendly	Significant absorption of moisture
No environmental impact	Form clusters when mixed with resins

Source: Ramamoorthy et al. (2015), Sydow et al. (2019)

Table 3 Comparison of natural and synthetic fibers

Natural fibers	Synthetic fibers
These are produced naturally	These are manufactured
These are usually high in strength	These are average in strength
These are environment friendly	These are not environment friendly
There is no release of carbon dioxide	There is an emission of carbon dioxide
These are less costly	These are high in cost
These are biodegradable	Very few are biodegradable

Source: Trivedi et al. (2023)

The components of natural fibers are pectin, lignin, pectin, hemicellulose, and cellulose. These natural fibers differ from synthetic fibers and are now widely used. Natural fibers differ from synthetic fibers in the following ways. Table 3 shows the comparison of natural and synthetic fibers.

Because of their superior properties, composite materials are becoming high in demand. These are drawing the attention of researchers due to the expanding range of utilization. Composite materials have a high strength-to-weight ratio compared to typical bulk materials. Typically, petroleum-based compounds are used to manufacture synthetic polymers, which are used to create composite materials.

Environmental concerns have resulted in the widespread rejection of these polymers as they are not biodegradable, sustainable, and are noxious when broken down. However, the rapid depletion of petroleum sources and the possibility of their total exhaustion is due to the widespread utilization. One of the best possible substitutes for nondegradable and unsuitable composites is the biocomposite based on PLA (polylactic acid) (Sanjay et al., 2018). Polylactic acid is sustainable and biodegradable and has no negative impact on the environment and the economy. Biopolymers manufactured from renewable sources, like starch, are used as matrices in biocomposites, whereas natural fibers are used as reinforcement materials. Applications of composites are numerous and expanding especially in the industries of sports, furniture, medicines, and automobiles (Bax et al., 2008).

7 Fabrication Techniques of PLA

Typically, conventional techniques for producing synthetic fiber-reinforced polymer composites are used in creating eco-friendly biocomposite. Closed molding processes, open molding procedures, and additive manufacturing processes are the three fabricating techniques for producing biocomposite. Figure 3 shows the fabrication techniques of polylactic acid.

7.1 Closed Mold Process

Closed mold processes in the fabrication of polylactic acid consist of injection molding and compression molding.

7.1.1 Injection Molding

The technology of injection modeling is widely used in the manufacturing of composite materials. This process involves injecting a mixture of molten thermoplastic polymer and fiber above the melting point into the mold's hollow form. Once the material has solidified, the component is extracted from the mold. The injection molding process is the widely used technique among researchers to fabricate polylactic acid biocomposite. This process is cost-effective and efficient to produce challenging components on a wide scale for commercial use (Thompson et al., 2010).

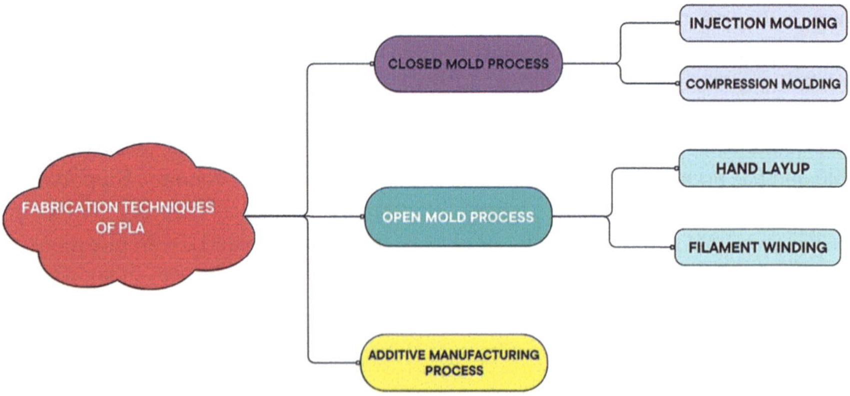

Fig. 3 Fabrication techniques of polylactic acid

7.1.2 Compression Molding

An additional recognized technique for creating PLA-based composite is compression molding. For materials that can be thermosets and thermoplastics, this technique is preferred. Utilizing this technique, polylactic acid and reinforcement or filler materials are brought to the lower mold. A mat of fiber is formed by the placement of the reinforcing material. High heat is used to melt PLA and uniformly combine it with fiber mats. Following this, the mold is subjected to high pressure to produce a composite's product final shape (Shumigin et al., 2011). High-rate production, highly strengthened goods, and exceptional dimension precision are all achieved using compression molding. By compression molding, Bax and Müssig created biocomposites based on PLA reinforced with 11% flax (Bax et al., 2008).

7.2 Open Mold Process

Hand layup and filament winding are the techniques that can be used in open mold processes to manufacture PLA.

7.2.1 Hand Layup

This is the earliest and least complicated open mold method for creating PLA-based biocomposites. During this procedure, dried fibers were inserted into the mold, capped with matrix resin, and forced through rollers. For the easy removal of composite from mold, it is sprayed with a reagent. Fiber, fiber matting, and short fibers can be altered using this technique. Low-volume manufacturing techniques are the applications of hand layup (Billah et al., 2021; Milenkovic et al., 2021).

7.2.2 Filament Winding

The first step of this process involves passing the fibers across a resin and thermoplastic matrix bath. In the filament winding process, the impregnated composite material is rolled onto the rotating mandrel until the required thickness is attained. A hollowed composite material is made by the filament winding procedure. The stacking order of the deposited fibers around the mandrel dictates the mechanical properties of the fabricated composite part. The high tensile strength can be achieved by lowering the angle of filament deposition, whereas the higher crushing strength is achieved by a higher angle (Preethikaharshini et al., 2022).

7.3 Additive Manufacturing Process

The more modern method of additive manufacturing, also known as 3D printing, is used to create biocomposites. Fused deposition modeling (FDM) is the most common and economical method for employing thermoplastic polymers among several additive manufacturing processes (Alturkestany et al., 2018; Ambone et al., 2020). This approach involves employing a screw extruder to generate fiber-reinforced PLA biocomposite filaments. The filament is then fed via the roller mechanism to the liquefier. When the filament comes across the heating block, it melts and emits in the form of thread, which is then placed onto the mobile substrate.

The additive manufacturing process uses nanosized fibers. Although there has been less research on fused deposition modeling than traditional approaches, recently it is becoming more common for fabricating biocomposites. The components made by 3D printing had noticeably less mechanical strength. The parameters of fused deposition modeling impact attributes of the 3D-printed part (Roxas et al., 2008; Dey et al., 2019).

8 Evaluation of Biocomposites Containing PLA

8.1 Mechanical Behavior

Materials called PLA biocomposites are created by mixing polylactic acid (PLA) with other biobased fillers or natural fibers. The desired mechanical qualities and sustainability are balanced in these composites. The chemical structure of PLA determines its mechanical, physical, and biodegradation rates. Compared to amorphous PLA, semicrystalline PLA has better mechanical properties. The elongation at the material's break reduces as PLA crystallinity increases and molecular chain mobility decreases, but the tensile strength and modulus rise. The D-content is a crucial characteristic that allows PLA's properties to change in several ways. For example, a higher D-content in PLA can lower the melting point by slowing down the process of crystallization.

The manufacturing process, the characteristics of the fibers and matrix, the amount of component materials, the size and shape of the fibers, the temperature, the interfacial bonding, etc., all affect the mechanical properties of the biocomposites. A PLA-based biocomposite reinforced with 20-weight percent sisal fibers was created. Observations revealed that mechanical characteristics like with the addition of fibers, the tensile, flexural, and impact characteristics increased dramatically to 80.6 MPa, 249.8 MPa, and 106.06 kJ/m^2, respectively. With the increase in reinforcement from 20 to 40 weight percent, it was observed. Significant increases in modulus to 13.9 MPa and tensile strength to 352 MPa were observed (Orue et al., 2016). Adding 30-weight percent ramie fiber to PLA biocomposites improved their tensile, flexural, and impact properties, increasing them to 55.3 MPa, 105.2 MPa, and 10.2 kJ/m^2, respectively.

Furthermore, Yu and Li (2014) and Singh et al. (2020) looked into the impact and flexural strength after adding the same percentage of ramie fiber (30 wt%). The results showed a rise in flexural strength to 132.5 MPa and impact strength to 11.3 kJ/m^2. The PLA matrix showed a maximum tensile strength of 76.2 MPa and an impact strength of 21.6 kJ/m^2 upon the addition of 20-weight percent flax fibers. A maximum tensile modulus of 8.1 GPa was demonstrated by PLA composite reinforced with 30-weight percent flax fibers. In a similar vein, adding additional fibers to the PLA matrix, such as kenaf, hemp, bamboo, and jute fibers, greatly improved the biocomposites'mechanical qualities.

When oil seed fibers were added to the PLA for reinforcement, their mechanical characteristics were reduced by 50%. The mechanical properties of PLA were reduced as a result of the improper reinforcing distribution brought on by the subpar manufacturing process. Nonetheless, the mechanical qualities might be improved by altering the production method. By blending PLA with aluminum triflate and then heating the mixture in a microwave for 6 hours at 170 °C, biocomposites with 20-weight percent reinforcement would be created. Furthermore, using a torque rheometer during manufacture would increase the PLA composites' tensile and impact strength by a factor of 1.5 (Ranakoti et al., 2022).

The mechanical characteristics of the biocomposites were significantly altered by thermal annealing. Three cellulosic fibers—agave, coir, and pine—were used as reinforcement to create PLA biocomposites. After conducting this experiment, they concluded that the annealed treatment and reinforcing greatly improved the mechanical properties of biocomposites. Petroleum-based polymers and PLA-based biocomposites have similar mechanical qualities. Table 4 represents the mechanical properties of PLA-based biocomposites.

8.2 *Thermal Behavior*

PLA degrades at 300 °C, whereas fibers typically begin to disintegrate at a lower temperature of 200–210 °C. Because natural fibers break down at a lower temperature than PLA biocomposites, this degradation is the primary reason for the PLA

Table 4 Mechanical properties of PLA

Mechanical parameter	PLA
Maximum tensile strength (MPa)	21–60
Melting temperature (°C)	150–162
Specific tensile modulus (kNm/g)	0.28–2.8
Maximum tensile strain (%)	2.5–6
Amorphous glass temperature (°C)	45–60

Source: Mehmood et al. (2023)

biocomposites'decreased thermal stability (Ngaowthong et al., 2019a, b). Thus, it is imperative to study the thermal characteristics of fiber-reinforced PLA biocomposites. Thermogravimetric analysis (TGA) and differential thermogravimetric (DTG) characterization are used to assess the thermal stability of PLA and fiber ingredients in biocomposites. Three main locations are involved in the heat degradation of PLA biocomposites. Around 100 °C is when the first thermal degradation zone appears (Sun et al., 2017a, b).

According to the literature, the thermal degradation of 30-weight percent PLA biocomposites reinforced with pineapple fibers started at 325 °C and reached its maximum at 415 °C, leaving an 18% residue and an 80% weight loss (Kaewpirom et al., 2014). The hemp fiber-reinforced (20 weight percent) PLA biocomposites showed a maximum degradation temperature of 398 °C and a beginning temperature of 303 °C, resulting in 95% weight loss and 1.2% residues. After conducting their experiment, it is concluded that the thermal characteristics of PLA biocomposites are not significantly affected by the inclusion of waste wood flour. However, the biocomposites'higher crystallinity upon reinforcing suggests that wood flour serves as a nucleating agent (Masirek et al., 2007).

9 Characterization and Performance of PLA Biocomposites

PLA (polylactic acid) is a bioplastic derived from renewable resources like corn starch. Fungal biomass, the body material of fungi cultivated on organic materials, can be incorporated into PLA to create biocomposites with unique properties and performance characteristics. PLA possesses the fundamental properties of universal polymer materials and is a member of the aliphatic polyester family. PLA has a tensile strength of roughly 54 MPa, which is comparable to polyethylene terephthalate (PET), and a tensile modulus of 3.4 GPa, which is marginally more than PET (Henton et al., 2005; Scaffaro et al., 2018).

PLA is ideal for a variety of applications due to its exceptional qualities, which include its outstanding flexibility, biocompatibility, and gas barrier capabilities. Comparing PLA's characteristics to those of conventional polymers, we find that PLA has a greater tensile modulus than nylon, PVC, and polypropylene (PP), and a higher flexural strength than PP. These results highlight PLA's potential, which can be further investigated in a variety of technical domains. PLA has a great deal of room to improve its mechanical and physical qualities, as previous research from the past 20 years has shown.

9.1 PLA-Based Biocomposites

Adding nanocellulose reinforcement to PLA-based bio-nanocomposites can significantly improve its properties, making it a more appealing option (Gond et al., 2021). Since nanocellulose and PLA are both sourced from renewable resources, the PLA-based bio-nanocomposites are fully biodegradable.

Because of the exceptional qualities of nanocellulose, which include low density, high aspect ratio, high specific modulus, high stiffness, low coefficient of thermal expansion, and nontoxicity, PLA composites reinforced with nanocellulose offer improved mechanical and thermal properties. Because PLA and nanocellulose have different hydrophobic and hydrophilic properties, there are some challenges in maintaining the compatibility of the nanocellulose within the PLA matrix (Zaaba et al., 2021).

9.2 PLA-Based Films Biocomposites

Another technique for fabricating PLA film biocomposites with reinforcement from biofillers is the PLA-based film. Gond and Gupta (2022) created PLA bio-nanocomposite films by solvent-casting different weight percentages of nanocellulose as reinforcement. Through chemical processing, the nanocellulose from the sugarcane bagasse was recovered. The use of nanocellulose enhanced the PLA bio-nanocomposite films' crystallinity and barrier qualities. Through the use of an extrusion technique and a dicumyl peroxide (DCP) cross-linking agent, Dhar et al. (2016) produced cellulose nanocrystals (CNC)-reinforced PLA biocomposite films. There was a 40% and 490% increase in the tensile strength and modulus, respectively. They discovered that PLA/CNC biocomposite film made with the cross-linking agent DCP had greater thermal stability (Masirek et al., 2007).

The PLA grafting on the cellulose nanocrystals (CNC) serves as a nucleating agent, which also increases the crystallinity. Biodegradable biocomposite mulch film was created by Finkenstadt and Tisserat et al. (2010), by reinforcing osage orange wood fibers with varying weight percentages (10 and 25 wt%), resulting in fiber lengths of 140 nm, 200 nm, and 400 nm. They looked at how reinforcement affected the mechanical qualities and found that while the tensile strength fell when osage orange wood fibers were added, the tensile modulus increased.

It was observed that adding nanocellulose up to 0.3 weight percent boosted the tensile strength and tensile modulus. By inhibiting the propagation of cracks, the inclusion of CNC enhanced the interaction between the PLA matrix and CNC, leading to an improvement in strength and modulus. However, strength and modulus may decrease with additional addition above 0.3 weight percent of nanocellulose aggregates in the matrix. Using the solvent casting process, Liu et al. (2016) created films based on bagasse hemicellulose and added glycerol and chitosan. They then examined the films' mechanical and barrier qualities.

They stated that maximum tensile strength and elongation percentage of 22 MPa and 65% were noted at 4 wt%, respectively, and that mechanical parameters increase with hemicellulose percentage (1.6–4 wt%). Because of the poor compatibility between the molecular chains, mechanical characteristics decline as the percentage of hemicellulose increases. The reason chitosan is added to films is because it has a high molecular weight and excellent film-forming ability. Similar to this, increasing the solution's glycerol content increases the film's flexibility throughout the production process.

9.3 3D/4D-Printed PLA Biocomposites

Among scientists and researchers, 3D printing PLA biocomposites and nanocomposites have proven to be a highly popular technology. One method of 3D printing that has been utilized to produce thermoplastic materials is fused deposition modeling or FDM. Numerous studies on the 3D printing of PLA biocomposites reinforced with natural fibers—such as microcrystalline cellulose (MCC), nanocrystalline cellulose (CNC), and cellulose nanofibrils (CNF)—extracted from diverse biomass in multiscale settings have been conducted. With these reinforcements, it was shown that PLA biocomposites'mechanical and thermal properties greatly improved. For 3D printing, Murphy et al. (2016) created PLA biocomposite filament reinforced with MCC.

Lignin-reinforced PLA biocomposites were made by Hong et al. (2021), who also evaluated how well they performed in 3D-printed parts. It was discovered that altering the lignin's surface might greatly enhance its mechanical qualities. Compared to pristine lignin-reinforced PLA biocomposites, the tensile strength of the PLA biocomposites reinforced with surface-modified lignin by carboxyl functional groups performed better.

According to Duigou et al. (2020), the deposition properties are significantly influenced by the materials' rheological behavior. To aid in the extrusion process, the material should have low viscosity at high shear rates. It should also have high zero-shear viscosity to maintain its shape.Because 3D-printed PLA biocomposites are biocompatible and biodegradable, they are recommended for use in advanced biomedical applications. It can be applied to cellular structure, tissue engineering, and medication delivery. Additionally, various organs and cellular structures can be prepared using it (Liu et al., 2020).

9.4 Biodegradability

PLA is biodegraded by a multistep process that combines microbial and chemical reactions. Comprehending this mechanism is essential for controlling PLA-based materials' biodegradability in a variety of applications. Ester bond cleavage produced by alkaline and enzymatic hydrolysis is the main mechanism of PLA breakdown. Alcohols and carboxylic acids are formed as a breakdown product when the ester linkages within the PLA polymer chains are disrupted. Microorganisms then use these shorter PLA units as substrates. Enzymes found in microorganisms can further break down PLA. The extracellular and intracellular depolymerizing enzymes play a role in this microbial depolymerization process.

The natural biodegradation of PLA occurs when bacteria, fungi, and algae are present. In the meantime, some biomass that can be used as manure is also produced during the process. When the residue produced is invisible and nonpoisonous, the terms biodegradability and compostability become synonymous (Wang et al., 2021). While the degradation behavior of PLA in an anaerobic environment has not been well researched by researchers, the PLA degradation phenomena in an aerobic environment have. Anaerobic environments cause the PLA to break down more slowly, according to observations.

On the other hand, the rate of degradation can be accelerated by raising the temperature above PLA's glass transition temperature (Tg). Aerobic degradation is comparable to composing when different microorganisms are involved, where humus is the result of the breakdown process (Ghosal et al., 2016). In industrial settings, where a significant volume of garbage is collected daily, composting is primarily employed to break down waste. To compare the rate of PLA degradation in industry and at home, a comparison study was conducted.

When compared to home composting, it was shown that industrial composting caused PLA to degrade much more quickly (60–80 days) due to several conditions, including high temperature, high pressure, and appropriate humidification. PLA is susceptible to hydrolysis because it has a hydrolyzable functional group. There are two phases to the aerobic PLA biodegradation process. Water diffuses in the hydrolysis chain during the first stage. The weak molecular chain is attacked by the microorganisms in the second stage, which results in the production of CO_2, H_2O, and microbial biomass.

Note that the PLA can only be attacked by an enzymatic reaction if the molecular weight drops to 10,000 Da or lower. Lipase, cutinase, alkaline proteases, subtilisin, trypsin, and elastase are among the several enzymes that are essential to the quick breakdown of PLA. The bodies of bacteria, fungi, and algae all contain these enzymes (Ahmed et al., 2018).

10 Impact of Fillers on PLA/Nanocomposite Properties

The development of PLA biocomposites appears to mark the conclusion of the search for a completely biodegradable composite with acceptable mechanical characteristics. Many biocomposites were created using different production techniques to explore PLA's potential as a polymer matrix. A mixture of 7 to 40-weight percent sugar beet pulp and PLA was injected into an injection molding machine in an attempt to create biocomposites. According to the results, the inclusion of sugar beet pulp caused a 29% fall in tensile strength and a 45% increase in PLA's Young's modulus.

The length of the short fiber, which resulted in an ineffective stress transfer mechanism between the fibers and matrix, was blamed for the bad outcome.

Thus far, there have been no positive mechanical performance outcomes from adding biofilters to PLA at varying loading situations. Flax fibers were reinforced in PLA as a bidirectional mat to achieve superior mechanical qualities that could rival the strength of composites made of synthetic fibers (Suryanegara et al., 2010). The findings showed a 20% and 32% increase in elongation at the break and absorption energy, respectively. Because of the bidirectional flax fibers' mat's reinforcing effect, the produced composites also displayed increased ductility. Following the reinforcement of kenaf fibers in the PLA matrix, which was created via a melt mixing process, encouraging mechanical properties were also attained (Ibrahim et al., 2009). The composites' tensile strength and tensile modulus rose by up to 30% relative to the weight of the fibers. Triacetin was shown to be responsible for the improvement since it gave the PLA stiffness and strengthened the interfacial connection between the fibers and matrix.

The lower mechanical properties of the composites at the higher reinforcement fibers were caused by improper wetting of the kenaf fiber as a result of an inadequate matrix. Because the nod in the PLA impeded the flow of the resin, adding a bamboo stick linked with a nod in the PLA produced an exceptional increase in the mechanical characteristics of the composites. PLA's relatively poor mechanical strength can be increased by adding 5% benzylated treated rice straw to a PLA and acetylation compound solution. PLA's very low crystallinity, which is about 10%, is the cause of this (Chaboki et al., 2019).

PLA-based nanocomposites have been developed using a variety of nanoparticles. The most prevalent ones among them include clays, carbon, bioglass, metals, and natural nanomaterials (Alikarami et al., 2022).

10.1 PLA/Metal Oxides

Because of their many characteristics, PLA/metal oxide-containing nanoparticles have attracted a lot of attention in research in recent years. In particular, because of their significant potential for practical use of antibacterial and antiviral materials, the construction of metal-containing nanocomposites is a current trend in scientific study (Demchenko et al., 2022).

Metal oxide nanoparticles have been created in a variety of forms, such as aluminum oxide (Al_2O_3), magnesium oxide (MgO), zinc oxide (ZnO), silica dioxide (SiO_2), and titania dioxide (TiO_2). They display a wide range of morphologies, including star, triangle, sphere, nanotube, nanorod, and nanowire shapes. The creation of polymer-metal oxide nanocomposites with a nanophase-separated structure necessitates both a decrease in the size of the polymer-metal oxide interface and the uniform dispersion of the metal oxide NPs (Shifrina et al., 2020).

Metal oxide NPs have intriguing findings in terms of their chemical and physical properties because of their high density and small size. The production of such nanocomposites using PLA is generally found to improve attributes such as optical, tensile strength, toughness, glass transition temperature, thermal stability, etc. As a result, these nanocomposites find extensive usage in a variety of fields, such as active surface coating and biomedical applications such as structural materials (Prasanna et al., 2019).

10.2 PLA Nanocomposites with Natural Nanoadditives

The commercial and academic sectors currently rank replacing dangerous and petroleum-derived complexes with natural, eco-friendly compounds derived from a variety of renewable resources as one of their top goals. Over the years, environmentally friendly and economically viable extraction techniques have been developed to produce encouraging results. After being separated, these environmentally friendly substitutes are successfully used in a wide range of industries with wildly different application goals, including biomolecules, adhesives, coagulants, dyes, and additives. The major challenge is creating high-performance thermoplastic composites reinforced with polyphenols, where nontoxic natural fillers are used in place of typical chemical additives to increase the desired multifunctionalities (antibacterial, structural, and antioxidant) and improve overall performance.

Esters and glycosides are just two examples of the vast family of secondary metabolites called polyphenols that are kept in the vacuoles of vegetative cells. Despite the large size of this family of compounds, they have certain characteristics in common, like the ability to form colored complexes with iron salts, be easily electrophilically substituted aromatically with diazonium salts and aldehydes, and be easily oxidized by potassium permanganate in alkaline environments (Smeriglio et al., 2017). High molecular weight polyphenols include tannins and lignin.

As a result, in addition to having characteristics common to the polyphenols group, they also have a high hydroxyl content, which gives them the capacity to form bonds and form a stable cross-linked association with a variety of molecules, including proteins or carbohydrates. Their distinct feature sets them apart from the other members of the polyphenolic group.

10.3 PLA/Ceramic Nanocomposites

Research on creating ceramic nanoparticles with better characteristics has been conducted successfully in several fields, including surface science and synthesis. Ceramics are described as solid compounds made of at least two elements, one of which must be a nonmetal or a metalloid, and which are created by the application of heat and occasionally pressure. One or more metals or other metalloids could be the other substance(s).

These multipurpose materials have good mechanical strength and hardness, good chemical and thermal stability, and feasible electrical, magnetic, optical, and thermal performance. Ceramic components are often shaped into the appropriate shapes using standard methods like injection molding, die pressing, tape casting, gel casting, etc. from a mixture of powder with or without binders and other additives. Densification can also only be achieved by sintering the green portions at high temperatures. One type of biomaterial that is often used in biomedical devices is ceramics.

Ceramics are commonly used as implant materials because of their high compressive strength, variable porosity, and bioactive qualities in the body. They can also be formed into a wide range of forms. Certain ceramics, such as calcium phosphate, have a chemical makeup that is quite close to that of human bone minerals, which makes them appropriate for usage as dental materials and orthopedic implants (the human skeleton, bones, and joints). These materials have outstanding osteo conduction properties, good biocompatibility, and excellent bioactivity (Chen et al., 2019).

11 Applications of PLA-Based Biocomposites

PLA composites reinforced with natural fibers are widely employed in a variety of applications, including infrastructure, furniture, cars, and food packaging (Sanjay et al., 2016). Reinforced polymer composites are employed in the construction of many automotive parts, including dashboards, seat backs, and door panels. Building elements include the floor, wall dividers, sliding windows, and so forth (Bax et al., 2008). It has been briefly mentioned how PLA may be used in tissue engineering, cardiovascular, drug carrier, epithelial cell regeneration, cancer therapy, tendon and skin regeneration, dental implants, orthopedics, antibacterial packaging, surgical instruments, respiratory equipment, and contemporary healthcare equipment.

PLA and its copolymers including glycolic acid and polyethylene glycol (PEG) have the necessary qualities to be used in the production of various medication carriers (Lassalle & Ferreira, 2007). The PLA copolymers for drug delivery carriers were identified by Tyler et al. (2016) who also covered the problems with these PLA-based materials. Liposomes, dendrimers, polymeric nanoparticles, micelles, and toxic antitumor medications can all be transported using PLA-based drug carriers. According to Mundel et al. (2022), PLA is a crucial biomaterial for micelles, and its copolymer with PEG can be utilized as a potent drug carrier for cancer-tumor targeting. Vert et al. (2015) used protein adhesion monitoring to examine the function of a PLA-based bioresorbable stent in the human body (globulins, fibrinogen, and albumin).

Using a PLA stent strut enhanced the hemocompatibility of surfaces in contact with blood. LA is appropriate for the cosmetic industry due to its antiaging, anti-acne, and controlled release of active substances for the skin. The suitability of PLA-based drug delivery vehicles for dermatotherapy was examined by Rancan et al. (2009). Fluorescent dyes were loaded into PLA nanoparticles, and the release of colors from the particles was examined. The outcomes demonstrated that hair follicle and sebaceous gland targeting are appropriate uses for PLA-based carriers. Melt electrospraying was used by Nam et al. (2020) to create the microbeads for cosmetics.

Comparing PLA microbeads to non-biodegradable microbeads, the former show potential as cosmetic ingredients. Hajleh et al. (2020) Covered a variety of approaches for creating PLA micro- and nanoparticles for use in cosmetics, including emulsion-based, microfluidic, direct compositing, and in situ forming procedures. Testing of PLA-based materials for cosmetic packaging was done by Connolly et al. (2019). A dermal danger was not observed in the skin cell assay.

The most popular 3D printing substance is PLA, which is utilized in the plastics sector to create a variety of engineering parts. According to Trivedi et al. (2023), PLA-based materials can be substituted to create environmentally friendly engineered products at a reasonable cost. Gere & Czigany (2020) looked into bottle-grade PLA's capacity for recycling. The outcomes demonstrated that compatibilizers are a useful tool for improving PLA's thermal stability when it comes to bottle production. The influence of various printing parameters on the 3D printing of PLA-based engineering items was examined by Roy et al. (2021).

The performance of PLA and polyethylene terephthalate (PET) drinking water bottles was compared by Tamburini et al. (2021). The findings demonstrated that drinking hot water and soap from PLA bottles is safe. The PLA was changed by Knoch et al. (2020) from hydrophilic to hydrophobic to produce exclusion nets that would provide a consistent pesticide supply in crops. Due to the increased phytosanitary effect of PLA cloth, PLA nets demonstrated molecular absorption for the integration of active ingredients. Films made of PLA were developed for use as agricultural mulch by Franca et al. 2019.

These PLA-based films are inexpensive and safe for the environment. Mochizuki (2022) talked about PLA's characteristics for nonwoven materials. The potential for PLA weathering resilience, thermal stability, resistance to hydrolysis, and electrical qualities demonstrated PLA fibers' suitability for use in industrial settings. PLA/ramie weave was created by Yang et al. (2021) for the automotive sector.

A PLA-based hybrid nanogenerator electronic skin device was created by Gong et al. (2020). The light-emitting diode and output signals can be turned on and off with this PLA-based gadget. Numerous other researchers documented the employment of PLA-based polymers in numerous further applications. It is possible to build various orthopedic and prosthetic devices using PLA-based materials (Büyük et al., 2023; Gendviliene et al., 2020).

PLA-based polymers are utilized in implants to create plates, screws, and other bone-application devices (Li et al., 2019). Degradable scaffolds and stents for the treatment of coronary artery disease are made from PLA. Table 5 lists many biomedical devices composed of PLA-based polymers. For applications like orthopedic screws and plates, the long-term mechanical strength preservation of PLA is

Table 5 Applications of PLA composites

Application area	Specific applications	References
Automotive	Dashboards, seat backs, door panels	Yang et al. (2021)
Construction	Flooring, wall dividers, sliding windows	Franca et al. (2019)
Packaging	Food packaging, antibacterial packaging	Connolly et al. (2019)
Medical and pharma	Tissue engineering, drug carriers, implants	Lassalle & Ferreira, (2007)
Cosmetics	Microbeads, dermal therapy, cosmetic packaging	Nam et al. (2020)
3D printing	Engineering parts, consumer products	Trivedi et al. (2023)
Agriculture	Mulching films, exclusion nets	Knoch et al. (2020)
Electronics	Nanogenerators, electronic skin devices	Gong et al. (2020)
Industrial	Nonwoven materials, industrial fibers	Mochizuki (2022)

ensured by its progressive deterioration (Feng et al., 2021). Table 5 summarizes the applications of PLA composites.

12 Future Prospects

As environmental consciousness has grown through the use of sustainable materials in engineering applications, PLA-based biocomposites and green composites have become more and more popular (Ilyas et al., 2021). Further investigation is needed in the following cases.

12.1 Enhanced Mechanical Properties

Investigate methods to improve the mechanical strength, toughness, and flexibility of PLA biocomposites. This could involve optimizing filler-matrix interactions, exploring novel reinforcements, or using advanced processing techniques like nanocomposite fabrication.

12.2 Functionalization and Compatibility

Study surface modification techniques to enhance the compatibility between PLA and natural fibers or other fillers. This could lead to improved adhesion at the interface, reducing potential weaknesses and enhancing overall material properties.

12.3 Durability and Aging

Assess the long-term durability and aging characteristics of PLA biocomposites, particularly under different environmental conditions (e.g., temperature, humidity, UV exposure). Understanding degradation mechanisms and developing strategies to mitigate degradation will be crucial.

PLA-based biocomposites are being used in a wide range of applications; however, more study and development are needed to reduce processing costs, enhance performance, and increase use in industrial settings. The most popular materials for 3D printing globally are PLA and its biocomposites; however, obtaining performance close to that of widely used traditional techniques (such as injection and compression molding) is a major difficulty. The benefits of complicated design manufacturing and the capacity to produce self-actuating items are provided by 4D printing. PLA has several beneficial properties, including a form memory effect, so PLA biocomposites could be highly helpful for 4D printing (Gurunathan et al., 2015).

13 Conclusion

Biocomposites, which blend natural fibers with biopolymers, are vital for sustainability. Polylactic acid (PLA) is notable for its renewable origins, such as corn starch and sugarcane, and its biodegradability. PLA production involves converting these resources into lactic acid, which is then polymerized. Despite its environmental benefits, PLA has good mechanical properties. This chapter explores the significance of biocomposites, provides an overview of PLA, outlines its sources and production, discusses its limitations, and analyzes the need for performance improvements to meet various industrial requirements. By fusing polylactic acid (PLA) with natural fibers, PLA biocomposites combine the biocompatibility and flexibility of PLA with increased mechanical strength to provide a sustainable substitute for conventional plastics. These characteristics are greatly influenced by the type, content, and processing techniques of the fiber, with efficient bonding and dispersion being essential. Heat stability can be increased by methods such as heat annealing, even though natural fibers reduce it. When compared to household composting, industrial composting speeds up the microbiological and chemical breakdown of PLA biocomposites. Even while some fillers could weaken the mechanical strength, they can improve performance when used properly. These composites find application in a variety of industries, such as construction, medicine, and automobiles. Future studies will concentrate on enhancing durability, compatibility with fillers, and mechanical qualities, with an eye toward possible improvements in manufacturing processes like 4D printing.

References

Ahmed, T., Shahid, M., Azeem, F., Rasul, I., Shah, A. A., Noman, M., et al. (2018). Biodegradation of plastics: Current scenario and prospects for environmental safety. *Environmental Science and Pollution Research, 25*, 7287–7298.

Akampumuza, O., Wambua, P. M., Ahmed, A., Li, W., & Qin, X. (2016). Review of the applications of biocomposites in the automotive industry. *Polymers and Polymer Composites, 38*(11), 1–17.

Alikarami, N., Abrisham, M., Huang, X., Panahi-Sarmad, M., Zhang, K., Dong, K., & Xiao, X. (2022). Compatibilization of PLA grafted maleic anhydrate through blending of thermoplastic starch (TPS) and nano clay nanocomposites for the reduction of gas permeability. *International Journal of Smart and Nano Materials, 13*(1), 130–151.

Alturkestany, M. T., Panchal, V., & Thompson, M. R. (2018). Improved part strength for the fused deposition 3D printing technique by chemical modification of polylactic acid. *Polymer Engineering and Science, 1*, 6. https://doi.org/10.1002/pen.24955

Ambone, T., Torris, A., & Shanmuganathan, K. (2020). Enhancing the mechanical properties of 3D printed polylactic acid using nanocellulose. *Polymer Engineering and Science, 60*, 1–14.

Appels, F. V. W., Dijksterhuis, J., Lucas, S. A., & Wösten, H. A. B. (2019). Hydrophobic surface properties of mycelium-based composites. *Fungal Biology and Biotechnology, 6*(1), 1–9.

Asokan, P., Firdoous, M., & Sonal, W. (2012). Properties and potential of bio fibers, bio binders, and biocomposites. *Reviews on Advanced Materials Science, 30*(3), 254–261.

Attias, N., Danai, O., Abitbol, T., Tarazi, E., & Ezov, N. (2021). Mycelium bio-composites in industrial design and architecture: Comparative review and experimental analysis. *Journal of Cleaner Production, 312*, 127733.

Auras, R., Harte, B., & Selke, S. (2004). An overview of polylactides as packaging materials. *Macromolecular Bioscience, 4*(9), 835–864.

Bax, J. M. (2008). Impact and tensile properties of PLA/Cordenka and PLA/flax composites. *Composites Science and Technology, 68*, 1601e1607. https://doi.org/10.1016/j.compscitech.2008.01.004

Bax, B., & Müssig, J. (2008). Impact and tensile properties of PLA/Cordenka and PLA/flax composites. *Composites Science and Technology, 68*(7–8), 1601–1607.

Billah, M. M., Rabbi, M. S., & Hasan, A. (2021). A review on developments in manufacturing process and mechanical properties of natural fiber composites. *Engineering Advances, 2*(1), 13–23. https://doi.org/10.38032/jea.2021.01.003

Bledzki, A. K., & Gassan, J. (1999). Composites reinforced with cellulose-based fibers. *Progress in Polymer Science, 24*(2), 221–274.

Büyük, N. İ., Aksu, D., & Torun Köse, G. (2023). Effect of different pore sizes of 3D printed PLA-based scaffold in bone tissue engineering. *International Journal of Polymeric Materials and Polymeric Biomaterials, 72*(13), 1021–1031.

Chaboki, M. G., Mohammadi-Rovshandeh, J., & Hemmati, F. (2019). Poly (lactic acid)/thermoplasticized rice straw biocomposites: Effects of benzylated lignocellulosic filler and nano clay. *Iranian Polymer Journal, 28*, 777–788. https://doi.org/10.1007/s13726-019-00743-1

Chen, G.-Q. (2010). Microbial lactic acid, its polymer poly(lactic acid), and their industrial applications. *Plastics from Bacteria Microbiology Monographs, 14*, 121–132. https://doi.org/10.1007/978-3-642-03287-5_6

Chen, Z., Li, Z., Li, J., Liu, C., Lao, C., Fu, Y., Liu, C., Li, Y., Wang, P., & He, Y. (2019). 3D printing of ceramics: A review. *Journal of the European Ceramic Society, 39*, 661–687.

Christian, S. J. (2019). *Natural fibre-reinforced noncementitious composites (bio composites)*. Elsevier.

Connolly, M., Zhang, Y., Brown, D. M., Ortuno, N., Jorda-Beneyto, M., Stone, V., et al. (2019). Novel polylactic acid (PLA)-organoclay nanocomposite bio-packaging for the cosmetic industry; migration studies and in vitro assessment of the dermal toxicity of migration extracts. *Polymer Degradation and Stability, 168*, 108938.

Demchenko, V., Kobylinskyi, S., Iurzhenko, M., Riabov, S., Vashchuk, A., Rybalchenko, N., Zahorodnia, S., Naumenko, K., Demchenko, O., Adamus, G., et al. (2022). Nanocomposites based on polylactide and silver nanoparticles and their antimicrobial and antiviral applications. *Reactive and Functional Polymers, 170*, 105096.

Dey, A., & Yodo, N. (2019). A systematic survey of FDM process parameter optimization and their influence on part characteristics. *Journal of Manufacturing Material Process, 3*(64), 1e30.

Dhar, P., Tarafder, D., Kumar, A., & Katiyar, V. (2016). Thermally recyclable polylactic acid/cellulose nanocrystal films through reactive extrusion process. *Polymer, 87*, 268–282.

Djukic-Vukovic, D., Mladenovic, J., Ivanovic, J., & Pejin, L. (2019). Mojovic, towards sustainability of lactic acid and poly-lactic acid polymers production. *Renewable and Sustainable Energy Reviews, 108*, 238e252.

Dorgan, J. R., Braun, B., Wegner, J. R., & Knauss, D. M. (2006). Poly(lactic acids): A brief review. *ACS Symposium Series, 939*, 102–125. https://doi.org/10.1021/bk-2006-0939.ch007

Drumright, R. E., Gruber, P. R., & Henton, D. E. (2000). Polylactic acid technology. *Advanced Materials, 12*(23), 1841–1846.

Duigou Le, A., Correa, D., Ueda, M., Matsuzaki, R., & Castro, M. (2020). A review of 3D and 4D printing of natural fibre biocomposites. *Materials & Design, 194*, 108911.

Elsacker, E., Peeters, E., De Laet, L., & Graeve, I. D. (2019). Mechanical, physical and chemical characterization of mycelium-based composites with different types of lignocellulosic substrates. *PLoS One, 14*(7), e0213954.

Farah, S., Anderson, D. G., & Langer, R. (2016). Physical and mechanical properties of PLA, and their functions in widespread applications—A comprehensive review. *Advanced Drug Delivery Reviews, 107*, 367–392.

Faruk, O., Bledzki, A. K., Fink, H. P., & Sain, M. (2012). Biocomposites reinforced with natural fibers: 2000–2010. *Progress in Polymer Science, 37*(11), 1552–1596.

Feng, P., Jia, J., Liu, M., Peng, S., Zhao, Z., & Shuai, C. (2021). Degradation mechanisms and acceleration strategies of poly (lactic acid) scaffold for bone regeneration. *Materials & Design, 210*, 110066.

Finkenstadt, V. L., & Tisserat, B. (2010). Poly(lactic acid) and Osage Orange wood fiber composites for agricultural mulch films. *Industrial Crops and Products, 31*(2), 316.

Franca, T.C., Kitagawa, D.A., Cavalcante, S.F.D.A., da Silva, J.A., Nepovimova, E., & Kuca, K. (2019). Novichoks: the dangerous fourth generation of chemical weapons. *International Journal of Molecular Sciences, 20*(5), 1222.

Garlotta, D. (2001). A literature review of poly(lactic acid). *Journal of Polymers and the Environment, 9*(2), 63–84.

Gendviliene, I., Simoliunas, E., Rekstyte, S., Malinauskas, M., Zaleckas, L., Jegelevicius, D., Bukelskiene, V., & Rutkunas, V. (2020). Assessment of the morphology and dimensional accuracy of 3D printed PLA and PLA/HAp scaffolds. *Journal of the Mechanical Behavior of Biomedical Materials, 104*, 103616.

Gere, D., & Czigany, T. (2020). Future trends of plastic bottle recycling: compatibilization of PET and PLA. *Polymer Testing, 81*, 106160.

Ghosal, D., Ghosh, S., Dutta, T. K., & Ahn, Y. (2016). Current state of knowledge in microbial degradation of polycyclic aromatic hydrocarbons (PAHs): A review. *Frontiers in Microbiology, 7*, 1369.

Gond, R. K., & Gupta, M. K. (2022). A novel approach for isolation of nanofibers from sugarcane bagasse and its characterization for packaging applications. *Polym. Compos, 41*, 5216.

Gond, R. K., Gupta, M. K., & Jawaid, M. (2021). Extraction of nanocellulose from sugarcane bagasse and its characterization for potential applications. *Polymer Composites, 42*(10), 1e13.

Gong, S., Zhang, B., Zhang, J., Wang, Z. L., & Ren, K. (2020). Biocompatible poly (lactic acid)-based hybrid piezoelectric and electret nanogenerator for electronic skin applications. *Advanced Functional Materials, 30*(14), 1908724.

Gupta, A. P., & Kumar, V. (2007). New emerging trends in synthetic biodegradable polymers–Polylactide: A critique. *European Polymer Journal, 43*(10), 4053–4074.

Gurunathan, T., Mohanty, S., & Nayak, S. K. (2015). A review of the recent developments in biocomposites based on natural fibres and their application perspectives. *Composites Part A: Applied Science and Manufacturing, 77*, 1–25.

Hajleh Abu, M.N., AL-Samydai, A., & Al-Dujaili, E.A. (2020). Nano, micro particulate and cosmetic delivery systems of polylactic acid: a mini review. *Journal of Cosmetic Dermatology, 19*(11), 2805–2811.

Haneef, M., Ceseracciu, L., Canale, C., Bayer, I. S., Heredia-Guerrero, J. A., & Athanassiou, A. (2017). Advanced materials from fungal mycelium: Fabrication and tuning of physical properties. *Scientific Reports, 7*, 41292. https://doi.org/10.1038/srep41292

Henton, D. E., Gruber, P., Lunt, J., & Randall, J. (2005). Polylactic acid technology. In *Natural fibers, biopolymers, and biocomposites* (pp. 559–607). CRC Press.

Hong, S., Park, J. H., Kim, O. Y., & Hwang, S. (2021). Preparation of chemically modified lignin-reinforced PLA biocomposites and their 3D printing performance. *Polymers, 13*(667), 1e10.

Ibrahim, N. A., Yunus, W. M. Z. W., Othman, M., Abdan, K., & Hadithon, K. A. (2009). Poly (lactic acid) (PLA)-reinforced kenaf bast fiber composites: The effect of triacetin. *Journal of Reinforced Plastics and Composites, 29*, 1099–1111. https://doi.org/10.1177/0731684409344651

Ilyas, R. A., Sapuan, S. M., Harussani, M. M., Hakimi, M. Y. A. Y., Haziq, M. Z. M., Atikah, M. S. N., et al. (2021). Polylactic acid (PLA) biocomposite: Processing, additive manufacturing and advanced applications. *Polymers, 13*(8), 1326.

Jawaid, M., Sapuan, S. M., & Alotman, O. Y. (2017). *Sustainable biocomposites: Challenges, potential and barriers for development* (p. 409). Green Biocomposites Green Energy Technology.

John, M. J., & Thomas, S. (2008a). Biofibres and biocomposites. *Carbohydrate Polymers, 71*(3), 343–364.

John, M. J., & Thomas, S. (2008b). Biofibers and biocomposites. *Carbohydrate Polymers, 71*(3), 343–364. https://doi.org/10.1016/j.carbpol.2007.05.040

Kaewpirom, S., & Worrarat, C. (2014). Preparation and properties of pineapple leaf fiber reinforced poly (lactic acid) green composites. *Fibers and Polymers, 15*, 1469–1477.

Kargarzadeh, H., et al. (2017). Recent developments on nanocellulose reinforced polymer nanocomposites: A review. *Polymer, 132*, 368–393.

Knoch, S., Pelletier, F., Larose, M., Chouinard, G., Dumont, M. J., & Tavares, J. R. (2020). Surface modification of PLA nets intended for agricultural applications. *Colloids and Surfaces A: Physicochemical and Engineering Aspects, 598*, 124787.

Lassalle, V., & Ferreira, M. L. (2007). Author proof author proof. *Macromolecular Bioscience, 7*, 767–783.

Li, J. W., Du, C. F., Yuchi, C. X., & Zhang, C. Q. (2019). Application of biodegradable materials in orthopedics. *Journal of Medical and Biological Engineering, 39*, 633–645. https://doi.org/10.1007/S40846-019-00469-8/TABLES/12

Lim, L. T., Auras, R., & Rubino, M. (2008). Processing technologies for poly(lactic acid). *Progress in Polymer Science, 33*(8), 820–852.

Liu, Y., Sun, B., Wang, Z., & Ni, Y. (2016). Mechanical and water vapor barrier properties of bagasse hemicellulose-based films. *BioResources, 11*(2), 4226.

Liu, S., Qin, S., He, M., Zhou, D., Qin, Q., & Wang, H. (2020). Current applications of poly (lactic acid) composites in tissue engineering and drug delivery. *Composites: Part B, 199*(108238), 1e25.

Masirek, R., Kulinski, Z., Chionna, D., Piorkowska, E., & Pracella, M. (2007). Composites of poly (L-lactide) with hemp fibers: Morphology and thermal and mechanical properties. *Journal of Applied Polymer Science, 105*(1), 255–268.

Mehmood, A., Raina, N., Phakeenuya, V., Wonganu, B., & Cheenkachorn, K. (2023). The current status and market trend of polylactic acid as biopolymer: Awareness and needs for sustainable development. *Materials Today: Proceedings, 72*, 3049–3055.

Milenkovic, S., Slavkovic, V., Fragassa, C., Grujovic, N., Palic, N., & Zivic, F. (2021). Effect of the raster orientation on strength of the continuous fiber reinforced PVDF/PLA composites, fabricated by hand-layup and fused deposition modeling. *Composite Structures, 270*, 114063.

Mochizuki, M. (2022). Textile applications. *Poly (Lactic Acid) Synthesis, Structures, Properties, Processing, Applications, and End of Life*, pp. 619–629.

Mohanty, A. K., Misra, M., & Drzal, L. T. (2002). Sustainable biocomposites from renewable resources: Opportunities and challenges in the green materials world. *Journal of Polymers and the Environment, 10*, 19–26.

Mundel, R., Thakur, T., & Chatterjee, M. (2022). Emerging uses of PLA–PEG copolymer in cancer drug delivery. 3. *Biotech, 12*(2), 41.

Murphy, C. A., & Collins, M. N. (2016). *Microcrystalline cellulose reinforced polylactic acid biocomposite filaments for 3D printing* (p. 1e10). Pymer Compos.

Nam, G. E., Kim, Y. H., Han, K., Jung, J. H., Rhee, E. J., Lee, S. S., Kim, D. J., Lee, K. W., & Lee, W. Y. (2020). Obesity fact sheet in Korea, 2019: prevalence of obesity and abdominal obesity from 2009 to 2018 and social factors. *Journal of Obesity & Metabolic Syndrome, 29*(2), 124.

Nampoothiri, K. M., Nair, N. R., & John, R. P. (2010). An overview of the recent developments in polylactide (PLA) research. *Bioresource Technology, 101*(22), 8493–8501.

Ngaowthong, C., et al. (2019a). Recycling of sisal fiber reinforced polypropylene and polylactic acid composites: thermo-mechanical properties, morphology, and water absorption behavior. *Waste Manag, 97*, 71–81. https://doi.org/10.1016/j.wasman.2019.07.038

Ngaowthong, C., Borůvka, M., Běhálek, L., Lenfeld, P., Švec, M., Dangtungee, R., et al. (2019b). Recycling of sisal fiber reinforced polypropylene and polylactic acid composites: Thermo-mechanical properties, morphology, and water absorption behavior. *Waste Management, 97*, 71–81.

Nguyen, P. T., et al. (2022). Fungal-derived biocomposites: A new approach in the enhancement of PLA-based materials. *Journal of Sustainable Materials and Technologies, 45*, 123–130.

Orue, A., Jauregi, A., Unsuain, U., Labidi, J., Eceiza, A., & Arbelaiz, A. (2016). The effect of alkaline and silane treatments on mechanical properties and breakage of sisal fibers and poly (lactic acid)/sisal fiber composites. *Composites Part A: Applied Science and Manufacturing, 84*, 186–195.

Pickering, K. L., Efendy, M. G. A., & Le, T. M. (2016). A review of recent developments in natural fibre composites and their mechanical performance. *Composites Part A: Applied Science and Manufacturing, 83*, 98–112.

Prasanna, S. R. V. S., Pandey, S., Balaji, K., & Rana, S. (2019). *Metal oxide based nanomaterials and their polymer nanocomposites*. In Nanomaterials and Polymer Nanocomposites; Elsevier.

Preethikaharshini, J., Naresh, K., Rajeshkumar, G., Arumugaprabu, V., Khan, M. A., & Khan, K. A. (2022). Review of advanced techniques for manufacturing biocomposites: Non-destructive evaluation and artificial intelligence assisted modeling. *Journal of Materials Science, 57*(34), 16091e16146. https://doi.org/10.1007/s10853-022-07558-1

Ramamoorthy, S. K., Skrifvars, M., & Persson, A. (2015). A review of natural fibers used in bio-composites: Plant, animal and regenerated cellulose fibers. *Polymer Reviews, 55*(1), 107.

Ranakoti, L., Gangil, B., Mishra, S. K., Singh, T., Sharma, S., Ilyas, R. A., & El-Khatib, S. (2022). Critical review on polylactic acid: Properties, structure, processing, biocomposites, and nano-composites. *Materials, 15*(12), 4312.

Rancan, F., Papakostas, D., Hadam, S., Hackbarth, S., Delair, T., Primard, C., et al. (2009). Investigation of polylactic acid (PLA) nanoparticles as drug delivery systems for local dermatotherapy. *Pharmaceutical Research, 26*, 2027–2036.

Rasal, R. M., Janorkar, A. V., & Hirt, D. E. (2010). Poly(lactic acid) modifications. *Progress in Polymer Science, 35*(3), 338–356. https://doi.org/10.1016/j.progpolymsci.2009.12.003

Robert, J. L., & Aubrecht, K. B. (2008). Ring-opening polymerization of lactide to form a biodegradable polymer. *Journal of Chemical Education, 85*(2), 258–260. https://doi.org/10.1021/ed085p258

Roxas, M., Ju, S., & Sullivan, P. E. (2008). *Fluid dynamics analysis of desktop-based fused deposition modeling rapid prototyping*. Thesis of Bachelor of Applied Science, Department of Mechanical and Industrial Engineering, University of Toronto.

Roy, A., Saffar, M., Vaswani, A., & Grangier, D. (2021). Efficient content-based sparse attention with routing transformers. *Transactions of the Association for Computational Linguistics, 9*, 53–68.

Sanjay, M. R., Arpitha, G. R., Naik, L. L., Gopalakrishna, K., & Yogesha, B. J. N. R. (2016). Applications of natural fibers and its composites: An overview. *Natural Resources, 7*(3), 108–114.

Sanjay, M. R., Madhu, P., Jawaid, M., Senthamaraikannan, P., Senthil, S., & Pradeep, S. (2018). Characterization and properties of natural fiber polymer composites: a comprehensive review. *Journal of Cleaner Production, 172*, 566e581.

Scaffaro, R., Lopresti, F., & Botta, L. (2018). PLA based biocomposites reinforced with Posidonia oceanica leaves. *Composites Part B: Engineering, 139*, 1–11.

Shifrina, Z. B., Matveeva, V. G., & Bronstein, L. M. (2020). Role of polymer structures in catalysis by transition metal and metal oxide nanoparticle composites. *Chemical Reviews, 120*, 1350–1396.

Shumigin, E., Tarasova, A., & Krumme, P. M. (2011). Rheological and mechanical properties of poly (lactic) acid/cellulose and LDPE/cellulose composites. *Materials Science, 17*(1), 32e37.

Singh, J. I. P., Singh, S., & Dhawan, V. (2020). Influence of fiber volume fraction and curing temperature on mechanical properties of jute/PLA green composites. *Polymers and Polymer Composites, 28*(4), 273–284.

Singh, R., et al. (2021). Advances in fungal biomass utilization for PLA biocomposites: A review of recent progress. *Materials Science and Engineering C, 135*, 111202.

Smeriglio, A., Barreca, D., Bellocco, E., & Trombetta, D. (2017). Proanthocyanidins and hydrolysable tannins: Occurrence, dietary intake and pharmacological effects. *British Journal of Pharmacology, 174*, 1244–1262.

Sun, Z., Zhang, L., Liang, D., Xiao, W., & Lin, J. (2017a). Mechanical and thermal properties of PLA biocomposites reinforced by coir fibers. *International Journal of Polymer Science, 2017*(1), 2178329.

Sun, Z., Zhang, L., Liang, D., Xiao, W., & Lin, J. (2017b). Mechanical and thermal properties of PLA biocomposites reinforced by coir fibers. *International Journal of Polymer Science, 1*, 9. https://doi.org/10.1155/2017/2178329

Suryanegara, L., Nakagaito, A. N., & Yano, H. (2010). Thermo-mechanical properties of microfibrillated cellulose-reinforced partially crystallized PLA composites. *Cellulose, 17*, 771–778. https://doi.org/10.1007/s10570-010-9419-5

Sydow, Z., & Bienczak, K. (2019). The overview on the use of natural fibers reinforced composites for food packaging. *Journal of Natural Fibers, 16*(8), 1189.

Tamburini, E., Costa, S., Summa, D., Battistella, L., Fano, E. A., & Castaldelli, G. (2021). Plastic (PET) vs bioplastic (PLA) or refillable aluminium bottles—What is the most sustainable choice for drinking water? A life-cycle (LCA) analysis. *Environmental Research, 196*, 110974.

Thompson, M. R., & Sun, J. (2010). Wet granulation: implications of screw design. *Journal of Pharmacological Sciences, 99*(4), 2090–2103. https://doi.org/10.1002/jps

Trivedi, A. K., Gupta, M. K., & Singh, H. (2023). PLA based biocomposites for sustainable products: A review. *Advanced Industrial and Engineering Polymer Research, 6*(4), 382–395.

Tyler, B., Gullotti, D., Mangraviti, A., Utsuki, T., & Brem, H. (2016). Polylactic acid (PLA) controlled delivery carriers for biomedical applications. *Advanced Drug Delivery Reviews, 107*, 163–175.

Vert, M. (2015). After soft tissues, bone, drug delivery and packaging, PLA aims at blood. *European Polymer Journal, 68*, 516–525.

Vink, E. T. H., Rabago, K. R., Glassner, D. A., & Gruber, P. R. (2003a). Applications of life cycle assessment to nature works™ polylactide (PLA) production. *Polym. Degrad. Stabil., 80*(3), 403–419.

Vink, E. T. H., Rabago, K. R., Glassner, D. A., & Gruber, P. R. (2003b). Applications of life cycle assessment to NatureWorks™ polylactide (PLA) production. *Polymer Degradation and Stability, 80*(3), 403–419.

Wang, J., Peng, C., Li, H., Zhang, P., & Liu, X. (2021). The impact of microplastic-microbe interactions on animal health and biogeochemical cycles: A mini-review. *Science of the Total Environment, 773*, 145697.

Yang, X., Fan, W., Ge, S., Gao, X., Wang, S., Zhang, Y., et al. (2021). Advanced textile technology for fabrication of ramie fiber PLA composites with enhanced mechanical properties. *Industrial Crops and Products, 162*, 113312.

Yu, T., Jiang, N., & Li, Y. (2014). Study on short ramie fiber/poly (lactic acid) composites compatibilized by maleic anhydride. *Composites Part A: Applied Science and Manufacturing, 64*, 139–146.

Zaaba, N. F., Jaafar, M., & Ismail, H. (2021). Tensile and morphological properties of nanocrystalline cellulose and nanofibrillated cellulose reinforced PLA bionanocomposites: A review. *Polymer Engineering and Science, 61*(1), 22e38.

Zhou, Y., Cui, Y., & Qu, X. (2019). Exopolysaccharides of lactic acid bacteria: Structure, bioactivity and associations: A review. *Carbohydrate Polymers, 207*, 317.

Zhou, Q., et al. (2023). Integrating fungal biomass into PLA composites for improved mechanical and thermal properties. *International Journal of Biological Macromolecules, 180*, 245–252.

Biochar from Biobased Materials: Applications, Clean Technologies, and Novel Production Methods

Amit Kumar Tiwari, Ashish Kapoor, Dan Bahadur Pal, and Amina Othmani

Abstract Biobased materials received significant attention as a potential and renewable resource for producing valuable products such as biochar due to their abundant availability and low cost. Biobased material (biomass) is characterized by its chemical properties like low ash content and ability to yield high amounts of liquid. Over the years, researchers have dedicated great efforts to this utility by using biobased materials and lignocellulosic biomasses as they have excellent ability for wastewater treatment and the removal of hazardous materials. Recently, a rise of interest has been oriented towards biochar derived from waste biomass due to its interesting properties. This highly porous carbonous material, which is rich in charcoal, is becoming very attractive for many applications due to its abundance. Biochar is a renewable fuel that is added to soil to improve its quality because of its higher proportion of aromatic C and condensed aromatic structures. In this chapter, we have highlighted the economical and environmentally friendly technologies recently opted for biochar production and their benefits. The pyrolysis techniques possess good abilities to produce biochar. The effect of the influencing parameters (time and temperature) on the pyrolysis are described. Finding urgent solutions for the daily environmental challenges is becoming primordial for all researchers. Presently, the integration of clean technologies for environmentally safe production and applications of biochar has gained great interest. In the present chapter, we discussed the main production methods of biochar, their properties, merits and demerits, and main applications. A particular interest was given to the methods used

A. K. Tiwari
Department of Chemical Engineering, Birla Institute of Technology, Mesra, Ranchi, Jharkhand, India

A. Kapoor · D. B. Pal
Department of Chemical Engineering, Harcourt Butler Technical University, Kanpur, Uttar Pradesh, India

A. Othmani (✉)
Faculty of Sciences of Monastir, University of Monastir, Avenue of the Environment, Monastir, Tunisia

D. B. Pal, S. L. Bansal (eds.), *Fungal Waste Biomass Management for Energy, Environment and Value-Added Products*, Fungal Biology,
https://doi.org/10.1007/978-3-031-82599-6_8

for the modification of biochar. A variety of factors are discussed including the main influencing parameters. Finally, a critical overview was provided, followed by some potential future solutions for better use of biochar in environmental challenges.

Keywords Biochar · Biomass · Clean technologies · Pyrolysis · Gasification · Carbonization · Modification methods

1 Introduction

Using cheap and plentiful agricultural wastes to address environmental challenges has been proven to be a successful strategy (Othmani et al., 2022). Organic biomass is pyrolyzed to produce biochar, a highly porous carbonous substance useful in many fields in particular for wastewater treatment (Xu et al., 2015). This is the last product, which is extremely refractory in nature and has half of the original carbon. After the thermal decomposition of organic biomass and agricultural residues, it is naturally produced and used as an agent for soil reclamation (Lu et al., 2018; Wang & Wang, 2019). It contains different amounts of C, H, O, N, S, and ash content (Ahmad et al., 2014). Biochar has unique properties that make it suitable for a wide range of applications, like large surface area, high porousness, different functional groups, high cation exchange capacity, and good stability (Agegnehu et al., 2015; Domingues et al., 2017). These properties include its quick and simple preparation, environmental friendliness, cost-effectiveness, and reusability, making it one of the best candidates for removing various types of contaminants (Cywar et al., 2022). Apart from its intended uses, biochar modification has garnered interest to expand its application domains or improve its application efficiency (Benavente & Fullana, 2015). Many biochar-based new materials, such as activated, functionalized, and tailored biochar, biochar composites, and porous carbon, have been produced based on distinct activation/synthesis techniques (Dawood et al., 2017). Furthermore, sorbents generated from biochar have come to light as potential solutions for removing and/or controlling pollutants like arsenic, cadmium, methylene blue, bisphenol-a, peroxymonosulfate, and persulfate (El-Naggara et al., 2019). Biochar is a highly promising material for mitigating soil remediation and air and water purification, given its environmental friendliness and carbon neutrality; researchers underlined the significance of standardizing biochar characterization and suggested the launch of an international forum for the characterization of biochar (Hamza et al., 2016). Basically, three methods are used to transform biomass into biochar: fast pyrolysis, slow pyrolysis, and gasification (Han et al., 2016). Based on the literature, slow pyrolysis is the most acceptable method which is commonly used for conversion because it provides the highest yield of biochar. The ultimate yields of liquid (bio-oil) and gas (syngas) are obtained through fast pyrolysis and gasification processes, respectively (Othmani et al., 2022). This article contributes to a better interpretation and knowledge of biochar's production methods and fundamental applications. The production of biochar by making use of miscellaneous techniques has also been explored. Many aspects are discussed, including the main influencing parameters. The modification methods of biochar including acidic, alkaline, metal salts or metal oxides modifications, and other methods are discussed. Additionally, we highlighted the main merits and demerits of each production method. Finally, a critical overview was given followed by some solutions that may help for better use of the biochar in environmental challenges.

2 Biochar: Characteristics and Primary Applications

It is a carbon-rich organic stuff formed as a by-product after thermal degradation of organic material without oxygen. This last has a noticeable enrichment in C as well as other elements like N, O, and H (Blanco-Canqui, 2017). It has both alkyl and aromatic structures. The contents of these elements may vary as it is subject to the nature of feedstock (raw material) utilized for the production of biochar, which is further used for the reclamation of soil and plant enhancement. The kind of raw material also affects the pH of biochar; pH typically ranges from neutral to alkaline, though acidic pH has been reported when specific feedstocks are used and under various pyrolysis conditions (Tomczyk et al., 2020). The pH and pyrolysis temperature have direct effects on biochar production; for example, increasing the temperature causes bionic acid breakdown and increases the concentration of mineral alkali elements (Huang et al., 2021). Also, temperature can affect the porosity and the specific area of biochar (Smith, 2016). These functional organic groups affected the hydrophobicity of the biochar used (Lu et al., 2018). An inversely proportional relationship exists between the pyrolysis temperature and the polar functional groups existing on the surface of biochar, which enhances the hydrophobicity of this last. From here, we can conclude the importance of the pyrolysis temperature as a main factor that may affect all the other parameters that affected in turn the quality of the biochar produced (Wu et al., 2017). Figure 1 highlights the primary attributes and applications of biochar, as reported by multiple researchers.

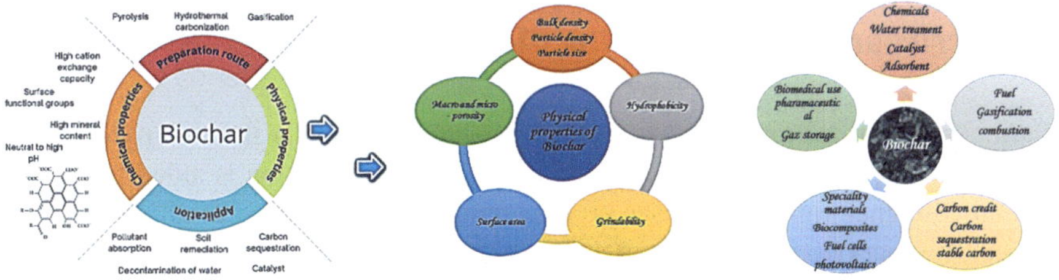

Fig. 1 Main properties and applications of biochar

Based on previous studies, biochar is a renewable fuel that can also be added to soil to improve its quality (Venegas et al., 2016; Zhu et al., 2017). The primary cause of the high stability is the nature of carbon structures (Bamdad et al., 2018). Compared to other soil organic matter like lignin, biochar has a significantly higher proportion of aromatic C and condensed aromatic structures (Evan Bassam et al., 2023). This is the obvious chemical distinction between biochar and other organic matter (Zhu et al., 2017). Biochars with condensed aromatic structures can take on different forms, such as graphite C, turbostratic C, and amorphous C, which are formed at different temperatures during pyrolysis. It is appropriate for a wide range of applications due to its interesting properties. A few benefits of biochar are its quick and simple preparation, environmental friendliness, cost-effectiveness, and reusability (Hamed et al., 2020). Biochar has shown benefits for human use due to its unique properties, which include its high biodegradability, healthy concentrations of organic and total carbon, and the optimal ratios of macro- and microelements like potassium, magnesium, calcium, sodium, copper, zinc, and iron (Chenhui et al., 2021). Moreover, its high specific area, high surface functional group content, pH, and propositivity have given further reasons for its use, especially in wastewater decontamination (Hamed et al., 2020). Biochar has been used as a soil amendment for several millennia. In addition to being a soil conditioner, biochar has other advantageous properties that make it useful for a variety of tasks (Chi et al., 2017).

A few significant uses for biochar are enumerated as a soil conditioner and a waste management tool, carbon sequestrant, climate change mitigant, and wastewater treating agent in various industries such as construction, cosmetics, metallurgy, food industry, and energy production sector as a means of advancing the creation of catalysts (Cheng et al., 2017). Even though biochar is widely used in many fields, one of its most effective uses is still for wastewater treatment and as a soil conditioner, energy production, and support for catalyst development (Sakhiya et al., 2020).

According to the literature, using biochar as a soil conditioner typically increases soil pH and nutrient acquisition (Melo et al., 2016). The presence of a variety of functional groups on the biochar helps to increase the number of available sites on the surface of this last, which leads to enhancing its capacity to uptake the target pollutants (Ippolito et al., 2020).

3 Overview of the Major Methods of Biochar Production: Merits and Demerits of the Method

Agricultural residue and biomass can be converted to biochar through pyrolysis in absence of oxygen (Sakhiya et al., 2020). Charcoal production is aided by longer vapour residence times and lower process temperatures (Hamed et al., 2020). The type of product obtained is dependent on the applied temperature; for instance, liquids are produced at low temperature and short residence time, whereas gases

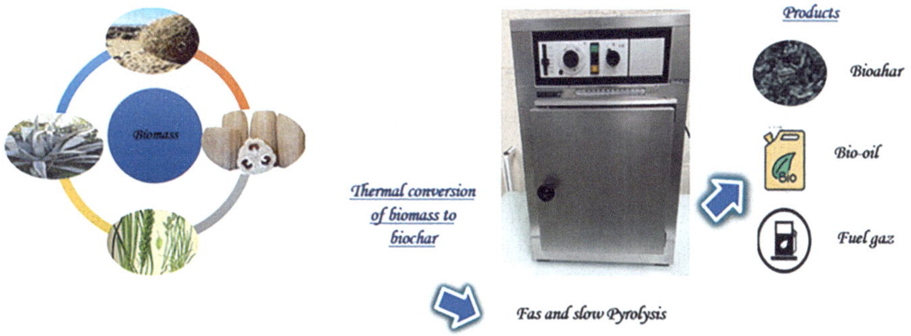

Fig. 2 Brief description of the concept for the production of biochar

require more time and high temperatures to be produced (Peiris et al., 2017). Therefore, the products are dependent on the initial applied conditions and, in particular, the temperature (Domingues et al., 2017). Figure 2 presents a brief description of the general concept for the production of biochar.

Currently, there is a lot of interest in fast pyrolysis for the production of liquids, this last can produce a good quality of chemicals that can be used in turn as a source of energy (Shaheen et al., 2018).

Biomass breaks down rapidly during fast pyrolysis, mostly producing vapours and aerosols along with a small amount of gas and charcoal. After cooling and condensation, a homogeneous, dark brown liquid with a heating value is produced (Qin et al., 2018). Biomass is known by its chemical compositions that contain feeds with a small amount of ash content yield and a huge high liquid yield. In order to protect the properties of the used biomass, specific attention must be given to the required temperature that may affect negatively the production of the desired product (Inyang et al., 2014).

As mentioned earlier, the desired product is strongly affected by the initial condition. In this context, the fast pyrolysis, which is well known for its speed, can take a few seconds to produce liquids. However, the products are still dependent on some parameters, among them the heat and mass transfer rate (Mishra & Upadhyay, 2021). As for slow pyrolysis, this method that produces more biochar than any other technology is employed in the chemical industries to produce coke, charcoal, methanol, activated carbon, and various wood-derived chemicals (Inyang et al., 2014). The duration of the heating presents the main influencing parameter in this method, it must be adequate to facilitate equilibration throughout the pyrolysis process (Othmani et al., 2022). The reactions can be affected by the residence time of the products in the reactor (Jafri et al., 2018).

Unlike slow pyrolysis, during fast pyrolysis, the reactor load is heated quickly with high heating rates, and the pyrolysis vapours are immediately swept out of the hot zone of the reactor and condensed. Time to retention of pyrolysis vapours is usually less than 2 s in fast pyrolysis for limited side reactions that break down heavy hydrocarbons (oils) into light hydrocarbons (gas) (Hamed et al., 2020). Fast decomposition at high temperatures (pyrolysis) is used for the production of liquid fuels to the detriment of coal and gas. The slow process of pyrolysis is distinguished by a slow heating rate, which involves heating the material to a temperature of around 400–500 °C at 0.1–1 °C/s for approximately 5–30 min. This method promotes carbon formation, but liquid products and gases are also produced in small quantities. A slow heating rate combined with a long residence time leads to secondary reactions, which ultimately result in increased tank formation (Han et al., 2016).

As for flash pyrolysis, it can also be taken as an improved way of fast pyrolysis. In fast pyrolysis, the heating speed is very high, around 1000 °C/s. The temperature reached in flash pyrolysis can exceed 900 °C and reaction times are significantly reduced (0.1–1 s) (El-Gamal et al., 2017). The high heating rate combined with an elevated temperature and less residence time of pyrolytic vapours yields a high amount of liquid and less formation of carbon. It can convert up to 70% of the biomass to bio-oil with a very good quality (Almeida & Marque, 2015). This method requires fast heating rates and high reaction temperatures, and the residence time is very short compared to the other methods used for biochar production (Angın, 2013). The efficiency of this method is dependent on the biomass particle size utilized as a feedstock which must be very small typically around 60–140 mesh size (Almeida & Marque, 2015).

Another important method used for the production of biochar is the gasification, this thermochemical conversion can be done under controlled oxygen conditions at high temperatures (>700 °C), producing producer gas or syngas, which are combustible gases (Mishra & Upadhyay, 2021).

The gasification process can convert plastics into syngas that can be used in the production of electricity or for the manufacture of chemicals, hydrogen, or transport fuels. Its main advantage is the probability of processing heterogeneous, mixed, and contaminated polymers with limited pretreatment is greater (Chowdhury et al., 2019). Carbonaceous material is converted into gaseous products, namely syngas, in the presence of gasification agents and at high temperatures. Syngas can then be burned to generate energy or used for other purposes. The gasification mechanism is divided into drying which is a separation process highly useful in case of a high moisture-containing feedstock; during the process, the entire moisture content of the biomass is evaporated without any energy recovery followed by an oxidation/combustion process (Özgün et al., 2022). Gasification is a more efficient method of producing syngas. Gasification is an additional use for biodegradable waste. Additionally, the elevated temperatures help to remove corrosive elements (chloride, potassium, etc.), and after that, the production of clean gas is achieved in place of vexed fuels (Matteo et al., 2023).

Furthermore, the process of carbonization was considered to help transform organic materials into high-carbon materials. It displays several pyrolytic reactions that are comparable to the conventional method of producing charcoal (Haider et al., 2022). Biochar can be produced using either large, stationary units or smaller, more portable ones. This is one of the main advantages of its production. Table 1 summarizes the main advantages and disadvantages of the methods used for biochar production.

Table 1 Benefits and drawbacks of biochar production (Ahmad et al., 2014; Domingues et al., 2017; Shaheen et al., 2018; Cywar et al., 2022; Matteo et al., 2023)

Process	Merits	Demerits
Fast pyrolysis	This direct thermochemical procedure converts solid biomass into a large amount of energy liquid bio-oil, resulting in a large amount of bio-oil yield. The main advantages of this process are its low temperature (500 °C) and short hot residence time (2 s).	Low yield of biochar. Very small particles of biomass feedstock (1–2 mm) is required. Prefer biomass with low moisture content (<10%).
Slow pyrolysis	The maximum biochar yield, the process's ability to handle a broad range of particle sizes, low to moderate temperatures, and extended residence times are its key benefits.	Further treatment of gases is needed due to high CO concentrations.
Flash pyrolysis	Hot vapour residence time at low-moderate temperature. Production of fuel with higher energy density. Higher yields of bio-oil.	This method has a number of drawbacks, including low thermal stability. The oil contains a solid phase. Oil corrosion and chemical instability. Viscosity increased over time. Pyrolytic water production. Biochar alkali dissolution in bio-oil.

4 Effects of Main Parameters Affecting the Biochar Production

The efficiency of the pyrolysis of biochar to the desired product is influenced by a set of factors (Matteo et al., 2023). The primary factors are described in Fig. 3.

4.1 Effect of Temperature

As seen in Fig. 3, the influencing parameters on the production of biochar are numerous; however, the significance is debatable. Let us begin with the basic factor, which is temperature. The importance of this parameter is related to its ability to control the main decompositional behaviour of the used polymers (Domingues et al., 2017). It also controls the cracking reaction of plastics that occurs due to loss of Van der Waals force (between the molecules), this loss is always generated due to the increased vibration of the molecules. This tends to evaporate the molecules and results in the breaking of the carbon chain (Al-Salem et al., 2017). Consequently, the temperature has a crucial effect on pyrolysis products and secondary reactions (Demirbas, 2004). At low temperatures, long-chain hydrocarbons are produced, while increasing temperature promotes the production of short-chain compound due to the cracking of C-C bonds (El-Gamal et al., 2017). Several researchers have studied the effect of temperature variations on the distribution of products obtained

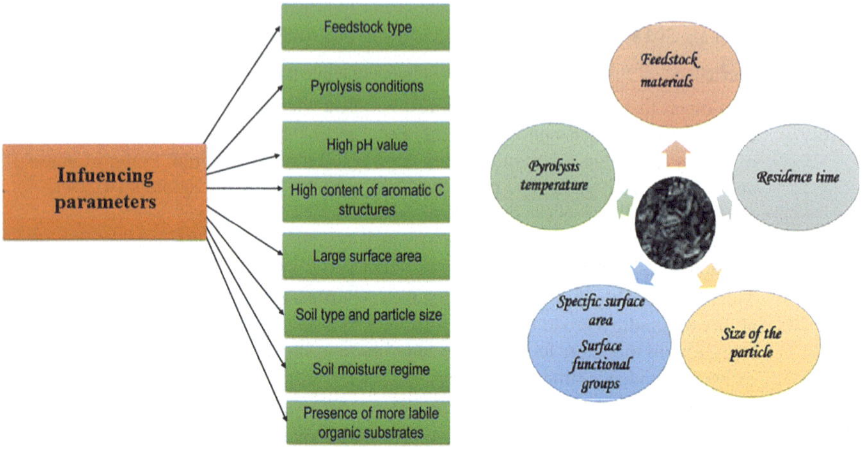

Fig. 3 Main influencing parameters on the biochar production

after pyrolysis of plastic waste (Domingues et al., 2017). They found that the properties of the products can vary at a specific temperature. According to Lopez et al. (2011), a large proportion of highly viscous liquids with a high content of long carbon chains is obtained at 460 °C, while a small proportion of liquids with high aromatic contents was noticed at 600 °C (Lopez et al., 2011).

4.2 *Effect of the Residence Time*

The definition of residence time varies from one study to another. However, its importance in biochar production is the same (Almeida & Marque, 2015). Some researchers consider it as the contact time of the material (e.g. plastic) with the hot surface of the reactor during the fast or continuous pyrolysis process (Ansah et al., 2016). On the other hand, residence time in slow pyrolysis or process batch refers to the time between the start of heating the raw material and the end of the reaction where all products are recovered. Longer residence time improves the secondary cracking conversion of primary products, resulting in more thermally stable products such as lighter molecular weight hydrocarbons and non-condensable gases (Chen et al., 2015). The researchers have succeeded to prove that the residence time had a significant impact on the composition of the pyrolysis products. Temperatures below 700 °C and a longer residence time resulted in a higher liquid yield. At temperatures above 700 °C, however, the residence time had less of an effect on the liquid and gas yield (Ansah et al., 2016; Lopez et al., 2018).

4.3 *Effect of Feedstock Properties*

The quality of the biochar produced is related to the biomass used; therefore, the properties of the final products can establish the performance of the process. The chemical composition of the biomass used may vary from one type to another, so the quality of the biochar produced will also vary. In other words, the ratio of lignin, hemicellulose, and cellulose content in the feedstock determines the percentage of volatile carbon and fixed carbon in pyrolytic products such as bio-oil, gas, and biochar (Chen et al., 2015). The higher amount of lignin in the feedstock is responsible for the higher yield of biochar. As a result, the feedstock chosen may be determined by the final type of the required product (Hamed et al., 2020). Table 2 collects some biochar properties derived from various feedstocks.

4.4 *Effect of Surface Area, Porosity, and Cation Exchange Capacity*

The porosity and surface area of biochar preferentially affects the ability of biochar to adsorb many pollutants, especially in adsorption processes. Therefore, special attention should be paid to this along with the applied temperature that can positively or negatively affect the performance of biochar. Another important parameter (cation exchange capacity) describes the total amount of exchangeable cations that are strongly affected and can be retained by biochar (Haider et al., 2022). Biochar is considered effective for recovering nutrients from wastewater due to its properties like high porosity and adsorption, which allow poisons to adhere to its surfaces,

Table 2 Biochar properties derived from various feedstocks

Feedstock	pH	Moisture	C (%)	O (%)	Pyrolysis temperature	CEC (C mol kg^{-1})
Coconut shell	9.18	4.40	50.20	43.40	–	–
Rice straw	64	–	41	–	700	127.5
Eucalyptus	6	–	37	–	700	129.75
Peanut shell	9.5	1.90	82.20	3.35	700	–
Corn stover	–	2.30	33.20	8.60	700	–
Soybean	11.3	–	82	–	700	–
Sugarcane bagasse	9.30	1.30	85.59	10.48	685	–
Cotton seed hull	8.50	6.53	87.50	7,.60	600	–
Pine cone	9.8	1.20	95.16	0.60	700	–
Tire rubber	–	–	86	–	800	–
Acacia	8.1	–	66	–	800	117
Green waste	6.2	–	70	–	680	129
Turkey litter	10	–	45	–	700	–
Bamboo	–	–	83.6	–	83.6	129

Modified from Chen et al. (2015), Lopez et al. (2018), Hamed et al. (2020), Haider et al. (2022)

resulting in clean wastewater and nutrient-rich biochar. Biochar and activated carbon are widely used to eliminate toxic pollutants from water (Marta et al., 2022).

5 Biochar Modification

One of the main limitations of the use of biochar in some applications is its cost and its fundamental properties, some researchers have proposed some modification methods to enhance its ability to absorb pollutants with better performance (Bo et al., 2023). In this chapter, we present some commonly used methods for biochar modification to improve its physicochemical properties for various environmental processes. A few important methods used to modify the biochar are shown in Fig. 4.

The acidic modification of biochar consists of introducing acid functional groups on its surface by removing any impurities. It can be done using any type of acid, strong or weak; however, the introduction of the acidic functional group will be different. For instance, for a weak acid, the introduction can be done via the esterification method (Rajapaksha et al., 2016). These methods are adopted by researchers to increase the surface area of biochar and enhance the hydrophobicity and the ability of biochar to absorb the targeted pollutant.

For alkali modification, the purpose of modification of biochar is the same, to increase the surface area and concentration of oxygen functional groups on the surface of the biochar. Generally, potassium hydroxide and sodium hydroxide are the main products used for this utility. Many researchers have succeeded in showing the ability of alkali modification to enhance the properties of their used biochar such as Jin et al. (2014) and Sun et al. (2015), who have used different methods for modification using biochar (municipal waste-derived) prepared by pyrolysis, which has shown good As (V) removal potential; another modification was also tried by using wheat straw-derived biochar prepared through hydrothermal carbonisation process, which was proven efficient in increasing the surface area. Furthermore, the

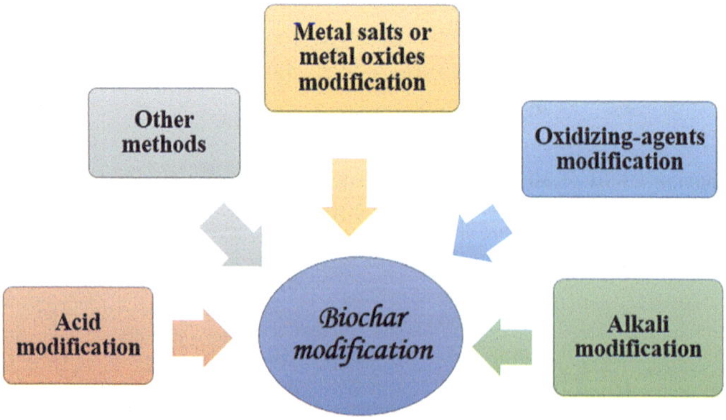

Fig. 4 Biochar's modification methods

modification of biochar by adding oxidizing agents has shown good potential to enhance the properties of used biochar by increasing the oxygenated functional groups, thereby increasing the adsorption capacity (Wang et al., 2015). Wood-derived biochar has been successfully modified by potassium permanganate to enhance its adsorption capacity for the adsorption of Cd (II), Cu (II), and Pb (II). Others have chosen the metal salts or metal oxide modification for enhancing the properties of the used biochar, like Samsuri et al. (2013) who have modified the biochar (derived from husk) by Fe (III) for the adsorption of As (III) and (V).

Some other methods have been used for the modification of biochar, such as carbonaceous material modification, carbon nanotubes, and graphene, which can successfully adsorb organic pollutants and heavy metals (Zhao et al., 2011). Many studies have discussed the modification of biochar by organic solvent, steam modification, modification using ammonia gas, etc. (Wang & Wang, 2018). It is classified into physical modification methods for steam activation and widely applied to escalate the porosity of biochar and for the removal of impurities such as incomplete combustion products. This treatment helps to raise the surface area available for adsorption process.

Furthermore, biochar-based composites are prepared using the impregnation process, where other materials such as organic compounds, metal oxides, clay, graphene oxide, or carbon nanotubes are impregnated with biochar and microbial inoculation. In this case, biochar mainly acts as a platform with a raised surface area to carry the materials deposition on it. Composites differ from chemical activation because they are concerned with the formation of absolutely new functional groups on the surface that were not present before on the surface of the feedstock or biochar. These can enhance the performance of the final modified biochar. Based on all these discussed methods, there is no doubt about the performance of the modification methods used to enhance the properties of the used biochar; however, the actual application of this modified biochar for agriculture, energy, and sustainable environment development needs to be thoroughly reviewed.

These emerging and effective applications will provide a solid foundation for future engineered applications of modified biochar. Extensive research is devoted to biochar modification; despite that, most of the work is in theoretical stages and concentrated in the laboratory (Li et al., 2019). There are few reports available on large-scale applications of modified biochar. Industrial-scale production of modified biochar is still in its early childhood stages and requires additional support for expansion. In addition, the cost of modification is also a critical factor in their promotion and use (Domingues et al., 2017). Chemical modification may be more advantageous than physical modification in terms of developed properties like greater specific surface area, good porous structure, and more oxygenated surface functional groups. However, the use of the chemical modifiers may create secondary contaminants and pollution risks to the environment. Therefore, the adoption of environmentally friendly organic acids must be used during the modification process. As a consequence, we have to seriously think about the choice of the added chemical to achieve both goals enhancing the performance of the used biochar and avoid the secondary effects of these last (Jafri et al., 2018).

Doing characterizations for modified biochar can help to better understand the main process and predict the proposed mechanism, especially during the adsorption of pollutants. This way can help to avoid wasting efforts to produce more hazardous materials when choosing the wrong chemicals that may interact during modification process and create a new source of environmental danger.

Some future research must be dedicated to track the rationality of modifications of biochar-based composite over the time to avoid the toxicity generated by the materials absorbed in the biochar matrix. As a solution, leaching tests are required to better control this issue. Another lacuna is also related to the ecotoxicology of these biochar formulations (Yang et al., 2019a, b; Ye et al., 2018).

6 Challenges for Biochar Applications: A Brief Description of Problems Encountered

Starting from the methods used for the biochar production, pyrolysis, a thermochemical breakdown of organic matter, requires lower processing temperatures and produces fewer air pollutants. Despite its main merits and when talking from an environmental standpoint, pyrolysis provides an alternative to landfill, does not pollute water, and reduces greenhouse gases (Luz et al., 2018). It also generates a high calorific value fuel that is easily commercialized and can be used in internal combustion engines to generate electricity and heat (Lyu et al., 2020). It presents a set of demerits that limit the performance of the by-product. Biochar has several utilizations in different fields such as in climate change, renewable electricity, food security, and waste control. The biochar production process also produces a mixture of fractions of liquids and gases as a by-product (Kamali et al., 2022). Its highly porous surface provides an excellent natural territory for the survival of microorganisms and enhances bioavailability. In some applications, it also forms a reservoir for nutrients, water, and pollutants. Adding it to the soil increases the growth rate of plants. This in turn provides an effective sink for oneself segregation of atmospheric CO_2 (Hamza et al. 2016). Biochar also has other benefits: reduced risk of reduced crop yields during dry seasons, reduced need for chemical fertilizers containing nitrogen and phosphorus, aids in retaining nitrogen and sulfur in the soil, thereby reducing emissions. In addition, it aids in the regrowth of vegetation on barren lands, it inhibits the growth of mould and mildew. Currently, many types of biochar derivatives are being produced, with different properties depending on the feedstock, pyrolysis conditions, residence time, and additives added. Standardization and classification of the different types is necessary if it is to be marketed for public use (Feng & Zhu, 2018). Furthermore, biochar can adjust climate versions by using sequestering carbon. The potential of this substance to enhance soil fertility and raise agricultural yield is another special quality; as a result, soils with biochar content material have the highest fertility when achieved nicely and below controlled conditions, pyrolysis can yield biochar with an extensive variety of uses (Thomas et al., 2020; Huff & Lee, 2016; El-Naggara et al., 2019; Hamza et al. 2016). However, considering the deliberations done by numerous investigators who have considered the various interactions between biochar production methods and waste reuse practices, it is necessary to develop new opportunities. A circular bio-economy has been established by utilizing waste from one agro-processing industry to another by exposing the toxic pollutant issues and the by-product going into soil application (Huang et al., 2019; Hamza et al. 2016). This has enabled new product and procedure advancement as well as the potential for the formation of new businesses. Significant items become fluids due to high moisture content, and if there is a low degree of water, there is a high risk that the procedure produces a higher amount of residue rather than oil. However, at higher temperatures, the warming rate is high, a greater division of debris and vaporous items is produced (Hamza et al., 2016). Bio-oil can be delivered in this manner, and constrained discharges could be incorporated into the local network for the feasible generation of biochar noting both specialized and financial viewpoints, recouping additionally the biochar and heat produced. Furthermore, more efforts should be made to increase the use of vegetable materials on a large scale and to meet the usual demand for these materials (Li et al., 2019).

Also, a specific interest must be dedicated to the price. Along with the widespread use of biochar, we must consider its negative impact on the environment. The stability of biochar related to the stability of carbon can present an environmental problem. Therefore, it is primordial to take into consideration these physicochemical properties during the search for the solutions that may enhance its full potential use (Daful & Chandraratne, 2018). To summarize, as with any other adsorbent, biochar has many advantages and disadvantages, including unintended environmental and health consequences. For example, we cannot ignore the fact that the use of biochar on soil biota and aquatic organisms presents an increasing toxicity, as demonstrated by the avoidance behaviour bioassay, which assessed survival rates, vertical distribution, weight changes, superoxide dismutase activity, and protein content for soil biota, as well as inhibition rates and lethality percentages for aquatic organisms. The toxicity of biochar can be related to its particle size. The selection of the best biochar is still one of the main challenges. This subject needs more attention to avoid the creation of a secondary source of pollution; instead of solving the problems, we could increase it.

Biochar's high alkalinity, caused by abundant minerals, may also reduce microbial activity. This toxic effect could be reduced by washing biochar with acids and water to remove minerals, or by selecting feedstocks that produce neutral biochar. Given the benefits and widespread application of engineered biochar in the fields of sustainable environment, agriculture, and energy development, its potential and unintended ecotoxicological consequences should be considered before its widespread use. Nanometal oxide/hydroxide-biochar composites are widely manufactured due to their numerous advantages in environmental pollution remediation over pure biochar. In addition, the presence of metal oxides loaded on biochar can affect the terrestrial and aquatic ecosystems, especially when it is released into the environment without any treatment. Furthermore, due to the fact that the modification of biochar may present more serious problems, the potential of this ecotoxicity should be considered (Cen et al., 2019).

7 Conclusion

A description of the merits and demerits of pyrolysis-based technologies, such as gasification, slow and fast pyrolysis, as efficient methods of converting lignocellulosic biomass into high-value products was given. The solutions proposed by researchers to improve clean techniques and methods of biochar production, biochar modification, etc. are briefly discussed in the different sections of this chapter. Several advantages and challenges faced during the production of biochar from biobased materials are also highlighted and discussed thoroughly. Despite all of the recommended merits of biochar, it is also noticed that there is some disagreement about its uses, especially the modified forms of biochar. From the cited, it can be concluded that biochar is good for environmental applications; in some cases, it may also contain toxic compounds and can have hazardous effects. Modifying or activating biochar causes chemical and physical changes that may increase toxicity (reactivity, presence of nanoparticles, metals, etc.). Long-term and careful assessments of unintended consequences are required before the extensive applications of modified biochar, especially in agriculture. Meanwhile, the costs of the modifications limit their promotion and extensive applications; therefore, in the future, more attention should be given to low-cost and environmentally friendly methods for the commercial production of biochar.

References

Agegnehu, G., Bass, A. M., Nelson, P. N., & Bird, M. I. (2015). Benefits of biochar, compost, and biochar–compost for soil quality, maize yield and greenhouse gas emissions in a tropical agricultural soil. *Science of the Total Environment, 543*, 295–305.

Ahmad, M., Moon, D. H., Vithanage, M., Koutsospyros, A., Sang, S. L., Yang, J. E., Lee, S. E., Jeon, C., & Yong, S. O. (2014). Production and use of biochar from buffalo-weed (*Ambrosia trifida* L.) for trichloroethylene removal from water. *Journal of Chemical Technology and Biotechnology, 89*, 150–157.

Almeida, D., & Marque, M. d. F. (2015). Thermal and catalytic pyrolysis of polyethylene plastic waste in semi. *Polimeros, 26*, 1–8.

Al-Salem, S. M., Antelava, A., Constantinou, A., Manos, G., & Dutta, A. (2017). A review on thermal and catalytic pyrolysis of plastic solid waste (PSW). *Journal of Environmental Management, 197*, 177–198.

Angın, D. (2013). Effect of pyrolysis temperature and heating rate on biochar obtained from pyrolysis of safflower seed press cake. *Bioresource Technology, 128*, 593–597.

Ansah, E., Wang, L., & Shahbazi, A. (2016). Thermogravimetric and calorimetric characteristics during co-pyrolysis of municipal solid waste components. *Waste Management, 56*, 196–206.

Bamdad, H., Hawboldt, K., & MacQuarrie, S. (2018). A review on common adsorbents for acid gases removal: Focus on biochar. *Renewable and Sustainable Energy Reviews, 81*, 1705–1720.

Benavente, V., & Fullana, A. (2015). Torrefaction of olive mill waste. *Biomass and Bioenergy, 73*, 186–194.

Blanco-Canqui, H. (2017). Biochar and soil physical properties. *Soil Science Society of America Journal, 81*, 687–711.

Bo, X., Zhang, Z., Wang, J., Guo, Z., Li, Z., Lin, H., Huang, Y., Han, Z., Kuzyakov, Y., & Zou, J. (2023). Benefits and limitations of biochar for climate-smart agriculture: A review and case study from China. *Biochar, 5*, 77.

Cen, K., Zhang, J., Ma, Z., Chen, D., Zhou, J., & Ma, H. (2019). Investigation of the relevance between biomass pyrolysis polygeneration and washing pretreatment under different severities: Water, dilute acid solution and aqueous phase bio-oil. *Bioresource Technology, 278*, 26–33.

Chen, D., Yin, L., Wang, H., & He, P. (2015). Reprint of: Pyrolysis technologies for municipal solid waste: A review. *Waste Management, 37*, 116–136.

Cheng, H., Hill, P. W., Bastami, M. S., & Jones, D. L. (2017). Biochar stimulates the decomposition of simple organic matter and suppresses the decomposition of complex organic matter in a sandy loam soil. *Global Change Biology Bioenergy, 9*, 1110–1121.

Chenhui, Z., Binbin, S., Ming, Y., Zhifeng, L., Qinghua, L., Qingyun, H., Ting, W., Yang, L., Yuan, P., Jing, H., Jiajia, W., Jie, L., & Lin, T. (2021). Activation of peroxymonosulfate by biochar-based catalysts and applications in the degradation of organic contaminants: A review. *Chemical Engineering Journal, 416*, 128829.

Chi, T., Zuo, J. E., & Liu, F. L. (2017). Performance and mechanism for cadmium and lead adsorption from water and soil by corn straw biochar. *Frontiers of Environmental Science and Engineering, 11*, 15.

Chowdhury, S., Sikder, J., Mandal, T., & Halder, G. (2019). Comprehensive analysis on sorptive uptake of enrofloxacin by activated carbon derived from industrial paper sludge. *Science of the Total Environment, 665*, 438–452.

Cywar, R. M., Rorrer, N. A., & Hoyt, C. B. (2022). Bio-based polymers with performance-advantaged properties. *Nature Reviews Materials, 7*, 83–103.

Daful, A. G., & Chandraratne, M. R. (2018). Biochar production from biomass waste-derived material. In *Reference module in materials science and materials engineering* (pp. 11249–11258). Elsevier.

Dawood, S., Sen, T. K., & Phan, C. (2017). Synthesis and characterization of slow pyrolysis pine-cone biochar in the removal of organic and inorganic pollutants from aqueous solution by adsorption: Kinetic, equilibrium, mechanism and thermodynamic. *Bioresource Technology, 246*, 76–81.

Demirbas, A. (2004). Effects of temperature and particle size on bio-char yield from pyrolysis of agricultural residues. *Journal of Analytical and Applied Pyrolysis, 72*, 243–248.

Domingues, R. R., Trugilho, P. F., Silva, C. A., de Melo, I. C. N. A., Melo, L. C. A., Magriotis, Z. M., & Sanchez-Monedero, M. A. (2017). Properties of biochar derived from wood and high-nutrient biomasses with the aim of agronomic and environmental benefits. *PLoS One, 12*, 0176884.

El-Gamal, E., Saleh, M., Elsokkary, I., Rashad, M., & Abd El-Latif, M. M. (2017). Comparison between properties of biochar produced by traditional and controlled pyrolysis. *Alexandria Science Exchange Journal, 38*, 413–424.

El-Naggara, A., Sang, B., Sang, S. L., Rinklebed, J., Farooq, M., Songe, H., Sarmahh, A. K., Zimmerman, A. R., Ahmad, M., Shaheend, S. M., & Ok, Y. S. (2019). Biochar application to low fertility soils: A review of status, and future prospects. *Geoderma, 337*, 536–554.

Evan Bassam, D., Zoltan, T., & Angela, A. (2023). Biochar and its effects on soil properties and evapotranspiration: A sustainable solution for plant growth. *Cogent Food & Agriculture, 9*, 1.

Feng, Z. J., & Zhu, L. Z. (2018). Sorption of phenanthrene to biochar modified by base. *Frontiers of Environmental Science and Engineering, 12*, 1.

Haider, C. F. U., Coultier, J. A., Liqun, C. A. I., Hussain, S. S., Cheema, A., & Renzhi, J. W. U. (2022). An overview on biochar production, its implications, and mechanisms of biochar-induced amelioration of soil and plant characteristics. *Pedosphere, 32*, 107–130.

Hamed, K., Shariat, P., Hamed, K., Shariat, P., Mona, D., Mona, D., Ok, Y. S., Abdul-Sattar, N., Benyamin, K., Solange, I., Mussatto, A., Meisam, T., & Lam, S. S. (2020). A comprehensive review of engineered biochar: Production, characteristics, and environmental applications. *Journal of Cleaner Production, 270*, 122462.

Hamza, U. D., Nasri, N. S., Amin, N. S., Mohammed, J., & Zain, H. M. (2016). Characteristics of oil palm shell biochar and activated carbon prepared at different carbonization times. *Desalination and Water Treatment, 57*, 7999–8006.

Han, Y., Cao, X., Ouyang, X., Sohi, S. P., & Chen, J. (2016). Adsorption kinetics of magnetic biochar derived from peanut hull on removal of Cr (VI) from aqueous solution: Effects of production conditions and particle size. *Chemosphere, 145*, 336–341.

Huang, M., Li, Z., Luo, N., Yang, R. Y., Wen, J., Huang, B., & Zeng, G. (2019). Application potential of biochar in environment: Insight from degradation of biochar-derived DOM and complexation of DOM with heavy metals. *Science of the Total Environment, 646*, 220–228.

Huang, H., Reddy, N. G., & Huang, X. (2021). Effects of pyrolysis temperature, feedstock type and compaction on water retention of biochar amended soil. *Scientific Reports, 11*, 7419.

Huff, M. D., & Lee, J. W. (2016). Biochar-surface oxygenation with hydrogen peroxide. *Journal of Environmental Management, 165*, 17–21.

Inyang, M., Gao, B., Zimmerman, A., Zhang, M., & Chen, H. (2014). Synthesis, characterization, and dye sorption ability of carbon nanotube–biochar nanocomposites. *Chemical Engineering Journal, 236*, 39–46.

Ippolito, J. A., Cui, L., Kammann, C., Wrage-Monnig, N., Mendizabal, T., Sigua, G., Novack, J., Spokas, K., & Borchard, N. (2020). Feedstock choice, pyrolysis temperature and type influence biochar characteristics: A comprehensive meta-data analysis review. *Biochar, 2*, 421–438.

Jafri, N., Wong, W. Y., Doshi, Y., Yoon, L. W., & Cheah, K. H. (2018). A review on production and characterization of biochars for application in direct carbon fuel cells. *Process Safety and Environmental Protection, 118*, 152–166.

Jin, H., Capareda, S., Chang, Z., Gao, J., Xu, Y., & Zhang, J. (2014). Biochar pyrolytically produced from municipal solid wastes for aqueous As(V) removal: Adsorption property and its improvement with KOH activation. *Bioresource Technology, 169*, 622–629.

Kamali, M., Sweygers, N., Al-Salem, S., Appels, L., Aminabhavi, T. M., & Dewil, R. (2022). Biochar for soil applications-sustainability aspects, challenges and future prospects. *Chemical Engineering Journal, 428*, 131189.

Li, C., Zhu, X., He, H., Fang, Y., Dong, H., Lü, J., Li, J., & Li, Y. (2019). Adsorption of two antibiotics on biochar prepared in air containing atmosphere: Influence of biochar porosity and molecular size of antibiotics. *Journal of Molecular Liquid, 274*, 353–361.

Lopez, A., de Marco, I., Caballero, B. M., Laresgoiti, M. F., & Adrados, A. (2011). Influence of time and temperature on pyrolysis of plastic wastes in a semibatch reactor. *Chemical Engineering Journal, 173*, 62–67.

Lopez, G., Artetxe, M., Amutio, M., Alvarez, J., Bilbao, J., & Olazar, M. (2018). Recent advances in the gasification of waste plastics. A critical overview. *Renewable and Sustainable Energy Reviews, 82*, 576–596.

Lu, H. P., Li, Z. A., Gascó, G., Méndez, A., Shen, Y., & Paz-Ferreiro, J. (2018). Use of magnetic biochars for the immobilization of heavy metals in a multi-contaminated soil. *Science of the Total Environment, 622-623*, 892–899.

Luz, F. C., Cordiner, S., Manni, A., Mulone, V., & Rocco, V. (2018). Biochar characteristics and early applications in anaerobic digestion-a review. *Journal of Environmental Chemical Engineering, 6*, 2892–2909.

Lyu, H., Zhang, Q., & Shen, B. (2020). Application of biochar and its composites in catalysis. *Chemosphere, 240*, 124842.

Marta, M., Ok, Y. S., & Patryk, O. (2022). From waste to fertilizer: Nutrient recovery from wastewater by pristine and engineered biochars. *Chemosphere, 306*, 135310.

Matteo, P., Peter, B., Theo, T., Pierre-Alexandre, G., Anthony, D., & Olivier. (2023). Study of flash pyrolysis and combustion of biomass powders using the Godbert-Greenwald furnace: An

essential step to better understand organic dust explosions. *Process Safety and Environmental Protection, 169*, 458–471.
Melo, L. C. A., Puga, A. P., Coscione, A. R., Beesley, L., Abreu, C. A., & Camargo, O. A. (2016). Sorption and desorption of cadmium and zinc in two tropical soils amended with sugarcane-straw-derived biochar. *Journal of Soils and Sediments, 16*, 226–234.
Mishra, S., & Upadhyay, R. K. (2021). Review on biomass gasification: Gasifiers, gasifying mediums, and operational parameters. *Materials Science for Energy Technologies, 4*, 329–340.
Othmani, A., John, J., Rajendran, H., Mansouri, A., Sillanpää, M., & Velayudhaperumal Chellam, P. (2022). Biochar and activated carbon derivatives of lignocellulosic fibers towards adsorptive removal of pollutants from aqueous systems: Critical study and future insight. *Separation and Purification Technology, 274*, 119062.
Özgün, T., Nazlıcan, K., Atakan, Ö., Can Özgür, Ç., & Azize, A. (2022). Biomass gasification for sustainable energy production: A review. *International Journal of Hydrogen Energy, 47*, 15419–15433.
Peiris, C., Gunatilake, S. R., Mlsna, T. E., Mohan, D., & Vithanage, M. (2017). Biochar based removal of antibiotic sulfonamides and tetracyclines in aquatic environments: A critical review. *Bioresource Technology, 246*, 150–159.
Qin, P., Wang, H., Yang, X., He, L., Müller, K., Shaheen, S. M., Xu, S., Rinklebe, J., Tsang, D., & Ok, Y. S. (2018). Bamboo- and pig-derived biochars reduce leaching losses of dibutyl phthalate, cadmium, and lead from co-contaminated soils. *Chemosphere, 198*, 450–459.
Rajapaksha, A. U., Chen, S. S., Tsang, D. C., Zhang, M., Vithanage, M., Mandal, S., Gao, B., Bolan, N. S. & Ok, Y.S. (2016). Engineered/designer biochar for contaminant removal/immobilization from soil and water: potential and implication of biochar modification. *Chemosphere, 148*, 276–291.
Sakhiya, A. K., Anand, A., & Kaushal, P. (2020). Production, activation, and applications of biochar in recent times. *Biochar, 2*, 253–285.
Samsuri, A. W., Sadegh-Zadeh, F., & Seh-Bardan, B. J. (2013). Adsorption of As(III) and As(V) by Fe coated biochars and biochars produced from empty fruit bunch and rice husk. *Journal of Environmental Chemical Engineering, 1*, 981–988.
Shaheen, S. M., Niazi, N. K., Hassan, N. E., Bibi, I., Wang, H., Tsang, D. C., Ok, Y. S., Bolan, N., & Rinklebe, J. (2018). Wood-based biochar for the removal of potentially toxic elements in water and wastewater: A critical review. *International Materials Reviews*, 1–32.
Smith, P. (2016). Soil carbon sequestration and biochar as negative emission technologies. *Global Change Biology, 22*, 1315–1324.
Sun, K., Tang, J., Gong, Y., & Zhang, H. (2015). Characterization of potassium hydroxide (KOH) modified hydrochars from different feedstocks for enhanced removal of heavy metals from water. *Environmental Science and Pollution Research, 22*, 16640–16651.
Thomas, D., Fernandez, N. B., Mullassery, M. D., & Surya, R. (2020). Iron oxide loaded biochar/polyaniline nanocomposite: Synthesis, characterization and electrochemical analysis. *Inorganic Chemistry Communications, 119*, 108097.
Tomczyk, A., Sokołowska, Z., & Boguta, P. (2020). Biochar physicochemical properties: Pyrolysis temperature and feedstock kind effects. *Reviews in Environmental Science and Biotechnology, 19*, 191–215.
Venegas, A., Rigol, A., & Vidal, M. (2016). Changes in heavy metal extractability from contaminated soils remediated with organic waste or biochar. *Geoderma, 279*, 132–140.
Wang, J. L., & Wang, S. Z. (2018). Activation of persulfate (PS) and peroxymonosulfate (PMS) and application for the degradation of emerging contaminants. *Chemical Engineering Journal, 334*, 1502–1517.
Wang, S. Z., & Wang, J. L. (2019). Activation of peroxymonosulfate by sludge-derived biochar for the degradation of triclosan in water and wastewater. *Chemical Engineering Journal, 356*, 350–358.

Wang, H., Gao, B., Wang, S., Fang, J., Xue, Y., & Yang, K. (2015). Removal of Pb(II), Cu(II), and Cd(II) from aqueous solutions by biochar derived from $KMnO_4$ treated hickory wood. *Bioresource Technology, 197*, 356–362.

Wu, H., Lai, C., Zeng, G., Liang, J., Chen, J., Xu, J., Dai, J., Li, X., Liu, J., Chen, M., Lu, L., Hu, L., & Wan, J. (2017). The interactions of composting and biochar and their implications for soil amendment and pollution remediation: A review. *Critical Reviews in Biotechnology, 37*(6), 754–764.

Xu, G., Yang, X., & Spinosa, L. (2015). Development of sludge-based adsorbents: Preparation, characterization, utilization and its feasibility assessment. *Journal of Environmental Management, 151*, 221–232.

Yang, F., Zhang, S., Sun, Y., Tsang, D. C. W., Cheng, K., & Ok, Y. S. (2019a). Assembling bio- char with various layered double hydroxides for enhancement of phosphorus recovery. *Journal of Hazardous Materials, 365*, 665–673.

Yang, X., Zhang, S., Ju, M., & Liu, L. (2019b). Preparation and modification of biochar materials and their application in soil remediation. *Applied Sciences, 9*, 1365.

Ye, M., Liu, J., Ma, C., Li, Y. Y., Zou, L., Qian, G., & Xu, Z. P. (2018). Improving the stability and efficiency of anaerobic digestion of food waste using additives: A critical review. *Journal of Cleaner Production, 192*, 316–326.

Zhao, G., Li, J., Ren, X., Chen, C., & Wang, X. (2011). Few-layered graphene oxide nanosheets as superior sorbents for heavy metal ion pollution management. *Environmental Science and Technology, 45*, 10454–10462.

Zhu, X., Chen, B., Zhu, L., & Xing, B. (2017). Effects and mechanisms of biochar-microbe interactions in soil improvement and pollution remediation: A review. *Environmental Pollution, 227*, 98–115.

Advanced Materials from Fungal Mycelium: Fabrication and Tuning of Physical Properties

Praveen Kumar, Ashish Kapoor, Dan Bahadur Pal, and Muthukumar Raghunathan

Abstract Fungal mycelium, the vegetative structure of fungi, is emerging as a promising biological material for sustainable fabrication due to its biodegradability, low energy production requirements, and tuneable physical properties. Mycelium-based composites are engineered by growing fungal networks on various organic substrates, which can be moulded and processed into materials with diverse applications such as packaging, insulation, construction, and even bio-textiles. These materials are highly considered for their low environmental footprint and potential to replace synthetic, non-biodegradable products. This chapter provides a comprehensive overview of the cultivation of mycelium on various agricultural waste substrates, highlighting the eco-friendly nature of this approach. The ability to tune the physical properties of mycelium-based materials, including mechanical strength, density, thermal conductivity, and hydrophobicity, is crucial for their application in different industries. This chapter also highlights the primary methods used to modulate these properties, such as adjusting the type of fungal strain, substrate composition, growth conditions (temperature, humidity, nutrients), and post-processing techniques (e.g. drying, heat treatment, and chemical cross-linking).

Keywords Fungal mycelium · Cultivation · Myco-composite · Tuning properties

---

P. Kumar · M. Raghunathan (✉)
Department of Biotechnology, School of Bioengineering, Chennai, Tamil Nadu, India
e-mail: muthukur@srmist.edu.in

A. Kapoor · D. B. Pal
Department of Chemical Engineering, Harcourt Butler Technical University, Kanpur, Uttar Pradesh, India

D. B. Pal, S. L. Bansal (eds.), *Fungal Waste Biomass Management for Energy, Environment and Value-Added Products*, Fungal Biology,
https://doi.org/10.1007/978-3-031-82599-6_9

1 Introduction

The exploration of advanced materials derived from biological sources has gained significant momentum in recent years. Among these, fungal mycelium has emerged as a promising candidate due to its unique structural and functional properties. Mycelium, the vegetative part of a fungus, consists of a network of thread-like structures called hyphae. These hyphae can intertwine to form a dense and robust material that can be manipulated to achieve various desirable physical properties (Alaneme et al., 2023). This chapter explores the fabrication methods and tuning strategies employed to harness the potential of fungal mycelium in developing advanced materials. Historically, the use of fungi in materials science has been limited, often confined to the realms of food production and traditional medicine. However, the recognition of mycelium's capabilities has shifted this perspective, bringing fungal mycelium to the forefront of sustainable material development. Mycelium-based materials are gaining attention due to their biodegradability low environmental impact and the ability to be cultivated using agricultural waste. This not only provides a sustainable alternative to conventional materials but also offers solutions to waste management challenges (Alaneme et al., 2023). The pursuit of sustainable materials has driven scientists and innovators to investigate the vast potential of mycelium, the vegetative component of fungi. Unlike traditional materials that often rely on finite resources and produce significant environmental waste, mycelium offers a renewable, biodegradable alternative.

Mycelium materials derived from the root structure of fungi have found diverse applications across multiple industries due to their sustainability, biodegradability, and versatility. In construction, mycelium composites serve as eco-friendly alternatives to traditional building materials like insulation and panelling, offering comparable thermal and acoustic properties while reducing environmental impact (Alemu et al., 2022; Van Wylick et al., 2022). The packaging industry benefits from mycelium's lightweight and mouldable nature producing protective packaging that can decompose naturally minimizing waste as shown in Fig. 1 (Bruscato et al., 2019). Additionally, mycelium-based leather substitutes are gaining traction in the fashion industry providing a sustainable option that mimics animal leather without the associated ethical and environmental concerns (Cartabia et al., 2021; Silverman et al., 2020). Innovations in mycelium technology are also expanding into biomedical fields, where its biocompatibility and natural healing properties are being explored for wound dressings and tissue engineering (Danninger et al., 2022; French et al., 2023). The versatility of mycelium materials continues to inspire new applications, reinforcing their potential as key players in advancing sustainable practices across various sectors. This chapter delves into the fascinating world of mycelium-based materials, exploring their fabrication processes, the tuning of their physical properties, and their future scope and challenges across various industries.

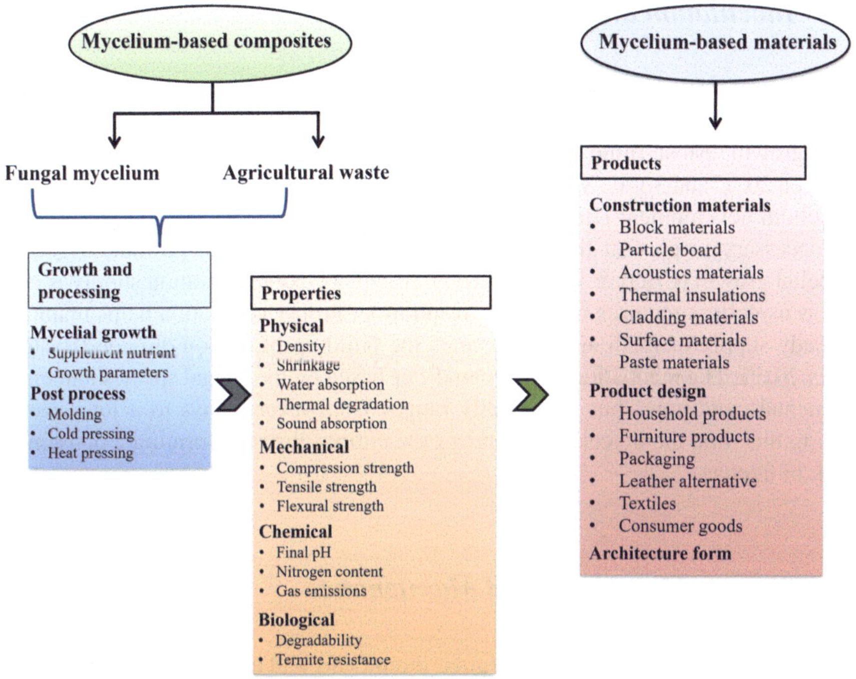

Fig. 1 Illustrative representation of the mycelium-based composites synthesis process, including key stages, potential variations, and the design of mycelium-based materials

2 Selection and Preparation of Substrate

The first step in cultivating mycelium for material production is selecting the appropriate fungal species. Different fungi produce mycelium with varying characteristics such as growth rate texture and mechanical properties. Commonly used species include *Ganoderma lucidum*, *Pleurotus ostreatus* (oyster mushroom), and *Trametes versicolor*. The choice of species depends on the specific requirements of the intended application, such as strength, flexibility, or biodegradability (Dessi-Olive, 2022).

The substrate serves as the nutrient base for mycelium growth. It is typically composed of agricultural by-products like sawdust, straw, or other lignocellulosic materials. The substrate must be sterilized to eliminate any competing microorganisms that could inhibit mycelium growth. This is usually achieved through autoclaving or pasteurization (González et al., 2022). After sterilization the substrate is inoculated with fungal spores or mycelial fragments. This inoculation can be done in sterile conditions to prevent contamination. The choice of substrate and its preparation are crucial as they directly influence the growth rate and properties of the mycelium (Chulikavit et al., 2022).

2.1 Inoculation and Incubation

Once the substrate is prepared and inoculated, it is placed in a controlled environment for incubation. The conditions during this stage are critical for optimal mycelium growth. Most fungi have an optimal growth temperature range typically between 20 °C and 30 °C. Maintaining this temperature range is essential for healthy mycelium development (Bazli et al., 2020). High humidity levels around 80–90% are necessary to prevent the substrate from drying out and to promote vigorous mycelial growth (Cricrì & Perrella, 2017). Adequate oxygen/aeration supply is crucial as mycelium requires oxygen for respiration. Proper ventilation helps maintain a steady supply of fresh air and prevents the build-up of carbon dioxide (Rastogi et al., 2017). The incubation period can differ based on the fungal species and environmental conditions, but it generally ranges from several days to a few weeks. During this time the mycelium colonizes the entire substrate, forming a dense network of hyphae.

2.2 Growth Monitoring and Maintenance

Throughout the incubation period regular monitoring and maintenance are essential to ensure optimal growth. This involves checking for signs of contamination, such as unwanted mould or bacteria, and maintaining consistent environmental conditions. Any deviation from the optimal conditions can significantly affect the quality and properties of the mycelium (Yang et al., 2021). If contamination is detected, affected areas must be removed or treated promptly to prevent it from spreading. In some cases, adding beneficial microorganisms that can outcompete contaminants may be a viable strategy.

2.3 Harvesting and Post-harvest Processing

Once the mycelium has fully colonized the substrate, it is ready for harvesting. The timing of the harvest is crucial; harvesting too early can result in incomplete colonization, while waiting too long can lead to overgrowth and degradation of the substrate. The harvested mycelium-substrate composite is then subjected to further processing depending on the intended application (Madusanka et al., 2024).

Post-harvest processing typically involves a heat treatment step to stop further fungal growth and stabilize the material. This is done by drying the mycelium at a controlled temperature to ensure it retains its desired properties while preventing decomposition. Depending on the final application, the mycelium material may undergo additional treatments such as compression to increase density coating to enhance durability or chemical treatments to modify specific properties (Lurie & Pedreschi, 2014).

3 Fabrication Techniques

3.1 Mould-Based Fabrication

Mycelium-based materials can be produced in a variety of ways depending on the application and end product. Mould design is the most significant step because mycelium grows into complicated shapes by being confined within pre-designed moulds. This kind of fabrication is commonly used to create panels furniture, bricks, architectural patterns, and other structural objects. Mycelium-Miscanthus composites as bio-composite building insulation materials to lessen the environmental impact of construction materials as shown in Fig. 2. This research revealed that the thermal conductivity of Mycelium-Miscanthus composites varied between 0.0882 and 0.104 Wm^{-1} K^{-1}, highlighting their potential as environmentally friendly insulating materials (Dias et al., 2021). Mycelium bricks are used as environmentally friendly building materials by using agricultural waste such as sawdust, rice bran, and clean clay as a substrate to promote the development of mycelium. The compressive strength of bricks made from rice bran and sawdust, when combined with mycelium, exhibited a significant improvement compared to bricks without

Fig. 2 Schematic representation of (**a**) mycelium mat and (**b**) mycoblock (mycelium bricks). (Reproduced from [Cartabia et al., 2021; Alemu et al., 2022] under CC-BY license)

mycelium. The rise in compressive strength was remarkable, with a respective enhancement of 38.5% and 31.0% (Ghazvinian & Gürsoy, 2022). Another study developed mycoblocks by combining sawdust, bagasse, and coffee husk with the fungus *Pleurotus ostreatus.* The mycelium growth on coffee husk exhibited the highest rate, while the growth on sawdust shown the lowest rate. The results showed that the block produced using sawdust (after 21 days of moulding) had higher compressive strength and density (measuring 750 kPa and 343.44 Kg/m^3, respectively), with bagasse and coffee husk blocks following suit (Alemu et al., 2022). Mycelium composite bricks were cultivated naturally using the edible fungus Pleurotus ostreatus grown on rye berrygrains as biodegradable substrates. Their moisture-dependent mechanical properties were examined, and theenergy performance of buildings insulated with mycelium composites was analyzed using EnergyPlus, comparingthe results to those of conventional natural insulation materials.

The study discovered that increasing the density of these bricks enhanced their flexural strength making them appropriate for building insulation and lowering indoor temperature variations and energy consumption (Zhang et al., 2022). In another study, reported the feasibility of *Fusarium oxysporum*, a filamentous fungi strain, for the development of self-healing concrete. *Fusarium oxysporum* can effectively mediate the self-healing of concrete cracks. The fungi's rapid growth, hydrophobic nature, and ability to precipitate calcium minerals make it an ideal candidate for improving concrete sustainability and durability (Zhang et al., 2021). Elsacker et al. reported the use of *Ganoderma resinaceum* and *Trametes versicolor* mycelium for the self-healing construction materials. This paper also concluded that the wire cutting of living and dried mycelium blocks can be performed outside controlled bio-lab environments (Elsacker et al., 2021a, b). Mycelium-based composite materials have been examined using various fabrication methods, with a focus on their properties including compressive and flexural strength, as well as fire resistance (Chen & Poon, 2009). A bio-composite was developed as a substitute for expanded polystyrene (EPS) by combining sawdust with coir pith in a 3:2 ratio. Growing *Pleurotus ostreatus* on a sawdust-coir pith substrate produces a material that is not only cost-effective and sustainable but also exhibits strong mechanical and thermal properties (Sivaprasad et al., 2021). Figure 2 showed the schematic representation of (a) mycelium mat and (b) mycoblock (mycelium bricks) (Fig. 3).

3.2 Layer-by-Layer Growth (Panel and Sheet Production)

Another novel technique is layer-by-layer development which enables the production of increasingly complex shapes and structures. This approach involves growing mycelium in thin layers, which solidify before the next layer is added, resulting in a composite material with improved qualities. The porosity and filter-like qualities of *P. ostreatus* can be utilized to create a panel for adsorbing atmospheric particulates. The study used substrates such as hemp, rice, lacquer tree wood chips, and oak wood chips. It was found that hemp-based mycelium composite panels had the

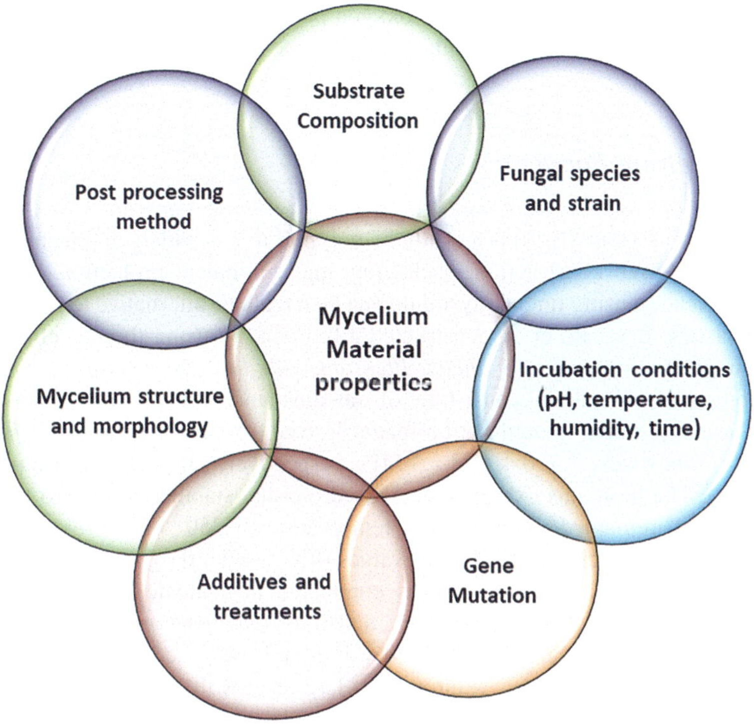

Fig. 3 Illustrative representation of the factors responsible for tuning the properties of mycelium material

highest NO_3^- adsorption, followed by panels made from rice straw and lacquer tree wood chips.

Furthermore, mycelium composite panels with better particulate matter adsorption capacity showed a higher rate of water absorption (Lee & Choi, 2021). Cartabia et al. used wood decay fungi, *Fomitopsis iberica*, to create mycelium mats of leather-like materials because of their ease of cultivation, absence of sporification, and mycelial structure. Sterilized millet grains were employed as a substrate to promote mycelium colonization. The colonized millet was placed in flat moulds and incubated for 21 days to restart the colonization process. The resulting upper layer of the mycelial mat was removed (Cartabia et al., 2021). In another study, Raman et al. (2022) investigated 14 Polyporales species for mycelial leather production on oak sawdust and rice bran substrate. The physical-chemical and mechanical parameters, along with the morphological images, indicated that *Fomitella fraxinea* is an appropriate strain for producing mycelial leather (Raman et al., 2022). On the other hand, using *P. ostreatus* mycelium as an inexpensive bio-matrix for fabricating bio-composite boards proved challenging. The small particle size of the RWS hindered the mycelium's ability to infiltrate the board and effectively bond the fibres. The small particle size of RWS prevented oxygen penetration in the board, which is a critical element for mycelium inoculation (Alaneme et al., 2023).

3.3 Mycelium Foam

Expanded polystyrene (EPS) is a synthetic polymer that is widely applicable in the fields of food civil construction, packaging, and equipment protection. The biofoams fabrication using fungal mycelium can be a realistic alternative to petroleum-based plastics. Bruscato et al. constructed a biofoam by using the mycelia of the macrofungi *Pycnoporus sanguineus*, *Pleurotus albidus*, and *Lentinus velutinus*. They prepared a medium consisting of sawdust and wheat bran to substitute EPS. The compressive strengths of biofoams derived from *P. sanguineus*, *P. albidus*, and *L. velutinus* were 1.3, 0.4, and 1.3 MPa, respectively, surpassing those of EPS (0.4 MPa). The biofoams exhibited decreased thermal stability compared to EPS, however, they remained stable up to a temperature of 350 °C. Biofoams have a higher density compared to EPS, with values of 0.3 and 0.03 g cm^{-3}, respectively (Bruscato et al., 2019). Research was conducted to evaluate the suitability of *Pleurotus ostreatus*, *Volvariella volvacea*, and *Schizophyllum commune* as potential species for biofoam generation. Blended wastepaper was used as the substrate in this study. *Pleurotus ostreatus* exhibited the largest growth area (0.76 cm^2) on the wastepaper after 7 days of inoculation, surpassing *V. volvacea* and *S. commune*. The mycelium biofoam produced from *P. ostreatus* exhibited the greatest dry density at a moisture percentage of 60% (w/w) (0.451 g/cm^3) and the best compressive strength at a temperature of 30 °C (2.343 MPa) (Nashiruddin et al., 2021). Fungal mycelium (*G. lucidum*) skins were used as a biodegradable substrate material for electronics. Using laser patterning and physical vapour deposition method, the developed mycelium skins can be integrated with standard electronic processing techniques to create electronic traces with high conductivity (up to $9.75 \pm 1.44 \times 10^4$ S/cm). The mycelium-based electronics can withstand over 2000 bending cycles and multiple folds with only a moderate increase in resistance (Danninger et al., 2022).

Sustainable mycelium-based insulation material with low thermal conductivity was developed using spent mushroom substrate (SMS) and beech sawdust as a substrate. *Trametes versicolor* shown vigorous development on all surfaces, with a mycelial growth rate ranging from 7.4 to 11.8 mm per day and a density ranging from 190 to 200 kg per cubic meter. *Ganoderma lucidum* exhibited robust development on substrates composed of sawdust, with a growth rate ranging from 5.9 to 6.9 mm per day. However, the growth of *Ganoderma lucidum* was impeded by the discarded mushroom substrate. The mycelium-based materials exhibited a low thermal conductivity range of 0.06 to 0.07 W/m•K, rendering them well-suited for insulation applications (Schritt et al., 2021). Natural nanofibrils, known as nanopaper, with similarities to chitosan and chitin, are derived from fungal mycelium (specifically *Artemisia arbuscula*, *Mucor genevensis.*, and *T. versicolor*) that has been extracted from liquid waste. Achieving polymer yields of approximately 10–26%, similar to those of chitin derived from crustaceans, resulted in the production of nanopapers with significantly higher tensile strengths compared to existing mycelium materials. The tensile strengths reached up to approximately 25 MPa for mycelium and 98 MPa for mushroom. Additionally, the nanopapers exhibited useful hydrophobic surface properties due to the presence of organic lipid residues (Nawawi et al., 2020).

3.4 Extrusion and 3D Printing

3D printing with mycelium is a new technology that creates printable pastes by combining mycelium and a binding agent. These pastes may be extruded using a 3D printer to create complex shapes and structures providing a high level of customization and precision. An extrudable paste using a low-cost energy and simple extrusion approach that has been underutilized for mycelium-bound composites. Soh et al. (2020) used a bamboo fibres and sawdust as a substrate material to provide nutrients for the mycelium (*Ganoderma lucidum*) growth and chitosan as a rheological modifier to increase workability and allow extrusion with the mycelium binding the entire component together. A hydrogel solution containing *P. ostreatus* liquid culture is used for 3D printing in environments that are not sterile, leading to the creation of complex and operational bio-composites based on mycelium (Lin et al., 2024). *Ganoderma lucidum* is used in 3D bioprinting to create lattice patterns that facilitate growth and colonization. This approach led to the development of materials that could autonomously mend, regenerate, and adapt, all while carrying out engineering tasks. This work developed robotic skin that is mechanically durable, self-cleaning, and capable of self-regeneration following damage (Gantenbein et al., 2023). The potential of *Pleurotus ostreatus* and *Ganoderma lucidum* mycelia as self-growing, natural and biodegradable biomedical scaffolding materials was explored. Mycelial bio-composites provide a new and sustainable alternative to standard tissue engineering technologies as shown in Table 1 (Antinori et al., 2021).

3.5 Textile Infusion

The incorporation of fungal mycelium into textiles is a novel method to developing sustainable biodegradable fabrics. This method includes growing mycelium directly onto textile surfaces yielding products with the desirable qualities of both mycelium and the textile basis. *Pleurotus ostreatus*, a fungal species, is highly biocompatible and non-pathogenic. French et al. used *P. ostreatus* to fabricate a monolayer mask that has the functionalities of a conventional three-layer mask. *Pleurotus ostreatus* has colonized the surface of a protective layer, prompting the mycelium to penetrate the filter template and extend over the malt extract agar plate. Additionally, it was found that the mask treated with mycelium exhibits asymmetrical hydrophobicity,

Table 1 This table illustrates the advance materials fabricated from fungal mycelium using various waste substrates and fungal strain

Prototype	Fungi strain	Substrate	Mould size/cultivation condition	Reference
Mycelial foam	*Pycnoporus sanguineus*, *Pleurotus albidus*, *Lentinus velutinus*	Sawdust and wheat bran	Mould: 10 cm × 6 cm, incubated at 24 ± 2 °C, 66% humidity until complete colonization of the medium occurred	Bruscato et al. (2019)
Biofoam	*Pleurotus ostreatus*, *Volvariella volvacea*, *Schizophyllum commune*	Wastepaper	–	Nashiruddin et al. (2021)
Extrudable paste	*Ganoderma lucidum*	Bamboo fibres and sawdust	Incubated at 25–28 °C, relative humidity of 65–80% and at pH of 5–6 for 1–4 weeks	Soh et al. (2020)
Single layer mask	*Pleurotus ostreatus*	Agar	Incubated at 27 °C for 10, 15, or 20 days	French et al. (2023)
Mycelium-composite panel	*Pleurotus ostreatus*	Hemp, rice straw, lacquer tree wood chips, and oak wood chips	Mould: 10 cm × 10 cm × 1 cm, the mould is kept at 22 °C (±2 °C), relative humidity of 65% (±5%) for 21–25 days, with ventilation and no light	Lee and Choi (2021)
Shoe sole	*P. ostreatus*, *Pleurotus citrinopileatus*, *Pleurotus eryngii*, *G. lucidum*	Husk psyllium, flour, chicken feathers, sawdust	23 °C and 50% relative humidity for more than 40 h	Silverman et al. (2020)
Mycelium bricks	Mycelium serum: rice and sugarcane molasses	Rice bran, sawdust, and pure clay	Mould: 200 mm length × 90 mm width × 60 mm height, moulds are stored inside a room with no sunlight at a constant temperature for a period of 25 days incubation	Maximino et al. (2020)
Mycelial mat	*Fomitopsis iberica*	Millet grains	Flat moulds: 9 cm diameter Petri plates, incubated at 23–28 °C for 21 days.	Cartabia et al. (2021)
Mycoblock	*P. ostreatus*	Coffee husk, bagasse, and sawdust	Mould: 11 cm × 8 cm × 4 cm, incubation temperature: 22 + 1 °C, for 7–21 days in aseptic condition	Shakir et al. (2020)
Bio-composite board	*P. ostreatus*	Rubberwood sawdust, corn grain, and glucose	Incubate at room temperature (~25 °C) for 45 days with 80% humidity	

(continued)

Table 1 (continued)

Prototype	Fungi strain	Substrate	Mould size/cultivation condition	Reference
Mycelial pellets	*Aspergillus niger* spores	Potato dextrose agar	–	Li et al. (2023)
Mycelium-based leather	*Fomitella fraxinea*, *Ganoderma lucidum*, *Elfvingia applanata*, *Bjerkandera adusta*, *Microporus affinis*, *Trametes versicolor*, *Fomitopsis pinicola*, *Wolfiporia extensa*, and *Postia balsamea*	Oak sawdust and rice bran	Incubated at 28 °C and 80–90% humidity at a high CO_2 level (1500 ppm) for 43 days in dark.	Raman et al. (2022)
Mycelium construction composite	*Trametes versicolor*	Rice hull and wheat grain	Mould: 100 mm long × 100 mm wide × 20 mm thick, incubated at 25 °C, 50% relative humidity for 12 days	Jones et al. (2018a, b)
Mycelium block	*Trametes versicolor* (M9912) and *Ganoderma resinaceum* (M9726)	Hemp hurds, bench sawdust	Mould: 185 × 185 × 78 mm, incubated at 26 °C, 60% relative humidity for 3–4 weeks	Elsacker et al. (2021a, b)
Engineered mycelium-composite	*P. ostreatus*	Coconut husk and bagasse	–	Lingam et al. (2023)
Concrete surface	*Fusarium oxysporum*	Type I portland cement and fine sand	Incubated at 29 ± 1.2 °C	Zhang et al. (2021)
Hydrogel	*Ganoderma lucidum*	Agar-based hydrogel substrate	Grown at 23 °C and relative humidity of 95%	Gantenbein et al. (2023)
Mycelium bio-composite	*P. ostreatus*	Sawdust, coir pith	Incubated at 27 °C and 80% relative humidity for 14 days	Sivaprasad et al. (2021)
Mycelium skin	*Ganoderma lucidum*	Beech wood shaving	Incubated at 25 °C in CO_2-rich atmosphere and dark room	Danninger et al. (2022)
Scaffold	*Ganoderma lucidum* and *Pleurotus ostreatus*	Potato dextrose broth	Incubated at 27 °C and 78% relative humidity	Antinori et al. (2021)
Thermal insulation	*Ganoderma lucidum* or *Trametes versicolor*	Beech sawdust and spent mushroom substrate	Mould: outer diameter: 87 mm, wall thickness: 0.7 mm, height: 60.5 mm, incubated in dark at 24.3 °C ± 3.5 °C	Schritt et al. (2021)

with very high hydrophobicity at the interface between the composite and air, and reduced hydrophobicity at the interface between the composite and the medium (French et al., 2023). A shoe sole that is capable of decomposing naturally, using the fungus species *Pleurotus eryngii* and *Pleurotus citrinopileatus*, together with composites made from waste chicken feathers. It has been shown that mycelium can infiltrate and interlace with feather fibres, resulting in the integration of feathers into the composite material. The mycelium composites of *P. eryngii* exhibited exceptional compressive strength, suggesting their potential use as sustainable and biodegradable materials in the manufacturing of footwear (Silverman et al., 2020).

3.6 Mycelial Millets and Microspheres

Mycelium pellets are compact clusters of mycelia, the vegetative part of a fungus consisting of a thick network of hyphae. These pellets are created by culturing fungal mycelium under precise circumstances that encourage the clustering of hyphae into pellet-shaped formations. A simple filtering procedure is used to generate a self-regenerating living membrane using *Aspergillus niger* pellets, which have the ability to manufacture melanin under normal environmental circumstances. A responsive living membrane, capable of changing colours, was created by manipulating the expression of genes involved in melanin production in response to the quantity of xylose in the environment. This membrane has the potential to be used as a biosensor for detecting xylose levels in industrial effluent (Li et al., 2023).

This study investigated Fe_2O_3 and polydopamine-functionalized *A. niger* microspheres (FPFMs) as promising adsorbents for thorium separation and removal from aqueous solution. The adsorption investigation found that FPFMs performed best with thorium, with maximal adsorption capacities of 304.276 ($q_{e,exp}$) and 326.346 ($q_{e,cal}$) mg•g^{-1} (Ding et al., 2019). A 3D porous mycelium-derived activated carbon (3D-MAC) was developed by activating mycelium pellets with $ZnCl_2$ and carbonizing them at high temperatures. The N-doped 3D-MAC demonstrated superior performance as a supercapacitor electrode, with a specific capacitance of 237.2 F g^{-1} at 10 mV s^{-1} and stability at high scan rates, indicating a substantial potential for high-performance energy storage applications (Hao et al., 2018).

4 Tuning Physical Properties of Fungal Mycelium

4.1 Fungal Species and Substrate Composition

The chemical composition and hydrodynamic properties of mycelium materials can be fine-tuned by altering the growth substrate. This leads to differences in morphology and functional characteristics allowing for the development of materials with

Table 2 Summary of the roles of various substrate and processing methods influencing material properties

Substrate/processing method	Fungus type/strain	Improved material property	Reference
Cotton straw	*G. lucidum*	Enhanced elasticity	Liu et al. (2019)
Cotton biomass	*G. lucidum*	High bending strength	Ziegler et al. (2016)
Rapeseed straw	*Trametes multicolor*	Flexible and soft skin	Appels et al. (2019)
Rapeseed straw	*P. ostreatus*	Firm and rough skin	
Cotton straw	*T. multicolor*	Enhanced water retention ability	
Heat pressing	*T. multicolor*	Uniform thickness, dark brown material	
Cold pressing	*P. ostreatus*	A twofold increase in density	
Heat pressing	*P. ostreatus*	Threefold increase in density, light brown material	
Cellulose	*P. ostreatus*	Firm	Haneef et al. (2017)
Cellulose and dextrose	*G. lucidum*	Enhanced elasticity	
Cellulose and dextrose	*P. ostreatus*	Enhanced elasticity, high water holding capacity	
Cellulose	*G. lucidum*	Loose structure	
Heat pressing	*G. lucidum*	Enhanced elasticity	Liu et al. (2019)
Heat pressing	*A. niger*	Reduced solubility	Solodovnik et al. (2017)

specific desired properties (Antinori et al., 2020). The effects of different chemical treatments on the mechanical properties and durability of six different mycelium strains, i.e. *Lentinus crinitus*, *Panus* aff. *conchatus, Ganoderma curtisii*, *P. Ostreatus*, *Ganoderma mexicanum*, and *Aurantiporus* sp. *G. mexicanum* and *G. curtisii* showed better results in mechanical test. Meanwhile treatments with glutaraldehyde and formaldehyde improved tensile strength and water resistance as shown in Table 2 (César et al., 2021).

The substrate composition (cotton, sawdust, straw), processing methods (no processing, cold and heat processing), and fungus species (*Trametes multicolor* vs. *P. ostreatus*) altered the characteristics of mycelium-based composites. Sawdust exhibited a greater density compared to cotton fibres or straw when used as a substrate. However, the density was further increased by a factor of 2 and more than 3 times by cold- and heat-pressing, respectively. Moreover, it was discovered that heat pressing enhances the uniformity, durability, and rigidity of the materials, transforming their characteristics from resembling foam to resembling wood. Furthermore, it was found that the water content of samples cultivated on rapeseed straw (7.6–9.6%) was greater than that on cotton (5.8–7.2%), and the first degradation temperature of rapeseed straw-based materials was lower (about ±225 °C) compared to cotton-based materials (approximately ±242 °C) (Appels et al., 2019). *Trichoderma asperellum* and *Agaricus bisporus* were cultivated on oat husk and rapeseed cake following oil extraction. The results showed that the mycelium composites had hydrophobic properties and shown significant strength especially when cultivated on rapeseed cake. An experiment shown that the rigidity of *A. bisporus* grown on rapeseed cake increased when the humidity was systematically raised and then lowered (Tacer-Caba et al., 2020).

In another study, molasses, a by-product of sugarcane, was used as a growth substrate. This approach led to polymer yields ranging from 10% to 26% (Jones et al., 2019a, b). To vary the physical and mechanical qualities of the bio-based material to match EPS standards, the particle size of the substrate (*Triticum* sp.) can be altered, substrate blends can be created, and other fungal species can be inoculated. Waxes like candelilla wax can also be applied to boost water resistant properties (López Nava et al., 2016). Jones et al. investigated the appropriateness of agricultural by-products (wheat straw, rice husks, and sawdust) for the formation of chitinous composites and nanofibers by mycelial growth of *Trametes versicolor* and *Polyporus brumalis*. As compared to other substrates, mycelium composites grown on wheat straw produced the most biomass and had a tensile strength of 2.8 MPa (Jones et al., 2019a, b). When approached through a biorefinery perspective, an agroindustrial byproduct like wheat bran can gain a newrole in the production of mycelium-based materials (Sisti et al., 2021). A urease-positive fungal strain, *Penicillium chrysogenum* CS1, was utilized in the biocementation of sand to create self-healing sandstone. This fungal strain effectively promoted the precipitation of calcite, which served as a cementing agent for the sand granules. The sandstone produced through this fungal-mediated process showed considerable compressive strength, highlighting the potential of fungi in construction engineering applications (Fang et al., 2018). The mechanical properties of the materials were assessed by varying the substrate particle size, fungus species (*Coprinus comatus* and *Pleurotus ostreatus*), and post-growth treatments, including baking, drying, compacting followed by baking, and compacting followed by drying. They concluded that composites constructed from substrates containing microparticles have a high elastic module and mechanical strength. In both compression and flexural tests, compacted and baked composites formed on substrates with microparticles demonstrated the highest elastic moduli (Houette et al., 2022). A mycelium composite containing four fungal species (*Geastrum fornicatum, Ganoderma williamsianum, Schizophyllum commune,* and *Lentinus sajor-caju*) and three types of lignocellulosic wastes (cornhusk, rice straw, and sawdust) was optimized. Mycelium-Based Composites (MBCs) from rice straw absorbed the most water, followed by maize husk and sawdust. Corn husk-based MBCs, on the other hand, demonstrated the highest flexural, impact, and tensile strengths when coupled with all fungal species (Aiduang et al., 2022).

4.2 *Growth Medium*

Antinori et al. showed that even small modifications in the composition of a standard fungal growth medium can induce significant differences in the morphology, chemical, and hydrodynamic properties of *G. lucidum* mycelium. Mycelium materials grown in potato dextrose broth (PDB) enriched with d-glucose are highly porous, thicker, and more apt to adsorb moisture with respect to mycelium materials grown in PDB with a small quantity of lignin. The latter, on the other hand, grow very fast, following a concentric pattern, and are denser and less hydrophilic (Antinori et al., 2020). It was demonstrated that changing the agar concentration during in vitro culturing can impact the physical growth and structure of the mycelial sheet without changing its permeability or chemical composition. When the agar concentration was raised from 1.5% to 6.0%, the mycelium density and hyphal breadth rose by 32.3% and 63.6%, respectively (Hotz et al., 2023).

5 Fabrication Methods

A novel method for manufacturing biocomposite sandwich structures is presented, utilizing entirely naturallyderived materials: natural fiber textiles (such as jute, hemp, and cellulose) for the skin, mycelium-bound agriculturalwaste for the core, and a bioresin as the matrix (Jiang et al., 2017). Rigobello et al. give insight into how anisotropic designs, achieved by orienting fibres within the composite, can influence the compressive behaviour and ultimate strength of the material (Rigobello & Ayres, 2022). Heat-pressing at 200 °C improved the mechanical properties significantly. Together, the mycelium, bacterial cellulose (BC), fibres, and type of drying treatment all contribute to the overall mechanical anisotropy of the composites. A heat-press treatment of BC-mycelium composites at 200 °C resulted in better mechanical properties (Elsacker et al., 2021a, b).

5.1 *Additives and Growth Condition*

The toughening mechanisms and fracture behaviour of mycelium-based biocomposites that included different weight percentages of laterite particles mixed with hemicellulose hemp conduits for nutritive and reinforcing components were investigated. Crack-bridging was found to be the dominant toughening mechanism at lower laterite weight fractions (0–20 wt %). However, as laterite content increased (20–40 wt %), crack-bridging and crack-deflection were detected. Finally, at larger laterite weight fractions (>40 wt %), crack tip shielding occurred predominantly through crack deflection (Etinosa et al., 2024).

Mycelium bio-composites, particularly those reinforced with natural reinforcement particles (NRPs), have promising applications as lightweight backfill materials in geotechnical engineering due to their improved physio-mechanical properties and sustainable nature (Gou et al., 2021). Teeraphantuvat et al. concluded that incorporating 20% paper waste into corn husks significantly improves the compression, bending, and impact strength of MBCs by over 20%. It also enhances the density of mycelium composite (Teeraphantuvat et al., 2024). *G. lucidum* mycelium grown on a substrate of sawdust and empty fruit bunch was tested for flexural, tensile, and compressive strength following 35 days of exposure to tropical weathering conditions. A significant decrease in maximum stress was observed for all loading types post-weathering. Protective coatings improved weathering resistance, with statistically significant improvements noted only in tensile strength (Chan et al., 2021).

5.2 Gene Mutation

The influence of environmental growth conditions and the deletion of the hydrophobin gene sc3 on the mechanical properties of the mycelium from the mushroom-forming fungus *Schizophyllum commune* was studied. The study reveals that the deletion of the sc3 gene significantly enhances the mechanical properties of the mycelium (Appels et al., 2018).

5.3 Thermal and Acoustic Properties

Mycelium offers excellent thermal and acoustic insulation properties, thanks to its natural structure, which results in low thermal conductivity. This makes it suitable for insulation panels in buildings improving energy efficiency. Substrates with higher cellulose content tend to enhance the thermal insulation properties of mycelium materials. Strains with dense compact hyphae networks tend to offer better thermal insulation. For instance, integrating graphene oxide (GO) and reduced graphene oxide (rGO) nanoparticles into mycelium composites has been shown to improve thermal conductivity making these materials suitable for applications requiring efficient heat dissipation (Gul et al., 2023). Mycelium composites have lower average and peak heat release rates compared to traditional construction materials, making them safer and more effective for thermal insulation. By varying the substrate composition and fungal strains they achieved materials with low thermal conductivity suitable for insulating walls and roofs. The study emphasized the environmental benefits of using renewable biodegradable materials in the construction industry (Abera, 2024). Furthermore, the porous structure of mycelium can absorb sound waves, reducing noise levels, and improving acoustic comfort in various environments (Barta et al., 2024). Mycelium bricks, as building insulation materials, show promising thermal performance and aim for enhanced biodegradability compared to existing alternatives like expanded polystyrene (Xing et al., 2018). The effects of cultivation factors on the compressive strength of mycelium-based bio-composites for architectural applications demonstrate the impact of substrate type and added supplements (Ghazvinian et al., 2019).

Acoustic insulation is another significant benefit of mycelium. Its porous structure can absorb sound waves, reducing noise levels and improving acoustic comfort. Higher porosity in mycelium materials can improve sound absorption by trapping and dissipating sound waves. It was demonstrated that substrates rich in lignocellulosic materials produce mycelium composites with superior thermal and acoustic properties (Haneef et al., 2017). Additionally, the incorporation of waste materials like straw, sawdust, and agricultural residues not only enhances the material properties but also promotes sustainability. Mycelium composites with different lignocellulosic substrates, such as hemp, flax, and straw, exhibit varying mechanical and thermal properties. These substrates influence the material's insulation characteristics, making them a promising alternative to conventional insulation materials (Elsacker et al., 2019). Maintaining optimal temperature and humidity levels during mycelium growth is crucial. Studies have indicated that lower temperatures (20–25 °C) favour denser mycelium growth, which improves both thermal insulation and acoustic damping (Pelletier et al., 2013). Post-cultivation drying reduces the moisture content, preventing decay and improving thermal insulation. Pressing the mycelium material can increase its density, which is beneficial for both thermal and acoustic performance (Vasatko et al., 2022).

6 Challenges and Future Perspectives

Mycelium-based materials face challenges in scalability, particularly in large-scale production. Addressing issues related to cultivation, processing, and quality control is crucial for widespread adoption. Standardizing production processes and properties is essential for industry acceptance. Collaboration between researchers, industry stakeholders, and regulatory bodies is necessary to develop these standards. The cost of producing mycelium-based materials can be higher than traditional materials, especially during early development. Reducing production costs through optimization of cultivation techniques, substrate selection, and post-processing methods is essential. Continued research and development are crucial for unlocking the full potential of mycelium-based materials. Collaboration between academia, industry, and government agencies will drive innovation and support the growth of this emerging field.

7 Conclusion

In summary, Mycelium-based materials offer a promising solution for sustainable material science, offering eco-friendliness, versatility, and performance in various applications such as construction, packaging, consumer goods, and textiles. Their ability to be customized through cultivation, fabrication, and post-processing techniques further enhances their appeal. As we move towards a more sustainable future, the development and commercialization of mycelium-based materials will play a crucial role. By harnessing the power of fungal mycelium, we can create innovative solutions that reduce environmental impact, promote circular economies, and provide high-performance alternatives to traditional materials. However, challenges like variability in material properties and production scalability need to be overcome. Collaboration across disciplines and industries is essential for integrating mycelium materials into mainstream applications.

Disclosure Statement The authors declare no possible conflicts of interest.

References

Abera, Y. A. (2024). Sustainable building materials: A comprehensive study on eco-friendly alternatives for construction. *Composites and Advanced Materials, 33*, 26349833241255957.

Aiduang, W., Kumla, J., Srinuanpan, S., Thamjaree, W., Lumyong, S., & Suwannarach, N. (2022). Mechanical, physical, and chemical properties of mycelium-based composites produced from various lignocellulosic residues and fungal species. *Journal of Fungi, 8*(11), 1125.

Alaneme, K. K., Anaele, J. U., Oke, T. M., Kareem, S. A., Adediran, M., Ajibuwa, O. A., & Anabaranze, Y. O. (2023). Mycelium based composites: A review of their bio-fabrication procedures, material properties and potential for green building and construction applications. *Alexandria Engineering Journal, 83*, 234–250.

Alemu, D., Tafesse, M., & Gudetta Deressa, Y. (2022). Production of mycoblock from the mycelium of the fungus Pleurotus ostreatus for use as sustainable construction materials. *Advances in Materials Science and Engineering, 2022*(1), 2876643.

Antinori, M. E., Ceseracciu, L., Mancini, G., Heredia-Guerrero, J. A., & Athanassiou, A. (2020). Fine-tuning of physicochemical properties and growth dynamics of mycelium-based materials. *ACS Applied Bio Materials, 3*(2), 1044–1051.

Antinori, M. E., Contardi, M., Suarato, G., Armirotti, A., Bertorelli, R., Mancini, G., Debellis, D., & Athanassiou, A. (2021). Advanced mycelium materials as potential self-growing biomedical scaffolds. *Scientific Reports, 11*(1), 12630.

Appels, F. V., Dijksterhuis, J., Lukasiewicz, C. E., Jansen, K. M., Wösten, H. A., & Krijgsheld, P. (2018). Hydrophobin gene deletion and environmental growth conditions impact mechanical properties of mycelium by affecting the density of the material. *Scientific Reports, 8*(1), 4703.

Appels, F. V., Camere, S., Montalti, M., Karana, E., Jansen, K. M., Dijksterhuis, J., Krijgsheld, P., & Wösten, H. A. (2019). Fabrication factors influencing mechanical, moisture-and water-related properties of mycelium-based composites. *Materials & Design, 161*, 64–71.

Barta, D. G., Simion, I., Tiuc, A. E., & Vasile, O. (2024). Mycelium-based composites as a sustainable solution for waste management and circular economy. *Materials, 17*(2), 404.

Bazli, M., Jafari, A., Ashrafi, H., Zhao, X. L., Bai, Y., & Raman, R. S. (2020). Effects of UV radiation, moisture and elevated temperature on mechanical properties of GFRP pultruded profiles. *Construction and Building Materials, 231*, 117137.

Bruscato, C., Malvessi, E., Brandalise, R. N., & Camassola, M. (2019). High performance of macrofungi in the production of mycelium-based biofoams using sawdust—Sustainable technology for waste reduction. *Journal of Cleaner Production, 234*, 225–232.

Cartabia, M., Girometta, C. E., Milanese, C., Baiguera, R. M., Buratti, S., Branciforti, D. S., Vadivel, D., Girella, A., Babbini, S., Savino, E., & Dondi, D. (2021). Collection and characterization of wood decay fungal strains for developing pure mycelium mats. *Journal of Fungi, 7*(12), 1008.

César, E., Canche-Escamilla, G., Montoya, L., Ramos, A., Duarte-Aranda, S., & Bandala, V. M. (2021). Characterization and physical properties of mycelium films obtained from wild fungi: Natural materials for potential biotechnological applications. *Journal of Polymers and the Environment, 29*(12), 4098–4105.

Chan, X. Y., Saeidi, N., Javadian, A., Hebel, D. E., & Gupta, M. (2021). Mechanical properties of dense mycelium-bound composites under accelerated tropical weathering conditions. *Scientific Reports, 11*(1), 22112.

Chen, J., & Poon, C. S. (2009). Photocatalytic construction and building materials: From fundamentals to applications. *Building and Environment, 44*(9), 1899–1906.

Chulikavit, N., Huynh, T., Dekiwadia, C., Khatibi, A., Mouritz, A., & Kandare, E. (2022). Influence of growth rates, microstructural properties and biochemical composition on the thermal stability of mycelia fungi. *Scientific Reports, 12*(1), 15105.

Cricrì, G., & Perrella, M. (2017). Investigation of mode III fracture behaviour in bonded pultruded GFRP composite joints. *Composites Part B: Engineering, 112*, 176–184.

Danninger, D., Pruckner, R., Holzinger, L., Koeppe, R., & Kaltenbrunner, M. (2022). MycelioTronics: Fungal mycelium skin for sustainable electronics. *Science Advances, 8*(45), eadd7118.

Dessi-Olive, J. (2022). Strategies for growing large-scale mycelium structures. *Biomimetics, 7*(3), 129.

Dias, P. P., Jayasinghe, L. B., & Waldmann, D. (2021). Investigation of Mycelium-Miscanthus composites as building insulation material. *Results in Materials, 10*, 100189.

Ding, H., Zhang, X., Yang, H., Luo, X., & Lin, X. (2019). Highly efficient extraction of thorium from aqueous solution by fungal mycelium-based microspheres fabricated via immobilization. *Chemical Engineering Journal, 368*, 37–50.

Elsacker, E., Vandelook, S., Brancart, J., Peeters, E., & De Laet, L. (2019). Mechanical, physical and chemical characterisation of mycelium-based composites with different types of lignocellulosic substrates. *PLoS One, 14*(7), e0213954.

Elsacker, E., Søndergaard, A., Van Wylick, A., Peeters, E., & De Laet, L. (2021a). Growing living and multifunctional mycelium composites for large-scale formwork applications using robotic abrasive wire-cutting. *Construction and Building Materials, 283*, 122732.

Elsacker, E., Vandelook, S., Damsin, B., Van Wylick, A., Peeters, E., & De Laet, L. (2021b). Mechanical characteristics of bacterial cellulose-reinforced mycelium composite materials. *Fungal Biology and Biotechnology, 8*, 1–14.

Etinosa, P. O., Salifu, A. A., Osafo, S., Eluu, S. C., Obayemi, J. D., & Soboyejo, W. O. (2024). Fracture and Toughening of Mycelium-based Biocomposites. *Materials & Design, 237*, 112592.

Fang, C., Kumari, D., Zhu, X., & Achal, V. (2018). Role of fungal-mediated mineralization in biocementation of sand and its improved compressive strength. *International Biodeterioration & Biodegradation, 133*, 216–220.

French, V., Du, C., & Foster, E. J. (2023). Mycelium as a self-growing biobased material for the fabrication of single-layer masks. *Journal of Bioresources and Bioproducts, 8*(4), 399–407.

Gantenbein, S., Colucci, E., Käch, J., Trachsel, E., Coulter, F. B., Rühs, P. A., Masania, K., & Studart, A. R. (2023). Three-dimensional printing of mycelium hydrogels into living complex materials. *Nature Materials, 22*(1), 128–134.

Ghazvinian, A., & Gürsoy, B. (2022). Mycelium-based composite graded materials: Assessing the effects of time and substrate mixture on mechanical properties. *Biomimetics, 7*(2), 48.

Ghazvinian, A., Farrokhsiar, P., Vieira, F., Pecchia, J., & Gursoy, B. (2019). Mycelium-based bio-composites for architecture: Assessing the effects of cultivation factors on compressive strength. *Materials Research Innovations, 2*, 505–514.

González, A. A. M., Zafra, L. M. C., Bordons, A., & RodrÃguez-Porrata, B. (2022). Pasteurization of agricultural substrates for edible mushroom production. *Journal of Microbiology, Biotechnology and Food Sciences, 12*(1), e5729–e5729.

Gou, L., Li, S., Yin, J., Li, T., & Liu, X. (2021). Morphological and physico-mechanical properties of mycelium biocomposites with natural reinforcement particles. *Construction and Building Materials, 304*, 124656.

Gul, W., Akbar Shah, S. R., Khan, A., Ahmad, N., Ahmed, S., Ain, N., Mehmood, A., Salah, B., Ullah, S. S., & Khan, R. (2023). Synthesis of graphene oxide (GO) and reduced graphene oxide (rGO) and their application as nano-fillers to improve the physical and mechanical properties of medium density fiberboard. *Frontiers in Materials, 10*, 1206918.

Haneef, M., Ceseracciu, L., Canale, C., Bayer, I. S., Heredia-Guerrero, J. A., & Athanassiou, A. (2017). Advanced materials from fungal mycelium: Fabrication and tuning of physical properties. *Scientific Reports, 7*(1), 41292.

Hao, Z., Hao, F., Singh, V. P., & Zhang, X. (2018). Changes in the severity of compound drought and hot extremes over global land areas. *Environmental Research Letters, 13*(12), 124022.

Hotz, E. C., Bradshaw, A. J., Elliott, C., Carlson, K., Dentinger, B. T., & Naleway, S. E. (2023). Effect of agar concentration on structure and physiology of fungal hyphal systems. *Journal of Materials Research and Technology, 24*, 7614–7623.

Houette, T., Maurer, C., Niewiarowski, R., & Gruber, P. (2022). Growth and mechanical characterization of mycelium-based composites towards future bioremediation and food production in the material manufacturing cycle. *Biomimetics, 7*(3), 103.

Jiang, L., Walczyk, D., McIntyre, G., Bucinell, R., & Tudryn, G. (2017). Manufacturing of biocomposite sandwich structures using mycelium-bound cores and preforms. *Journal of Manufacturing Processes, 28*, 50–59.

Jones, M., Bhat, T., Huynh, T., Kandare, E., Yuen, R., Wang, C. H., & John, S. (2018a). Waste-derived low-cost mycelium composite construction materials with improved fire safety. *Fire and Materials, 42*(7), 816–825.

Jones, M., Bhat, T., Kandare, E., Thomas, A., Joseph, P., Dekiwadia, C., Yuen, R., John, S., Ma, J., & Wang, C. H. (2018b). Thermal degradation and fire properties of fungal mycelium and mycelium-biomass composite materials. *Scientific Reports, 8*(1), 17583.

Jones, M., Weiland, K., Kujundzic, M., Theiner, J., Kählig, H., Kontturi, E., John, S., Bismarck, A., & Mautner, A. (2019a). Waste-derived low-cost mycelium nanopapers with tunable mechanical and surface properties. *Biomacromolecules, 20*(9), 3513–3523.

Jones, M. P., Lawrie, A. C., Huynh, T. T., Morrison, P. D., Mautner, A., Bismarck, A., & John, S. (2019b). Agricultural by-product suitability for the production of chitinous composites and nanofibers utilising Trametes versicolor and Polyporus brumalis mycelial growth. *Process Biochemistry, 80*, 95–102.

Lee, T., & Choi, J. (2021). Mycelium-composite panels for atmospheric particulate matter adsorption. *Results in Materials, 11*, 100208.

Li, K., Wei, Z., Jia, J., Xu, Q., Liu, H., Zhong, C., & Huang, H. (2023). Engineered living materials grown from programmable Aspergillus niger mycelial pellets. *Materials Today Bio, 19*, 100545.

Lin, N., Taghizadehmakoei, A., Polovina, L., McLean, I., Santana-Martínez, J. C., Naese, C., Moraes, C., Hallam, S. J., & Dahmen, J. (2024). 3D Bioprinting of Food Grade Hydrogel Infused with Living Pleurotus ostreatus Mycelium in Non-sterile Conditions. *ACS Applied Bio Materials, 7*(5), 2982–2992.

Lingam, D., Narayan, S., Mamun, K., & Charan, D. (2023). Engineered mycelium-based composite materials: Comprehensive study of various properties and applications. *Construction and Building Materials, 391*, 131841.

Liu, R., Long, L., Sheng, Y., Xu, J., Qiu, H., Li, X., Wang, Y., & Wu, H. (2019). Preparation of a kind of novel sustainable mycelium/cotton stalk composites and effects of pressing temperature on the properties. *Industrial Crops and Products, 141*, 111732.

López Nava, J. A., Méndez González, J., Ruelas Chacón, X., & Nájera Luna, J. A. (2016). Assessment of edible fungi and films bio-based material simulating expanded polystyrene. *Materials and Manufacturing Processes, 31*(8), 1085–1090.

Lurie, S., & Pedreschi, R. (2014). Fundamental aspects of postharvest heat treatments. *Horticulture Research, 1*, 14030.

Madusanka, C., Udayanga, D., Nilmini, R., Rajapaksha, S., Hewawasam, C., Manamgoda, D., & Vasco-Correa, J. (2024). A review of recent advances in fungal mycelium based composites. *Discover Materials, 4*(1), 13.

Maximino, C., Ongpeng, J., Inciong, E., Sendo, V., Soliman, C., & Siggaoat, A. (2020). Using waste in producing bio-composite mycelium bricks. *Applied Sciences, 10*(15), 5303.

Nashiruddin, N. I., May, K. W., Mansor, A. F., Rahman, R. A., Zaidel, D. N. A., & Shaarani, S. M. (2021, June). Fabrication of Mycelium Biofoam from Wastepaper. In *Euro Mediterranean Conference for Environmental Integration* (pp. 31–34). Springer Nature Switzerland.

Nawawi, W. M., Jones, M. P., Kontturi, E., Mautner, A., & Bismarck, A. (2020). Plastic to elastic: Fungi-derived composite nanopapers with tunable tensile properties. *Composites Science and Technology, 198*, 108327.

Pelletier, M. G., Holt, G. A., Wanjura, J. D., Bayer, E., & McIntyre, G. (2013). An evaluation study of mycelium based acoustic absorbers grown on agricultural by-product substrates. *Industrial Crops and Products, 51*, 480–485.

Raman, J., Kim, D. S., Kim, H. S., Oh, D. S., & Shin, H. J. (2022). Mycofabrication of mycelium-based leather from brown-rot fungi. *Journal of Fungi, 8*(3), 317.

Rastogi, A., Zivcak, M., Sytar, O., Kalaji, H. M., He, X., Mbarki, S., & Brestic, M. (2017). Impact of metal and metal oxide nanoparticles on plant: A critical review. *Frontiers in Chemistry, 5*, 78.

Rigobello, A., & Ayres, P. (2022). Compressive behaviour of anisotropic mycelium-based composites. *Scientific Reports, 12*(1), 6846.

Schritt, H., Vidi, S., & Pleissner, D. (2021). Spent mushroom substrate and sawdust to produce mycelium-based thermal insulation composites. *Journal of Cleaner Production, 313*, 127910.

Shakir, M. A., Azahari, B., Yusup, Y., Yhaya, M. F., Salehabadi, A., & Ahmad, M. I. (2020). Preparation and characterization of mycelium as a bio-matrix in fabrication of bio-composite. *Journal of Advanced Research in Fluid Mechanics and Thermal Sciences, 65*(2), 253–263.

Silverman, J., Cao, H., & Cobb, K. (2020). Development of mushroom mycelium composites for footwear products. *Clothing and Textiles Research Journal, 38*(2), 119–133.

Sisti, L., Gioia, C., Totaro, G., Verstichel, S., Cartabia, M., Camere, S., & Celli, A. (2021). Valorization of wheat bran agro-industrial byproduct as an upgrading filler for mycelium-based composite materials. *Industrial Crops and Products, 170*, 113742.

Sivaprasad, S., Byju, S. K., Prajith, C., Shaju, J., & Rejeesh, C. R. (2021). Development of a novel mycelium bio-composite material to substitute for polystyrene in packaging applications. *Materials Today: Proceedings, 47*, 5038–5044.

Soh, E., Chew, Z. Y., Saeidi, N., Javadian, A., Hebel, D., & Le Ferrand, H. (2020). Development of an extrudable paste to build mycelium-bound composites. *Materials & Design, 195*, 109058.

Solodovnik, T. V., Stolyarenko, H. S., Slis, A., & Kultenko, V. (2017). Study of heat treatment effect on structure and solubility of chitosan films. *Chemistry & Chemical Technology, 11*(2), 175–179.

Tacer-Caba, Z., Varis, J. J., Lankinen, P., & Mikkonen, K. S. (2020). Comparison of novel fungal mycelia strains and sustainable growth substrates to produce humidity-resistant biocomposites. *Materials & Design, 192*, 108728.

Teeraphantuvat, T., Jatuwong, K., Jinanukul, P., Thamjaree, W., Lumyong, S., & Aiduang, W. (2024). Improving the physical and mechanical properties of mycelium-based green composites using paper waste. *Polymers, 16*(2), 262.

Van Wylick, A., Elsacker, E., Yap, L. L., Peeters, E., & De Laet, L. (2022). Mycelium composites and their biodegradability: An exploration on the disintegration of mycelium-based materials in soil. *Construction Technologies and Architecture, 1*, 652–659.

Vasatko, H., Gosch, L., Jauk, J., & Stavric, M. (2022). Basic research of material properties of mycelium-based composites. *Biomimetics, 7*(2), 51.

Xing, Y., Brewer, M., El-Gharabawy, H., Griffith, G., & Jones, P. (2018). February. Growing and testing mycelium bricks as building insulation materials. In *IOP conference series: Earth and environmental science* (Vol. 121, p. 022032). IOP Publishing.

Yang, L., Park, D., & Qin, Z. (2021). Material function of mycelium-based bio-composite: A review. *Frontiers in Materials, 8*, 737377.

Zhang, X., Fan, X., Li, M., Samia, A., & Yu, X. B. (2021). Study on the behaviors of fungi-concrete surface interactions and theoretical assessment of its potentials for durable concrete with fungal-mediated self-healing. *Journal of Cleaner Production, 292*, 125870.

Zhang, X., Hu, J., Fan, X., & Yu, X. (2022). Naturally grown mycelium-composite as sustainable building insulation materials. *Journal of Cleaner Production, 342*, 130784.

Ziegler, A. R., Bajwa, S. G., Holt, G. A., McIntyre, G., & Bajwa, D. S. (2016). Evaluation of physico-mechanical properties of mycelium reinforced green biocomposites made from cellulosic fibers. *Applied Engineering in Agriculture, 32*(6), 931–938.

Rice Polishing as a Substrate for Fungal Biomass and Protein Production by *Trichoderma harzianum* and *Candida utilis*

Rajeshwari Chatterjee and Amit Kumar Tiwari

Abstract The waste bioconversion offers a natural avenue for resource recovery and facilitates the eco-friendly recycling process. As rice is a major staple food consumed globally, its cultivation leads to the generation of significant biomass residues. The primary solid wastes produced during the rice production process consist of straw, husk, ash, bran, and broken rice. Biotechnological methods can yield valuable end products such as microbial biomass protein, by purifying the waste stream specifically from food processing waste. Fungi emerge as particularly advantageous for microbial biomass protein production due to their pellet or filamentous morphology, enabling cost-effective isolation and recovery from culture media, thus reducing significant fractions of capital and operating costs associated with microbial biomass protein production. Fungal biomass boasts high levels of crude protein, reaching up to 45%, with a balanced array of essential amino acids. This chapter aimed to explore the potential of fungi like *Trichoderma harzianum, Candida utilis* in generating fungal biomass protein from rice polishings. The review investigates that the optimal biomass yield was achieved with 5% (w/v) rice polishings after 72 hours of incubation at 28 °C. Amino acid profiling revealed abundant quantities of all essential amino acids in the final fungal biomass protein. This protein source derived from *T. harzianum* and *C. utilis* demonstrates promising nutritional value for poultry supplementation. The present chapter elucidates a comprehensive review on the practical significance of producing fungal biomass protein from rice polishings.

Keywords Fungal biomass · Protein rich biomass · Rice polishing · Microbial biomass protein · *Trichoderma harzianum* · Rice husk · Rice husk ash

---

R. Chatterjee (✉) · A. K. Tiwari
Department of Chemical Engineering, Birla Institute of Technology, Ranchi, Jharkhand, India
e-mail: rajeshwari@bitmesra.ac.in

D. B. Pal, S. L. Bansal (eds.), *Fungal Waste Biomass Management for Energy, Environment and Value-Added Products*, Fungal Biology,
https://doi.org/10.1007/978-3-031-82599-6_10

1 Introduction

Biomass can be produced from a wide range of feedstocks that fall into three main categories: waste (like waste from cities and industries), agriculture (like rice husk, bran, straw, rice, corn stover), and forestry (like small trees from thinning and cleanings, and unmerchantable wood) (Lim et al., 2012). The forestry sector provided more than 87% of the biomass feedstock, with the agricultural sector contributing 10% and municipal solid waste and landfill gas providing the remaining 3% (Kummamuru, 2016). When it comes to the supply of energy crops needed to produce biofuels, the agricultural sector also plays a big role. Furthermore, agricultural waste products have the potential to be a significant source of energy production. It is estimated that the global theoretical energy potential from agricultural waste ranges from 17.8 EJ to 82.3 EJ. Based on estimates, the theoretical energy potential from rice residues ranges from 3.29 to 15.2 EJ, ranking second globally behind maize (estimates vary from 5.67 to 35.7 EJ) (FAOSTAT, 2019).

Microbial biomass protein (MBP) can be produced from agro-industrial waste, which not only minimizes pollution but also offers a high-quality, reasonably priced supply of MBP. This method tackles the problems of industrial waste and the environment. The manufacture of MBP from such waste is advantageous for feeds used by poultry and livestock, and these products can be consumed by humans and animals (Mondal et al., 2012). Both yeast and fungi are frequently used to produce MBP, however, fungi are especially useful because of their pellet-like or filamentous structures, which make it easier to separate and collect MBP from culture media and lower production costs. Fungi for MBP can be grown on a variety of substrates. Zhang et al. (2008) demonstrated the use of starch waste for MBP production. Ahmed et al. reported the production of fungal biomass protein (FBP) using rice polishings.

One of the most significant and widespread crops in the world, rice provides 60–70% of the calories for over two million people in Asia alone (Zhang et al., 2008). The most affordable source of protein and energy for feeding chickens is rice polishing, a byproduct of rice milling. Rice polishing, which is high in protein, vitamins, minerals, and energy, has a lot of potential applications as a component of chicken feed (Das et al., 2008). With the right hydrolysis, it can be converted into fermentable sugars and used as a culture substrate to support the growth of microorganisms.

Proper hydrolysis of rice polishing transforms it into fermentable sugars, creating an ideal substrate for microbial growth. This process not only enhances bioresource utilization but also provides an innovative approach to turning agricultural byproducts into valuable commodities. By converting rice polishing into fermentable sugars, it becomes a viable feedstock for microorganisms, which can then produce microbial biomass. This microbial biomass is rich in protein and can be sold as a value-added product, providing farmers with an additional revenue stream. This approach effectively addresses both economic and environmental concerns, as it utilizes waste materials that would otherwise be discarded (Zhang et al., 2008).

Microorganisms play a crucial role in this biomass conversion process due to their natural ability to produce enzymes that break down complex organic materials. Among these microorganisms, filamentous fungi, such as *Aspergillus*, *Penicillium*, *Chaetomium*, and *Trichoderma*, are particularly effective in biomass generation. These fungi, along with certain bacterial species and yeast genera, have been widely used in industrial biotechnology to produce microbial biomass from agricultural residues (Ahmed et al., 2009a, b).

From a biotechnological perspective, the filamentous fungus *Trichoderma harzianum* stands out for its efficiency in converting biomass into microbial protein. Its filamentous structure allows for easier separation and collection of the microbial biomass from the culture medium, reducing production costs and increasing the feasibility of large-scale operations (Ahmed et al., 2009a). To achieve optimal yields and ensure the economic viability of this process, it is essential to fine-tune various process parameters, such as pH, temperature, and nutrient availability.

This chapter provides an in-depth overview of the production of fungal biomass protein (FBP) from rice polishings using *T. harzianum* and *Candida utilis*. It discusses the process optimization required to maximize biomass protein yield, particularly focusing on the use of defatted rice polishings as a substrate. Defatted rice polishings, which contain 11.82% crude fiber and 13.9% crude protein, are an excellent substrate for microbial growth, offering a balanced nutrient profile that supports the robust production of microbial biomass. The chapter highlights the importance of optimizing these conditions to enhance biomass output and ensure the sustainability of the process.

2 Utilizing Rice Biomass as Feedstock

2.1 Global Rice Cultivation and Consumption

A significant segment of the global population, especially in Asia, rice is the most commonly consumed staple food and one of the most significant food crops in the world. From 167 million hectares of planting areas worldwide, 782 million t of paddy rice were harvested in 2018; Asia accounted for the majority of the world's rice production (FAOSTAT, 2019). China led the world in paddy rice output with an estimated 212 million t, followed by Indonesia and India. Asia produces 90% of the world's rice, along with Bangladesh, Vietnam, Thailand, Myanmar, Philippines, Pakistan, and Cambodia (FAOSTAT, 2019). Due to consumer preferences for rice and an increase in Asian immigration to the USA and Europe, rice consumption is still expanding steadily in sub-Saharan Africa, the USA, and the EU (Mohanty, 2013). The long-term sustainability of the rice sector will be ensured by this growing tendency, which will also lead to a significant increase in the amount of biomass produced by rice production.

2.2 Rice Byproduct Processing

The primary steps in the production and processing of rice include harvesting, drying, milling, and polishing. Bran, broken rice, husk, and rice straw are the main wastes produced by these operations. The hard protective shell that forms on a grain of rice during milling is called rice husk, or rice hull. For every kg of milled white rice, around 0.28 kg of rice husk are produced (Gummert et al., 2010). Table 1 displays the results of the biochemical and proximate examination of rice husk (Worasuwannarak et al., 2007). Due to their traditional status as trash, rice husk and straw would either be burned in the field or disposed of in rivers, which would result in pollution, contamination, and greenhouse gas (GHG) emissions. While burning rice straw in the open helps control weeds, pests, and illnesses and reduce a significant amount of biomass, research has shown that burning these wastes also depletes soil nutrients and organic matter. The productivity of the soil is significantly impacted by the open burning of rice straw, which also destroys beneficial soil bacteria and insects (Mandal et al., 2004).

One of the most prevalent agricultural residues containing lignocellulose worldwide is rice husk, with an estimated 156 million t produced annually worldwide in 2018 (calculated using a rice husk to paddy weight ratio of 5:1) (FAOSTAT, 2019). This quantity of rice straw and rice husk may be turned into MBP instead of being burned for disposal. Table 1 showed the biochemical and proximate examination of rice husk.

Table 1 Biochemical and proximate examination of rice husk

Composition (%)	Rice husk
Cellulose	28.6
Hemicellulose	28.6
Lignin	24.4
Proximate analysis (% dry fuel)	
Ash	20.26
Fixed carbon	16.22
Volatile matter	63.52
Ultimate analysis (% dry fuel)	
Ash	20.26
Carbon	38.33
Chlorine	0.12
Hydrogen	4.75
Nitrogen	0.52
Oxygen	35.47
Sulfur	0.05

2.2.1 RiceStraw

During the rice harvest, straw is separated from the grain. Straw makes up roughly 50% of the dry weight of a rice plant, while this percentage can range from 40% to 60% based on factors like growth techniques, field conditions, and harvesting techniques (Kadam et al., 2000). About 1.35 tons of rice straw are still in the field for every ton of rice grain that is harvested. The high SiO_2 content of rice straw, a fibrous lignocellulosic substance frequently found in agricultural waste, makes it noteworthy. Depending on postharvest conditions, rice straw's ash level can vary from 13% to 20%. Typically, it contains approximately 75% SiO_2, 10% K_2O, 3% P_2O_5, 3% Fe_2O_3, 1.3% CaO, and lesser amounts of magnesium and salt (Kadam et al., 2000). Season, location, and kind of plant all have a substantial impact on the composition of rice straw (Binod et al., 2010). Burning rice straw outside is a common way to get rid of it; it's an economical way to stop fungus from spreading over the field. About 23% of rice straw is burned publicly or left in the field in India; 48% of rice straw is burned in the fields in Thailand; and 95% of rice straw is burned in the Philippines (Kadam et al., 2000). Although burning rice straw in open fields is a convenient disposal method, it releases significant amounts of greenhouse gases such as CO, SOx, NOx, and particulate matter (Kadam et al., 2000).

2.2.2 Rice Husk

As rice grains mature, a protective outer layer called rice husk is created. It is made up of four different layers: structural, fibrous, spongy, or cellular, and has a low density and a high volume, accounting for about 20% of the grain's overall weight (Pouey, 2006). Rice husk's main constituents are cellulose (50%) followed by lignin (30%) and inorganic residues (20%). The majority of the inorganic residue is silica, which makes up between 95% and 98% of the residue's weight in its hydrated, amorphous state (Pouey, 2006). The husk is approximately 8–10 mm long, 2–3 mm wide, and 0.2 mm thick, with a packing density of 122 kg/m^3 (Fang et al., 2004). Despite being an inexpensive raw material, rice husk's abrasiveness, low nutritional value, resistance to deterioration, and bulkiness make it difficult to recycle (Evaldt, 2011). It is not suited for animal feed due to its high silica concentration and low nutritional value, making feed manufacturing unfeasible (Alfaro et al., 2013; Almeida, 2010).

Utilizing waste materials like rice husk in power reactors for energy production helps mitigate the environmental harm associated with waste disposal and reduces reliance on petroleum by promoting renewable energy sources. The benefits of using rice husk as fuel for generating steam and electricity have led several companies in southern Brazil to operate small thermal power plants powered by husk. Rice processing companies, in particular, are the primary users of husk for generating energy and providing heat for drying and parboiling cereal, offering an alternative to firewood (Kieling et al., 2009). Rice husk also has a significant calorific value, approximately 16.7×10^6 J kg^{-1} (Della et al., 2006).

Efforts have been extensively directed toward utilizing rice husk as an affordable and efficient adsorbent for eliminating heavy metals from aqueous environments. The most commonly studied metals include Cu, Cd, Co, Zn Ni, Pb, and Hg in wastewater. The adsorption capacity and efficiency of rice husk depend on several factors, including pH levels, initial metal concentration, agitation speed, adsorbent dosage, contact time, and temperature.

2.2.3 Rice Husk Ash

Rice husk ash refers to the ash generated from burning rice husks. The characteristics of this ash can differ significantly based on the conditions under which the husks are burned. Factors such as the burning process and equipment used (e.g., grate or fluidized bed) influence the structural form of silica in the ash, which can be either amorphous or crystalline. Additionally, the time and temperature of the burning process affect the composition and properties of the resulting ash. Rice husk ash is a material that is bulky, lightweight, and very porous, having a density of about 1800 kg m^{-3}. It is equivalent to about 4% of rice in husk and 20% of the husk bulk (Pouey, 2006). SiO_2, C, K_2O, P_2O_5, CaO, and trace amounts of Mg, Fe, and Na make up the majority of rice husk ash (Armesto et al., 2002). According to Della et al. (2002), the trace elements that are most frequently discovered in ashes include Na, K, Ca, Mg, Fe, Cu, Mn, and Zn. Variations in the composition of ashes may result from factors related to geography, the year of harvest, sample preparation, and analytical techniques. K_2O and Na_2O are the primary contaminants, and their concentrations are influenced by the type of soil and the quantity of fertilizer applied to the plant during growth (Moraes et al., 2014).

2.2.4 Rice Bran

Rice bran is the layer located between the husk and the endosperm of the grain, which is removed during the rice polishing process. This bran constitutes about 10% of the mass of rice in its husked form and contains oil content ranging from 15% to 20% (Rukmini, 1988). Rice bran is highly nutritious, rich in lipids, proteins, mineral salts, and vitamins. The specific composition of rice bran can vary depending on the type of rice, climatic conditions, and the processing methods used (Amarasinghe & Gangodavilage, 2004). Moongngarm et al. (2012) conducted an analysis of the chemical composition, phytochemicals, and antioxidant activity of rice bran from various rice cultivars sourced from mills in Maha Sarakham, Thailand. Their findings, which are detailed in Table 2, align with those of Lim et al. (2013), who noted that rice bran has traditionally been used as animal feed. Due to its high oil content, rice bran is also suitable for oil extraction. Fabian et al. (2011) pointed out that rice bran is an undervalued byproduct of rice processing. It is typically not consumed as food because of its high fiber content and the potential for contamination with rice husk (Fabian et al., 2011). Another drawback is the rapid onset of rancidity, which

Table 2 Physiochemical composition of rice bran

Composition	Non-waxy rice	Waxy rice
Ash	10.6	10.7
Carbohydrate	40.6	42.5
Fat	18.7	16.9
Fiber	12.4	11.7
Phenolic compounds (mg g^{-1})	1.5	1.9
Phytic acid (mg g^{-1})	50.6	48.1

occurs due to the high lipolytic activity that produces free fatty acids. Although stabilization techniques have been developed, allowing a small percentage of rice bran to be used in commercial food products, the majority of it continues to be uti lized as an ingredient in animal feed or as fuel in boilers (Fabian et al., 2011).

One significant challenge with rice bran is its tendency to develop rancidity quickly due to the high lipolytic activity, which produces free fatty acids. While stabilization techniques have enabled the use of a small portion of rice bran in commercial food products, most of it is still primarily utilized as an ingredient in animal feed or as fuel in boilers (Fabian et al., 2011).

Rice bran is rich in oil, protein, vitamins, and antioxidants. Various techniques, including solvent extraction, solid–liquid extraction, and supercritical CO_2, among others, can be employed to extract its valuable compounds (Pourali et al., 2009). Table 2 showed the physiochemical composition of rice bran.

2.2.5 Broken Rice

Rice experiences mechanical stress during processing, leading to the breakage of some grains, which results in what is commonly referred to as broken rice. This byproduct, also known as rice grits, consists of fragmented and defective grains. A significant challenge faced by the rice industry is reducing the quantity of broken rice and rice grits produced. During the milling of white rice, approximately 14% of the rice ends up as broken rice. This broken rice shares the same average composition as whole grain rice. Depending on the regulations in each country, broken rice can be blended with whole rice in specific ratios. Additionally, broken rice offers various opportunities for reuse, such as in the production of rice flour, as animal feed, or as a raw material for brewing beer. Like whole grain rice, broken rice is rich in starch, which is a significant energy source for human consumption (Lim et al., 2013; Limberger et al., 2008).

Limberger et al. (2008) conducted physical and chemical modifications on starch derived from brewer's rice, assessing the effectiveness of these alterations through the rheological properties of both native and modified starch. The findings revealed that phosphating and extrusion processes altered the rheological characteristics of native brewer's rice, thus broadening its applications within the food industry. Although broken rice contains only 6–8% protein, this protein is highly valued for being hypoallergenic, nutritious, and beneficial for human consumption compared to other cereals. Authors successfully extracted protein from broken rice by disrupting protein–starch complexes, achieving a protein recovery of around 82% and a purity level of nearly 88%. Additionally, broken rice serves as a valuable raw material for producing lactic acid and other functional materials. *Lactobacillus delbrueckii* was used to produce lactic acid from broken rice. Both broken rice and rice bran are ideally suited for human consumption. Although regulations in some countries permit specific proportions of broken rice to be mixed with whole rice, there remains a surplus of this byproduct. Given that broken rice has the same composition as whole rice, it can be utilized in various high-value applications, such as protein or starch extraction.

3 Experimental Methods and Materials

3.1 Organisms

Due to its versatility in consuming a variety of substrates, including wheat bran, and producing a range of industrial products suitable for both human and animal consumption, *Candida utilis* has been widely employed in the manufacture of single-cell proteins (Otero et al., 1998). When scaling up to large-scale bioreactors, factors like cell mass and product formation kinetic characteristics are crucial for assessing the economic viability of these biotechnological processes (Tobajas & Garcia-Calvo, 1999). There has been research on the impact of consuming rice polishings on rumen fermentation in vitro (Garcia et al., 1992), however, there is no information on its application to the synthesis of single-cell protein.

There has been research on the impact of consuming rice polishings on rumen fermentation in vitro (Garcia et al., 1992), however, there is no information on its application in the synthesis of single-cell protein. It was then assessed for crude protein, RNA, metabolizable energy, and amino acid content to determine whether the single cell protein product was suitable for replacing or fortifying standard poultry rations.

The review revealed fermentation was conducted in two fermentors: a 14 L fermentor with 9 L of optimized salt medium and a 50 L fermentor with 30 L of the same medium, both under optimized conditions. The medium was sterilized in situ within the fermentors. A 24-hour-old vegetative culture was used to inoculate the medium. The standard operational settings for the fermentors were temperature of 35 °C, agitation at 400 rpm, pH maintained at 6.0, and airflow rate of 2 v/v/m. The dissolved oxygen level was kept at 50% saturation throughout the process. The fermentation was carried out under four different oxygen concentrations: 30%, 40%, 50%, and 60%.

The literature review revealed that the agar slants preparation constituted 0.2% malt extract, 0.2% yeast extract, 2% glucose, and 2% agar, the fungus was kept at

28 °C. Slants that had just been injected were kept at 4 °C after 5 days of incubation at 28 °C. *T. harzianum* was prepared as an inoculum using Vogel's medium (Ahmed et al., 2012). The inoculum of fungus was grown on an orbital shaker at 28 °C (120 rpm for 24 h) using rice polishings as the substrate. To achieve the best production FBP, additional ingredients were added to the medium: 0.2% KH_2PO_4, 0.15% $MgSO_4.7H_2O$, and 0.1% $CaCl_2.2H_2O$.

3.2 *Enhancement of* T. harzianum

Several writers have extensively employed the conventional procedure of optimizing cultural conditions by optimizing one parameter at a time (Mondal et al., 2012). It was determined whether different concentrations of rice polishing might yield FBP from *T. harzianum*. The ideal pH ranges to generate FBP were tested at various pH levels. Additionally, the ideal temperature for FBP was established at various temperatures. In order to achieve the best FBP production by the fermentation of rice polishings with *T. harzianum,* several ionic concentrations of $CaCl_2.2H_2O$, $MgSO_4.7H_2O$, and KH_2PO_4 were tried.

3.3 *C:N Ratio Effect on FBP Production*

The growth medium's soluble carbohydrate content was calculated using glucose, which stands for "C." Using Kjeldahl's method, the growth medium's total nitrogen content was estimated. Urea was used as a supply of nitrogen and glucose was used as a source of carbon to construct various C:N ratios. A variety of C:N ratios, including 30:1, 25:1, 20:1, 15:1, 10:1, and 5:1, were investigated to determine the ideal C:N ratio for *T. harzianum* in order to produce the most FBP.

3.4 *Analytical Method*

Using AOAC procedures, the FBP obtained from *T. harzianum*'s fermentation of rice polishings was examined (AOAC, 1984). To measure the true and crude proteins as well as the protein content of the cell biomass, 200 mL aliquots were selected at different time periods. To remove the substrate and cell mass, the entire content (200 mL) was homogenized and centrifuged at 12,000 × g for 10 min at 4 °C. Low centrifugation (4000 × g for 15 min, 4 °C) was used to separate the substrate from the cell mass. After separating from the substrate and settling at the bottom, the cell mass formed the upper layer. The substrate was dried at 95 °C to a consistent mass after being cleaned three times with distilled water. In addition, the cell mass was rinsed three times and suspended in saline. Additionally, a 100 mL fraction of culture broth was centrifuged at 10,000 rpm for 10 min, and the cell pellet was cleaned with saline solution. The pellet was dried at 70 °C to a constant mass while suspended in 10 mL of concentrated H_2O. In order to ascertain the intracellular protein of the second fraction, it was centrifuged as well, twice washed in saline, and then suspended in saline, as previously mentioned (Rajoka et al., 2006).

In case of *Candida utilis*, samples and flasks were taken in triplicate at various intervals to evaluate different parameters. Paul et al. (2002) investigated these metrics comprised cell concentration, solid substrate, crude protein, true protein, crude fiber, nitrogen-free extract, and RNA. To separate the substrate, culture samples (100 mL) were centrifuged at 7000 g for 10 min at 10 °C. The substrate was then twice rinsed with saline and dried to measure the amount of solid material that was not used. A further 100 mL of the culture broth underwent a 10 min, 10,000 g centrifugation. The resulting cell mass was dried at 70 °C after being twice rinsed with saline and suspended in 10 mL of distilled water. To create the dried biomass, a different 100 mL sample that contained the cell mass and any leftover substrate was dried. Crude protein, true protein, and RNA content of this biomass were regularly measured (Paul et al., 2002).

The nitrogen concentration, as determined by the micro-Kjeldahl method, was multiplied by a factor of 6.25 to determine the crude protein content. True protein was measured by treating 0.5 g of homogenized dry biomass with 20 mL of 5% v/v trichloroacetic acid, shaking it for 5 minutes, and then heating it to 90 °C for 15 minutes while shaking it now and then. The residue was dried to a consistent weight following three rounds of hot water rinsing and hot filtering. Using the micro-Kjeldahl method, the nitrogen concentration of this residue was determined, and genuine protein was computed based on that value. Using the orcinol HCl reagent, the total RNA content was determined as previously mentioned (Paul et al., 2002). By applying the Parr method, a Parr oxygen bomb calorimeter was used to calculate the dry biomass's calorific value (Hill & Anderson, 1958). Based on the heat produced during the combustion of a known sample weight in the presence of 20 atmospheres of oxygen, the calorific value was determined. In addition, AOAC (1984) methods were used to analyze the following parameters: moisture, dry matter, ash, ether extract, crude fiber, nitrogen-free extract, carbon, and cellulose content.

3.5 FBP Biological Evaluation

According to earlier studies, broiler chicks that were 1 day old were bought from the neighborhood market and given maize feed for 10 days. Subsequently, 30 birds were chosen based on weight homogeneity, and they were assigned at random to rations A, B, C, and D. This ensured that six experimental birds were fed each ration. Every chick was housed in duplicate in its own space. To identify them, each bird's wing was tagged separately. The experimental birds' initial body weights varied from 180 to 190 g. For 10 experimental days, broiler chicks that were 10 days old were fed the FBP in order to conduct a biological evaluation. In order to assess the protein quality of the biomass, 30% and 60% of it was substituted with soybean meal. The four experimental meals, A, B, C, and D, were prepared for the aforementioned experiments. Diet "A" was a control diet with 3200 kcal of metabolizable energy and 23% crude protein. The formulation of rations B and C included 23% crude protein and 3200 kcal of metabolizable energy per kilogram. Thirty percent of the soybean meal in ration B and 60% of the soybean protein in ration C were substituted with FBP. A ration (D) free of nitrogen was developed, with a metabolizable energy content of 3200 kcal/kg. As previously mentioned, feed conversation ratio, net protein utilization, and protein efficiency ratio were studied taking into consideration initial body weight, feed offered and feed refused, and bird weight (Shahzad & Rajoka, 2011).

4 Results and Discussion

4.1 Optimum Conditions for FBP Production

In order to get a higher production of FBP, various fermentation conditions were refined. To manufacture FBP from *T. harzianum*, these tests were conducted in a 5 L fermenter with a 2.5 L working capacity to optimize the fermentation conditions. In case of *Candida utilis*, initial experiments in shake flasks were conducted to optimize fermentation conditions for crude protein production. A 9% concentration of rice polishings significantly enhanced protein productivity and other parameters, except cell yield ($P \leq 0.05$). Specifically, 90 g of rice polishings per liter supported optimal protein productivity at 0.88 g/l/h, with yields of 0.36 g/g crude protein, 0.84 g/g cell mass, and 0.29% crude protein, while reducing RNA accumulation. Higher concentrations led to decreased protein formation and cell mass synthesis. Enhanced substrate metabolism resulted in a maximum cell mass of 32.5 g/L, surpassing previously reported values (Nigam, 2000).

4.2 Effect of Rice Polishings on Biomass Production

The generation of FBP was found to be primarily higher when rice polishings at a concentration of 5% (w/v) were used (Fig. 1). As the concentration of substrate rose, FBP did not rise. The rates of protein synthesis, substrate consumption, and cell mass synthesis were all determined while the concentration of rice polishings was varied. Prior research revealed that rice polishings at a weight percentage of 9% (w/v) resulted in notably increased FBP production (Rajoka et al., 2006). As a result, *T. harzianum* is more cost-effective when it comes to production at an industrial scale and requires less substrate for FBP.

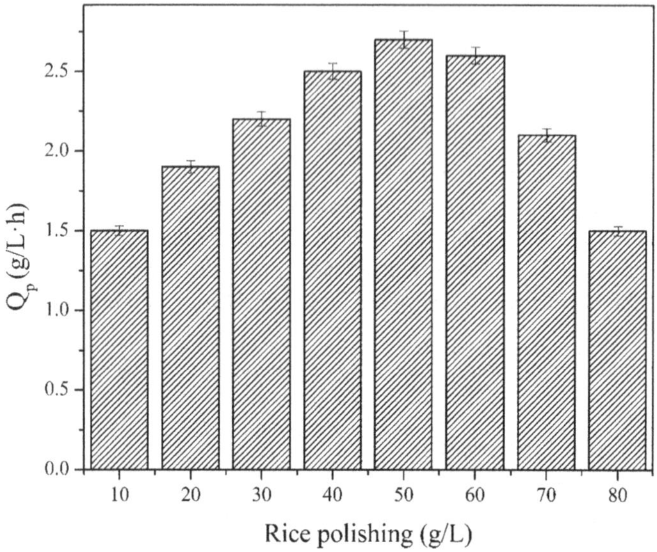

Fig. 1 Effect of rice polishings at different concentrations by *T. harzianum*

Fig. 2 Figure 2 shows a representative time course of *T. harzianum's* generation of FBP from rice polishing

4.3 *Impact on Production of Biomass*

The initial pH of the fermentation medium plays a crucial role in determining the yield of fructose-1,6-bisphosphate (FBP). Figure 2 highlights the relationship between pH and FBP synthesis, showing that the highest production of FBP occurs at pH 4 under a temperature of 280 °C. The impact of pH on FBP production is significant, as different pH levels can drastically influence the metabolic activities of the microorganisms involved. For example, *Candida utilis*, when cultivated on wheat bran, achieved its maximum biomass production at a pH of 6.5 (Arshad et al., 2014). Similarly, *Chaetomium* sp. was found to produce the highest protein content at a pH of 5.5 (Irfan et al., 2011). On the other hand, *Trichoderma viride* strain

WEBL0702 exhibited the greatest FBP production at a slightly more acidic pH of 4.5 (Zhang et al., 2008). Moreover, when the initial pH of the fermentation medium was adjusted from 5.5 to 6, both *Aspergillus niger* and *Rhizopus oryzae* showed a substantial increase in FBP yield (Ahmed et al., 2009a). These findings underscore the importance of optimizing pH conditions to maximize the production of FBP and other metabolites in various microbial strains.

4.4 Time Course of FBP Production

Using 5% (w/v) rice polishings as a substrate and Vogel's medium (beginning pH of 4) at 28 °C, the time course for FBP formation was examined. Figure 2 shows a representative time course of T. *harzianum's* generation of FBP from rice polishing. FBP reached its maximum after 72 h of fermentation. If the fermentation was continued, the FBP decreased after reaching its maximal value, which might be the result of cell autolysis. Therefore, in all subsequent trials, the ideal pH was 4, the ideal temperature was 28 °C, and the ideal incubation period was 72 h.

4.5 Effect of Ionic Concentration

For FBP to be produced on an industrial scale, *T. harzianum* must be produced economically. An investigation into the potential for enhancing microbial performance involved examining the impact of several sources of ionic concentrations. The variables' ideal concentrations were determined to be 0.2% KH_2PO_4, 0.15% $MgSO_4.7H_2O$, and 0.1% $CaCl_2.2H_2O$ (Fig. 3). According to previous review, *T. harzianum*'s low nutritional needs make it a potentially valuable strain for the manufacture of FBP. Therefore, using *T. harzianum*, we have created a medium with optimal ionic concentrations for the increased generation of FBP from rice polishings.

4.6 Effect of Carbon Ratios

The earlier study of *Candida utilis* showed carbon ratios significantly impact protein concentrate production. By varying urea concentration, it was found that a 10:1 ratio yielded the highest values for crude protein productivity (0.97 g/l/h), crude protein yield (0.35 g/g), cell mass (0.85 g/g), substrate yield (0.53 g/g), and RNA accumulation (0.29%), whereas in *T. harzianum,* the fermentation process's C:N ratio affects its ability to produce protein concentrates. By raising the urea concentration in the medium, this ratio was obtained in order to demonstrate its impact on the generation of biomass protein. A C:N ratio of 20:1 produced noticeably more

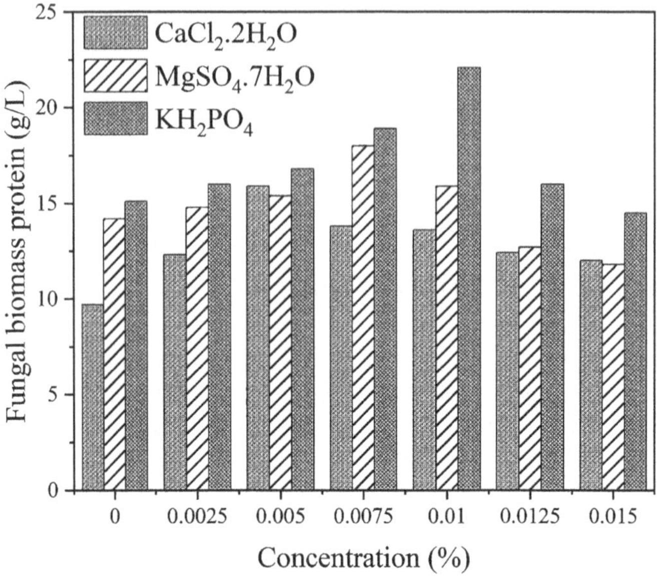

Fig. 3 Impact of different ionic concentration levels on *T. harzianum*'s generation of FBP from rice polishings

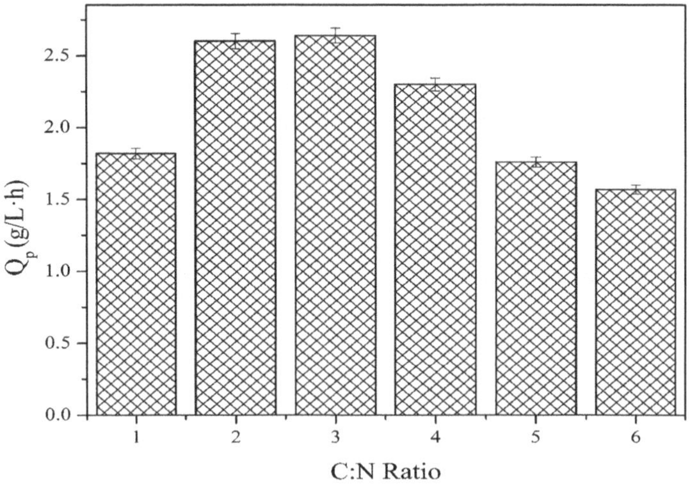

Fig. 4 Effect of varying C:N ratio on *T. harzianum* cultured on rice polishings on FBP protein production

FBP (Fig. 4). Harvesting MBP of superior quality requires a suitable C:N ratio (Yalemtesfa et al., 2010). For the formation of biomass, a C:N ratio of 8:10 has been recorded (Das et al., 2013). Overall, the data verified that maximum FBP production from *T. harzianum* is supported by urea, a lost-cost fertilizer.

Table 3 Chemical composition of FBP

Components	Rice polishings *T. harzianum*	Rice polishings *Candida utilis*	Biomass
Ash	9.50	3.01	9.62
Cellulose	10.76	9.76	11.5
Crude fat	11.73	–	21.25
Crude protein	11.25	13.9	49.5
Moisture	6.77	2.50	1.5
Nitrogen extract	54.5	57.37	28.3
True protein	1.0	0	32.0

4.7 Nutritional Value and Chemical Composition of FBP

The resulting FBP product was chemically examined, and an estimate of its amino acid composition was calculated. The ultimate biomass product produced by *T. harzianum* was analyzed chemically and determined to be dry weight based (Table 3). When *T. harzianum* was fermented in a 75 L fermenter, the actual protein content of the fungal biomass generated was 32.00%. As a result, the protein content was suitable for using FBP as fodder. According to Satari et al. (2016), the percentage of fungal biomass in *Mucor indicus* and *R. oryzae* was 40% and 51%, respectively. The composition of a protein's amino acids largely determines its potential nutritional value. *T harzianum* was cultivated in a 75 L fermenter for 72 h. The final FBP product's amino acid profile was estimated. The produced FBP has 16 amino acids in it. This FBP is a useful source of microbial protein. The *T. harzianum* fungal biomass protein has a higher percentage of lysine, which is a limiting amino acid in aquatic diets. In addition to addressing the low levels of phenylalanine and methionine in fungal biomass, cofeeding the fungal protein with commercially available protein can lower the cost of protein ingredients for feeds. Significant urea metabolism by the yeast is demonstrated by the rise in crude protein from 13.9% to 27.8%, with 22.3% of that amount being genuine protein. In case of *Candida utilis, a* large amount of nitrogen was used in the synthesis of actual amino acids (Table 2). Kluyveromyces fragilis biomass, grown on deproteinized whey supplemented with 0.8% diammonium hydrogen phosphate, contained 37% crude protein, whereas previous investigations, such as the single-cell protein product described by Singh and Kalamdhad (2014), contained just 30.4% crude protein (Paul et al., 2002).

4.8 FBP Assessment from a Biological Perspective

The study by Ahmed et al. (2017) revealed the chicks fed ration C had the most weight gain, with those fed ration B and control ration (A) following in decreasing order. It showed that FBP can substitute soybean meal to some extent in grill chick diets without negatively impacting the birds' ability to grow weight (results not shown). The current study's findings are consistent with those of Joshi et al. (2000). They utilized five yeasts and solid-state fermentation to generate animal feed on a big scale from apple pomace. The results of using fermented apple pomace in a 1:1 ratio with regular grill feed were comparable. The regular ration (A) had the lowest feed consumption. Chicks fed on rations consumed the maximum amount of feed. Chicks fed ration (C), which substituted 60% of the soybean protein with FBP, ate their maximum amount of feed. The results were similar to the production of single protein cell from *Arachniotus* sp. and *Brevibacterium flavum* and rice polishings were used as substrate for production of microbial protein.

5 Potential Applications

Rice husk presents a range of potential applications, especially as a biomass resource for generating power. It is often preferred over rice straw due to its status as an industrial byproduct, which mitigates logistical challenges associated with straw. Utilizing rice husk in power reactors for electricity generation effectively addresses waste disposal issues while reducing dependence on petroleum and promoting renewable energy sources.

The notable calorific value of rice husk, estimated to be around 16.7×10^6 J/kg^{-1} (Della et al., 2006), has made it an attractive fuel source. This high energy content has encouraged several companies in southern Brazil to incorporate rice husk into their operations, particularly in the form of small thermal power plants where it serves as a primary fuel for generating electricity and heat. These plants, primarily established by rice processing companies, use husk for generating power and heat, replacing firewood in processes like drying and parboiling rice. According to the National Electric Energy Agency (ANEEL, 2014), Brazil had 484 biomass-burning plants, of which nine exclusively used rice husk, with a combined capacity of 36.4 GW. This setup not only met the energy needs of the mills but also allowed for surplus electricity to be exported to the grid. Furthermore, rice husk has potential for generating other forms of energy, such as methane and hydrogen, through biomass conversion, and can be used to produce ethanol (Lim et al., 2012).

Beyond its application in power generation, rice husk has also been explored for its potential as a lignocellulosic feedstock for bioethanol production. Research conducted by Lim et al. (2012) highlighted the viability of utilizing rice husk in bioethanol production processes due to its rich content of cellulose and hemicellulose. These components are crucial for the production of bioethanol, making rice husk a promising alternative to other lignocellulosic materials in the development of sustainable and renewable energy sources. Prasara-A and Grant (2011) utilized a life-cycle analysis tool to compare the environmental impacts of using rice husk for power generation versus ethanol production. Their study revealed that electricity generation was more environmentally advantageous, showing lower impacts on human toxicity, acidification, and ecotoxicity compared to ethanol production.

Research has also investigated rice husk for converting carbohydrates into organic chemicals.

Suxia et al. (2012) conducted research aimed at optimizing the conditions necessary for efficient furfural production from rice husk. Their work involved fine-tuning various parameters to maximize the yield of furfural, a valuable chemical used in industrial applications. In a related study, Potumarthi et al. (2013) concentrated on improving the production of reducing sugars from rice husk through biological pretreatment methods. Their research sought to enhance the efficiency of sugar extraction by employing biological agents to pretreat the rice husk, thereby increasing the availability of reducing sugars, which are essential for numerous chemical and biofuel applications. Rice husk is used to produce levulinic acid, a valuable product in the tobacco and pharmaceutical industries. Although these applications are promising, they often involve complex processes, yield relatively small amounts, and utilize only a fraction of the available husk, failing to fully address the high volume of husk generated. The production of high-value products from rice biomass conversion, such as lactic acid, levulinic acid, xylose, activated carbon, and rice husk ash, along with their diverse applications in industries like pharmaceuticals, food, textiles, and construction, continues to drive advancements in rice biomass conversion technologies. Future efforts are focused on optimizing rice plant characteristics, increasing lignin content, reducing silica levels for improved thermochemical processing, implementing high-temperature ethanol fermentation, and developing biorefineries. There remains significant potential to fully harness this abundant bioenergy source to meet sustainable energy demands in the near future.

Rice husk has also been evaluated as a poultry litter material. The rice husk improved weight gain, feed intake, feed conversion, and productivity in poultry, making it a viable option for litter. However, the choice of litter material depends on factors such as availability, quality, cost, and intended use. In Brazil's climate, which supports open aviaries, rice husk's reuse is feasible and advantageous, as it generally has higher mineral content compared to wood shavings.

Numerous studies have investigated the potential of rice husk as a bulking agent in composting processes. Research conducted by Chowdhury et al. (2014) explored the effectiveness of rice husk in improving composting efficiency by enhancing aeration and facilitating the decomposition of organic materials. Similarly, Singh and Kalamdhad (2014) assessed the role of rice husk in composting, focusing on its ability to improve the physical properties of compost, such as its structure and porosity, which are crucial for maintaining adequate airflow and moisture levels. These studies collectively demonstrated that rice husk can significantly contribute to the composting process by acting as an effective bulking agent, thereby supporting more efficient and balanced decomposition of organic waste.

Moreover, research by Krishnani et al. (2008) has explored the production of activated charcoal from rice husk using chemical and physical activation methods. While these methods have shown satisfactory results, the high cost of production remains a significant barrier. Because it is a plentiful raw material, rice husk has garnered a lot of interest from researchers looking to create bio-composites. For artificial resin-based rice husk bio-composites, a number of surface modification methods have been suggested to enhance the adhesion between the fiber and the matrix. Nevertheless, using rice husk with naturally occurring resins still has a lot of potential. Natural resins have the potential to be essential in creating composite materials that are beneficial to the environment. The application of rice husk with recycled polymers is another unexplored field. Using currently available polymers could have a significant positive environmental impact, given the globe already has an oversupply of plastics that are harmful to the ecosystem. Numerous studies have demonstrated that with larger filler loadings, poor adhesion between the fiber and matrix decreases mechanical strength; however, this can be lessened by surface modification techniques. Many surface modification approaches have been studied, however, not all of them have been used to rice husk fibers. Hybrid bio-composites could also be developed with rice husk, and not much research has been done in this field. Given that hybrid composites have been demonstrated to greatly enhance characteristics, combining rice husk with other biofibers seems like a viable avenue for further research. Studies have looked into the use of rice husk ash, a result of rice husk thermal degradation that includes 92% silica. Further investigation into this area is anticipated in future studies. Crucially, a careful examination of the aging and weathering characteristics of these materials would be necessary for any commercial application of rice husk-based composites. To confirm the composites' suitability for commercial usage, it is also necessary to investigate the effects of microbes and fungi.

6 Conclusion

The large quantity of rice residues, coupled with advancements in conversion technologies, has enhanced the potential for bioenergy from rice byproducts, positioning it as a viable renewable energy source. As a protein supplement in animal feed, *T. harzianum*'s FBP has demonstrated considerable promise. Utilizing rice polishing at a 5% (w/v) concentration has shown to be an effective substrate for the generation of microbial biomass protein. A pH of 4, a temperature of 28 °C, and a 72-h incubation period were shown to be the ideal values for the formation of FBP. FBP output was greatly increased by a carbon to nitrogen ratio of 20:1. These results indicate that utilizing rice polishing as a substrate to produce FBP is a technically feasible way to turn agricultural waste into beneficial protein sources. The outcomes show promise for producing biomass protein from *T. harzianum* on a large scale, and they might also apply to other fungal fermentation processes. With a gross metabolizable energy value of 2678 kcal/kg, the dried biomass showed promise not just as a source of protein and amino acids but also as an energy source, especially for chicken feed. The only amino acid that was shown to be limiting was valine.

References

Ahmed, S., Bashir, A., Saleem, H., Saadia, M., & Jamil, A. (2009a). Production and purification of cellulose-degrading enzymes from a filamentous fungus *Trichoderma harzianum*. *Pakistan Journal of Botany, 41*(3), 1411–1419.

Ahmed, S., Imdad, S. S., & Jamil, A. (2012). Comparative study for the kinetics of extracellular xylanases from *Trichoderma harzianum* and *Chaetomium thermophilum*. *Electronic Journal of Biotechnology, 15*(3), 3–3.

Ahmed, S., Mustafa, G., Arshad, M., & Rajoka, M. I. (2017). Fungal biomass protein production from *Trichoderma harzianum* using rice polishing. *BioMed Research International, 2017*(1), 6232793.

Ahmed, S., Riaz, S., & Jamil, A. (2009b). Molecular cloning of fungal xylanases: An overview. *Applied Microbiology and Biotechnology, 84*, 19–35.

Alfaro, E. F., Dias, D. B., & Silva, L. G. (2013). The study of ionizing radiation effects on polypropylene and rice husk ash composite. *Radiation Physics and Chemistry, 84*, 163–165.

Almeida, S. R. (2010). *Pirólise rápida de casca de arroz: estudo de parâmetros e caracterização de produtos*.

Amarasinghe, B. M. W. P. K., & Gangodavilage, N. C. (2004). Rice bran oil extraction in Sri Lanka: Data for process equipment design. *Food and Bioproducts Processing, 82*(1), 54–59.

ANEEL—Agência Nacional de Energia Elétrica. (2014). *Matriz Energética do Brasil*. http://www.aneel.gov.br/aplicacoes/capacidadebrasil/OperacaoGeracaoTipo.asp?tipo=5&ger=Combustivel&principal=Biomassa. Accessed 17 July 2024.

Association of Official Analytical Chemists (AOAC). (1984). *The official methods of analysis of AOAC international*. AOAC. ISBN 0-93558424-2.

Armesto, L., Bahillo, A., Veijonen, K., Cabanillas, A., & Otero, J. (2002). Combustion behaviour of rice husk in a bubbling fluidised bed. *Biomass and Bioenergy, 23*(3), 171–179.

Arshad, M., Ahmed, S., Zia, M. A., & Rajoka, M. I. (2014). Kinetics and thermodynamics of ethanol production by *Saccharomyces cerevisiae* MLD10 using molasses. *Applied Biochemistry and Biotechnology, 172*, 2455–2464.

Binod, P., Sindhu, R., Singhania, R. R., Vikram, S., Devi, L., Nagalakshmi, S., Kurien, N., Sukumaran, R. K., & Pandey, A. (2010). Bioethanol production from rice straw: An overview. *Bioresource Technology, 101*(13), 4767–4774.

Chowdhury, A. K. M. M. B., Michailides, M. K., Akratos, C. S., Tekerlekopoulou, A. G., Pavlou, S., & Vayenas, D. V. (2014). Composting of three phase olive mill solid waste using different bulking agents. *International Biodeterioration & Biodegradation, 91*, 66–73.

Das, M., Gupta, S., Kapoor, V., Banerjee, R., & Bal, S. (2008). Enzymatic polishing of rice–a new processing technology. *LWT-Food Science and Technology, 41*(10), 2079–2084.

Das, A., Paul, T., Halder, S. K., Maity, C., Mohapatra, P. D., Pati, B. R., & Mondal, K. C. (2013). Study on regulation of growth and biosynthesis of cellulolytic enzymes from newly isolated Aspergillus fumigatus ABK9. *Polish Journal of Microbiology, 62*(1), 31.

Della, V. P., Hotza, D., Junkes, J. A., & Oliveira, A. P. N. D. (2006). Estudo comparativo entre sílica obtida por lixívia ácida da casca de arroz e sílica obtida por tratamento térmico da cinza de casca de arroz. *Química Nova, 29*, 1175–1179.

Della, V. P., Kühn, I., & Hotza, D. (2002). Rice husk ash as an alternate source for active silica production. *Materials Letters, 57*(4), 818–821.

Evaldt, D. C. (2011). *Influência do uso de aditivos na moagem de cinzas de casca de arroz para sua adequação como co-produto*.

Fabian, C., Ayucitra, A., Ismadji, S., & Ju, Y. H. (2011). Isolation and characterization of starch from defatted rice bran. *Journal of the Taiwan Institute of Chemical Engineers, 42*(1), 86–91.

Fang, M., Yang, L., Chen, G., Shi, Z., Luo, Z., & Cen, K. (2004). Experimental study on rice husk combustion in a circulating fluidized bed. *Fuel Processing Technology, 85*(11), 1273–1282.

Food and Agricultural Organization of the United Nations. (2019). *FAOSTAT database*. http://www.fao. org/faostat/en/?#home. Accessed 15 May 2020.

Garcia, D. C., Newbold, C. J., Galbraith, H., & Topps, J. H. (1992). The effect of including Colombian rice polishings in the diet on rumen fermentation in vitro. *Animal Science, 54*(2), 275–280.

Gummert, M., Hien, P. H., Pyseth, M., Rickman, J., Schmidley, A., & Pandey, S. (2010). Emerging technological and institutional opportunities for efficient postproduction operations. In *Rice in the global economy: strategic research and policy issues for food security* (p. 333).

Hill, F. W., & Anderson, D. L. (1958). Comparison of metabolizable energy and productive energy determinations with growing chicks. *J Nutr, 64*, 587.

Irfan, M., Nazir, M. I., Nadeem, M., Gulsher, M., Syed, Q., & Baig, S. (2011). Optimization of process parameters for the production of single cell biomass of *Candida utilis* in solid state fermentation. *American-Eurasian Journal of Agricultural and Environmental Sciences, 10*, 264–270.

Joshi, V. K., Gupta, K., Devrajan, A., Lal, B. B., & Arya, S. P. (2000). *Production and evaluation of fermented apple pomace in the feed of broilers*.

Kadam, K. L., Forrest, L. H., & Jacobson, W. A. (2000). Rice straw as a lignocellulosic resource: Collection, processing, transportation, and environmental aspects. *Biomass and Bioenergy, 18*(5), 369–389.

Kieling, A., Moraes, C. A., & Brehm, F. (2009). Utilização de cinza de casca de arroz na remoção de cromo hexavalente. *Estudos Tecnológicos em Engenharia, 5*(3), 351–362.

Krishnani, K. K., Meng, X., Christodoulatos, C., & Boddu, V. M. (2008). Biosorption mechanism of nine different heavy metals onto biomatrix from rice husk. *Journal of Hazardous Materials, 153*(3), 1222–1234.

Kummamuru, B. (2016). *WBA global bioenergy statistics 2017* (p. 436). World Bioenergy Association.

Limberger, V. M., Silva, L. P. D., Emanuelli, T., Comarela, C. G., & Patias, L. D. (2008). Modificação química e física do amido de quirera de arroz para aproveitamento na indústria de alimentos. *Química Nova, 31*, 84–88.

Lim, J. S., Manan, Z. A., Hashim, H., & Alwi, S. R. W. (2013). Towards an integrated, resource-efficient rice mill complex. *Resources, Conservation and Recycling, 75*, 41–51.

Lim, J. S., Manan, Z. A., Alwi, S. R. W., & Hashim, H. (2012). A review on utilisation of biomass from rice industry as a source of renewable energy. *Renewable and Sustainable Energy Reviews, 16*(5), 3084–3094.

Mohanty, S. (2013). Trends in global rice consumption. *Rice Today, 12*(1), 44–45.

Mondal, A. K., Sengupta, S., Bhowal, J., & Bhattacharya, D. K. (2012). Utilization of fruit wastes in producing single cell protein. *International Journal of Science, Environment and Technology, 1*(5), 430–438.

Mandal, K. G., Misra, A. K., Hati, K. M., Bandyopadhyay, K. K., Ghosh, P. K., & Mohanty, M. (2004). Rice residue-management options and effects on soil properties and crop productivity. *Journal of Food Agriculture and Environment, 2*, 224–231.

Moongngarm, A., Daomukda, N., & Khumpika, S. (2012). Chemical compositions, phytochemicals, and antioxidant capacity of rice bran, rice bran layer, and rice germ. *APCBEE Procedia, 2*, 73–79.

Moraes, C. A., Fernandes, I. J., Calheiro, D., Kieling, A. G., Brehm, F. A., Rigon, M. R., Berwanger Filho, J. A., Schneider, I. A., & Osorio, E. (2014). Review of the rice production cycle: By-products and the main applications focusing on rice husk combustion and ash recycling. *Waste Management & Research, 32*(11), 1034–1048.

Nigam, J. N. (2000). Cultivation of *Candida langeronii* in sugar cane bagasse hemicellulosic hydrolyzate for the production of single cell protein. *World Journal of Microbiology and Biotechnology, 16*, 367–372.

Otero, M. A., Cabello, A. J., Vasallo, M. C., Garcia, L., & Lopez, J. C. (1998). Preparation of an imitation soy sauce from hydrolyzed dried yeast *Candida utilis*. *Journal of Food Processing and Preservation, 22*(5), 419–432.

Paul, D., Mukhopadhyay, R., Chatterjee, B. P., & Guha, A. K. (2002). Nutritional profile of food yeast Kluyveromyces fragilis biomass grown on whey. *Applied Biochemistry and Biotechnology, 97*, 209–218.

Potumarthi, R., Baadhe, R. R., Nayak, P., & Jetty, A. (2013). Simultaneous pretreatment and sacchariffication of rice husk by *Phanerochete chrysosporium* for improved production of reducing sugars. *Bioresource Technology, 128*, 113–117.

Pouey, M. T. F. (2006). *Beneficiamentos da cinza de casca de arroz residual com vistas à produção de cimento composto e/ou pozolânico.*

Pourali, O., Asghari, F. S., & Yoshida, H. (2009). Sub-critical water treatment of rice bran to produce valuable materials. *Food Chemistry, 115*(1), 1–7.

Prasara-A, J., & Grant, T. (2011). Comparative life cycle assessment of uses of rice husk for energy purposes. *The International Journal of Life Cycle Assessment, 16*, 493–502.

Rajoka, M. I., Khan, S. H., Jabbar, M. A., Awan, M. S., & Hashmi, A. S. (2006). Kinetics of batch single cell protein production from rice polishings with Candida utilis in continuously aerated tank reactors. *Bioresource Technology, 97*(15), 1934–1941.

Rukmini, C. (1988). Chemical, nutritional and toxicological studies of rice bran oil. *Food Chemistry, 30*(4), 257–268.

Satari, B., Karimi, K., Taherzadeh, M. J., & Zamani, A. (2016). Co-production of fungal biomass derived constituents and ethanol from citrus wastes free sugars without auxiliary nutrients in airlift bioreactor. *International Journal of Molecular Sciences, 17*(3), 302.

Shahzad, M. A., & Rajoka, M. I. (2011). Single cell protein production from *Aspergillus terreus* and its evaluation in broiler chick. *International Journal of Bioscience, Biochemistry and Bioinformatics, 1*(2), 137.

Singh, W. R., & Kalamdhad, A. S. (2014). Potential for composting of green phumdi biomass of Loktak lake. *Ecological Engineering, 67*, 119–126.

Suxia, R., Haiyan, X., Jinling, Z., Shunqing, L., Xiaofeng, H., & Tingzhou, L. (2012). Furfural production from rice husk using sulfuric acid and a solid acid catalyst through a two-stage process. *Carbohydrate Research, 359*, 1–6.

Tobajas, M., & Garcia-Calvo, E. (1999). Determination of biomass yield of *Candida utilis* on glucose Black box and metabolic description. *World Journal of Microbiology and Biotechnology, 15*, 431–438.

Worasuwannarak, N., Sonobe, T., & Tanthapanichakoon, W. (2007). Pyrolysis behaviors of rice straw, rice husk, and corncob by TG-MS technique. *Journal of Analytical and Applied Pyrolysis, 78*(2), 265–271.

Yalemtesfa, B., Alemu, T., & Santhanam, A. (2010). Solid substrate fermentation and conversion of orange waste into fungal biomass using *Aspergillus niger* KA-06 and *Chaetomium* Spp KC-06. *African Journal of Microbiology Research, 4*(12), 1275–1281.

Zhang, Z. Y., Jin, B., Bai, Z. H., & Wang, X. Y. (2008). Production of fungal biomass protein using microfungi from winery wastewater treatment. *Bioresource Technology, 99*(9), 3871–3876.

Enzymes Playing Their Specific Roles in Biomass Waste Management

Muhammad Ammar, Nafeesa Yousaf, Marcelo Franco, and Muhammad Irfan

Abstract In this chapter, commercially available enzymes are discussed for bio-waste management. Sources and applications of laccases, chitinases, and cellulases are discussed briefly. Laccases are obtained from plants, bacteria, and fungi. These enzymes are used in agricultural residues, waste water treatment, biodegradation of dyes, xenobiotics, and pharmaceutical products. Chitin is present in arthropods, cell wall of fungi, nematodes, and in the shell of crab. Chitinases are used in shrimp waste treatment, sea shells waste treatment, plant diseases treatment, and also being used in the field of pharmaceutical. Cellulase is a multienzyme system that includes cellobiase, endoglucanases, and exoglucanases. In nature, a diverse range of microbes express cellulases. Cellulases are being used in the solid waste treatment, sewage and sludge water treatment, agricultural waste treatment, industrial waste treatment, and in the treatment of radioactive compounds.

Keywords Enzymes · Cellulases · Chitinases · Biomass · Waste management

1 Introduction

Organic stuff derived from live plants that is capable of being naturally regenerated or refilled is known as biomass. According to Pang (2016), cellulose, hemicellulose, and lignin are the basic elements that make up the structural makeup of biomass (Pang, 2016). Because biomass waste has a variety of dynamic waste streams and inconsistent data gathering techniques, it is difficult to measure. It is made up of food and lignocellulosic materials, the amount and content of which vary according to temperature, location, and social and economic factors. Organic waste, or

M. Ammar · N. Yousaf · M. Irfan (✉)
Department of Biotechnology, University of Sargodha, Sargodha, Pakistan

M. Franco
Department of Exact Science, State University of Santa Cruz, Ilheus, Brazil

D. B. Pal, S. L. Bansal (eds.), *Fungal Waste Biomass Management for Energy, Environment and Value-Added Products*, Fungal Biology,
https://doi.org/10.1007/978-3-031-82599-6_11

biomass, makes up around half of the world's trash (Kaltschmitt & Janczik, 2015). However, statistics indicate that biomass waste is growing quickly and produces over 140 Gt of trash year worldwide, posing serious disposal and governance challenges (Tripathi et al., 2019).

In recent decades, researchers have placed a strong emphasis on finding a sustainable method to handle the growing amount of biomass waste produced by the agri-food sector (Wancura et al., 2023). Because biomass waste has detrimental effects on the environment, the economy, and public health, its management and conversion are growing in popularity. Recent studies, however, have demonstrated that biomass waste may be economically and efficiently transformed into useful resources, saving money and protecting natural resources (Martin et al., 2023).

The process of transforming biological material from biomass into high-value goods and useful sources of energy is known as biomass conversion. Large amounts of agro-industrial waste are produced annually. The remaining agro-industrial food waste is either landfilled, burned, or composted; only 20% is used for animal feed (Palaniswamy et al., 2023). Nowadays, there is growing interest in utilizing agricultural leftovers to produce butanol in a sustainable manner. This is in reaction to worries that food costs may rise as a result of the rivalry between the production of food and fuel (Qureshi et al., 2014).

The oxidation reaction of a variety of substrates—which includes aromatic substances (ortho- and para-diphenols, methoxy phenols, amino phenols, and polyphenols), aliphatic amine, inorganic cation, and phenolic substances—is catalyzed by laccases, which are enzymes by means of copper molecules at cutting-edge of the dynamic site. Water is produced as a byproduct of molecular oxygen reduction (Dana et al., 2017). Laccases are appropriately referred to be "green catalysts" because of their properties (Arregui et al., 2019).

1.1 Industrial and Environmental Importance of Laccases Enzyme

The laccases are widely used in various industries such as in the synthesis of polymers and green nanoparticles, paper and pulp (delignification reaction, deinking procedure), textile-related (fiber bio bleaching, denim washing), wine evening out, fruitlet sap clearing up, roasting, in addition to sugar beetroot pectin manufacturing, biosensors, and lignocellulosic. Similar to hydrolytic enzymes, laccases are becoming more and more well-liked for the reason that of their astonishing abilities on the road to break down an inclusive range of compounds produced by both industry and agriculture as well as to remediate environmental contaminants and domestic waste (Arregui et al., 2019). Figure 1 shows the different uses of laccases enzyme.

Fig. 1 Different uses of laccases enzyme

1.2 Laccases Enzyme Sources

1.2.1 Plant Laccases

Yoshida identified laccases in the sumac in 1883 and made the first description of them (Singh & Arya 2019). Since then, plants, insects, lichens, bacteria, and fungi have been found to manufacture these enzymes. Vegetation that produces laccases enzyme includes asparagus, pears, apples, turnips, potatoes, cabbages, and others (Dana et al., 2017). Laccases enzyme sequestered from *Saccharum officinarum*, *Oryza sativa*, *Brassica napus*, *Acer psuedoplatanus*, *Liriodendron tulipifera*, *Nicotiana tabacum*, *Populus trichocarpa, Arabidopsis thaliana, Pinus taeda,* and *Brachypodium distachyon,* as well as *Zea mays, Lolium perenne*, have been studied extensively. Laccases play important roles in plant iron metabolism, resistance and shelter, lignification against tissue repair, and environmental stressors (Janusz et al., 2020).

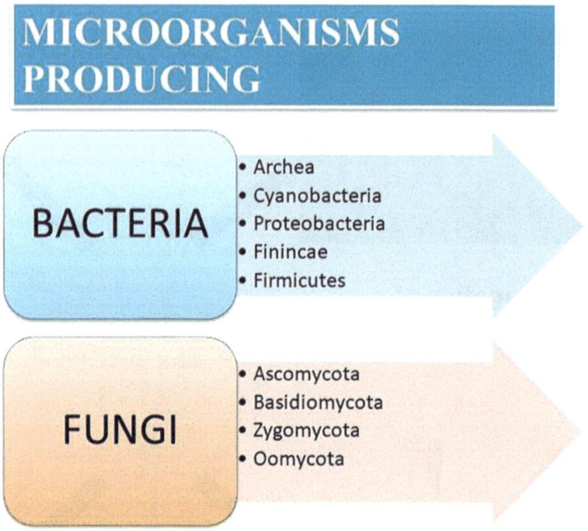

Fig. 2 Phyla of microorganisms producing laccases

1.2.2 Laccases from Bacteria

The members of this phylum comprise the most well-known bacteria that produce laccases. These include Aquificae; Deinococcus-Thermus (*Thermus thermophilus*); Cyanobacteria (*Oscillatoria boryana*); Proteobacteria (*Pseudomonas syringae*, *Sinorhizobium meliloti*, *Escherichia coli*, *Marinomonas mediterranea*, and *Azospirillum lipoferum*); Firmicutes (*Bacillus pumilus*, *Bacillus subtilis*, *Bacillus licheniformis*); Actinobacteria (*Streptomyces griseus*, *Pseudomonas syringae*, *Escherichia coli*); Archaea (*Haloferax volcanii*); and numerous humanoid pathogens (Janusz et al., 2020; Sondhi et al., 2021). The existence arising from a particular substrate induces the creation of bacterial laccases, which can stand either extracellular or intracellular and happen during the accelerated development stage. According to Janusz et al. (2020), bacterial laccases may be involved in sporulation, pigmentation, stress resistance, and homeostasis. Figure 2 shows the phyla of microorganisms producing laccases.

1.2.3 Laccases from Fungi

Similar to their bacterial counterparts, fungal laccases play a role in pigmentation, pathogenicity, sporogenesis, cell defense, and lignocellulosic substrate breakdown. Fungal laccases are mostly synthesized by fungus and are appropriate with the phylum Oomycota, Basidiomycota, Zygomycota, Chytridiomycota, and Ascomycota. They typically have a greater redox potential than bacteria, by means of standards of about 800 mV vs. normal hydrogen electrode (NHE). Fungi that produce laccase are also tangled in the humification procedure and the breakdown of trash. The most well-known fungiform species that produce laccases are the Ascomycetes

(*Trichoderma* sp., *Phlebia radiata*, *Pleurotus ostreatus*, *Lentinula* sp., *Chaetomium thermophilum*, *Trametes versicolor*) and the Basidiomycetes (*Agaricus bisporus*, *Lentinula edodes*, *Ganoderma* sp., *and Theiophora terrestris*) (Wang et al., 2021a, b).

2 Structure and Mechanism of Laccases

2.1 Structure of Laccases

The enzymatic base of the majority of laccases contains four copper atoms that are engaged in redox reactions. Three different structural kinds may be distinguished among these copper atoms:

(i). T1, also known as para-magnetic "blue" copper;
(ii). T2, also known as para-magnetic "non-blue" copper;
(iii). T3, also known as the diamagnetic spin-coupled center (Christopher et al., 2014; Wong, 2009).

The amino acids that these three forms of copper atoms bind to, as well as their electronic paramagnetic resonance (EPR) and ultraviolet and visible absorption spectra vary (Das et al., 2020).

2.2 Mechanism of Action

One of these different mechanisms can be used to explain how the copper cluster from the active site helps the laccases accomplish their function on the substrate. The most straightforward scenario is when an enzyme with the proper redox potential directly catalyzes the oxidation of substrates. In the event that this direct reaction is not possible, a chemical mediator may be used to cause the substrate to oxidize. This system, referred to as the "laccase-mediator-system," is present in biotechnological processes that include the degradation of certain chemicals as well as organic syntheses (Bassanini et al., 2020).

3 Microbial Biosynthesis of Laccases

Laccase-producing bacteria are extracted from a range of ordinary substrates, for instance decaying timber and tree bark, by means of a disperse-plate or smear-plate approach. Screening approaches employ chromogenic substrates, such as guaiacol, to determine which colonies are the most productive. Due to their relatively high cost, lignocellulosic substrates (sawdust, wheat dietary fiber, rice hay, maize fodder, in addition to supplementary cereal surplus) are among the many agricultural by-products that are now used in the production of laccases (Bassanini et al., 2020).

4 Laccases Enzyme Involved in Agro-Waste Treatment

Lignocellulose's material, which comprises agriculture and arboriculture waste, is the utmost copious in addition to reasonably priced replenishable raw material which is likely to be produced sustainably into ethanol globally (Qian, 2014). Many organic, thermodynamic, and biotic approaches are projected to convert Lignocellulosic Biomass into energy (Wang et al., 2021a, b). LB's biodegradability is greatly influenced by its composition, which is mostly prepared by cellulose, hemicellulose, and lignin (Xu and Li, 2017). Fermented carbohydrates of biological polysaccharides, which include both hemicellulose and cellulose, stay confined by the networking configuration of lignocellulose and are necessary for fermentation. Therefore, biomass needs to be processed at all times (Zheng et al., 2017).

Throughout the natural preprocessing procedure, the accomplishment of bacteria, fungus, and enzymes actually breaks down the lignocellulosic biomass. The fungi that are most frequently used in the delignification of biomass include white, brown, and soft-rotting fungus. The most eminent oxidative enzymes produced by white-rot fungi (WRF), which remain in charge of lignin biological degradation, include the laccases enzyme, lignin as well as manganese peroxidase; and out of this, two-lignin peroxidase is the chief enzyme engaged (Gunasekaran et al., 2021). The lignin-derived polymeric polyphenol compound oxidation procedure is accelerated by enzymes known as laccases, which contain copper in their location of activity. The laccase enzyme is essential for the breakdown of lignin and boosts the vintage of the dualistic progressions that follow: hydrolysis, as well as bacterial and yeast fermentation (Fillat et al., 2017).

Yadav and Vivekanand (2020) also conducted a genetic pretreatment of LB using pearl millet straw and wheat. They did this by using a fungal strain of *Curvularia lunata* that they had chosen, and they were able to improve the treatment parameters, which included moisture content (65%, 75%, and 85%), temperature (28 °C, 32 °C, and 36 °C), and treatment period (14, 28, and 42 days). The capability of the mycological straining to create laccases enzyme was evaluated using a colorimetric assay, along with the guaiacol serving similar to the substrate which is intended for the enzymatic reaction. The authors noted an enormous rise in biomethane (23% and 39%) and natural gas production (19% and 28%) for both substrates (flour and millet with pearls straw) when compared to unprocessed samples (Yadav & Vivekanand, 2020).

Fithri et al. (2020) also identified the fungal laccase activity in fungi residing in the empty fruit bunch of oil palms. Fungus laccase was used to investigate the biological management of two distinct waste products of agriculture (rice straw and maize cob). The structure of the waste fibers from cereal cultivation was also examined by the scientists using a scanning electron microscope (SEM) on both treated and untreated samples. After being treated with laccase, rice straw and corn cob waste were shown to have degrading activity by SEM pictures (Fithri et al., 2020).

LB was subjected to a fungal delignification pretreatment in other investigations to produce bioethanol. Salvachua, for instance, treated wheat straw with a combination of a mild alkali treatment and a fungal pretreatment in order to produce ethanol. According to published data, during a 21-day pretreatment period, the glucose fermentation was able to gather 66% to 69% of the cellulose that was accessible within the wheat chaff, producing 62% ethanol in both cases, which included *Irpex lacteus* and *Poria subvermispora* (Das et al., 2020).

Chang et al. (2021) examined the laccases enzyme (*T. versicolor*), which is present in the form of ionic liquescent, in an effort to increase rice straw's capacity to break down lignin and produce saccharification. When laccase + TritonX-100 was employed as a simultaneous pretreatment, the extreme cellulose hydrolysis was about 40.96%; when laccase + 1-allyl-3-methylimidazolium chloride was utilized, it was 38.24%, according to the authors' findings after 72 hours of enzymatic saccharification.

5 Degradation of Dyes

Laccases have demonstrated great importance in commercial and environmental applications recently, primarily in the decolorization of dyes. Studies show that synthetic paints are among the major pollutants in the world today. Younes and Sayadi (2013) demonstrated that after a single treatment day, using laccase enzyme made from *Scytalidium thermophilum*, the enzymatic degradation of indigo carmine dye yielded a decolorization rate of 98%. Moreover, Campos et al. (2001) provided proof for these conclusions. They found that utilizing the laccase that was obtained from the fungus *Trametes hirsuta*, instead of the laccases enzyme that was derived from the *Sclerotium rolfsii* bacterium, led to a more effective indigo degradation process. In a different study, Abadulla et al. (2000) investigated the decoloring and decontamination of fabric colors by using a laccase enzyme derived from *T. hirsuta* which is a fungus strain. They stated that anthraquinone, azo, indigoid, and triarylmethane colors could all be broken down by this laccase. An immobilized laccase treatment can cut dye toxicity by as much as 80%. An investigation by Rui et al. showed that textile dye enzymatic breakdown might be enhanced. The authors suggested a novel application of ionic liquids (ILs) to augment laccase's ability to break down indigo carmine (IC) dye. Figure 3 shows the degradation of dyes using surfactant ionic liquids.

It is well known that laccases may break down synthetic colors found in wastewater to produce fewer toxic byproducts. Studies in the literature have demonstrated that laccases are capable of dissolving the structures of different colors. In the study by Khelifi et al. (2009), indigo dye removal efficiency was found to be 98.6%. The

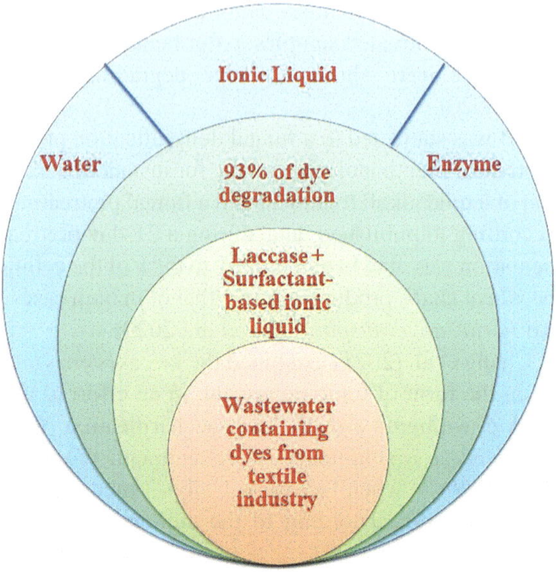

Fig. 3 Degradation of dyes using surfactant ionic liquids

study also utilized laccases' decolorization efficiency. Laccase yielded 90% decolorization for Procyon Red and 100% for Indole dye (Qu et al., 2015).

6 Degradation of Xenobiotic Compounds

The supreme prevalent mock substances are called xenobiotics, or extraneous substances in active custom, and they are used extensively in industry, agriculture, household products, and the environment. Scientists carried out a figure of readings to study the special effects of xenobiotics on the environment. These studies emphasized the special effects of xenobiotics on individuals, faunas, and the environment. Because xenobiotic contamination poses a threat, researchers have looked into a variety of physical and chemical techniques for the treatment of chemicals (which includes filtration, electrolysis, coagulation, ozonation, and precipitation) for the purpose of detoxification and degradation. Nevertheless, not entirely of these techniques remained effective or beneficial because of their high costs and the production of hazardous byproducts (Mishra et al., 2021). Bioremediation, which is actually called as a biological remediation technique that did not record any potentially hazardous compounds, was one approach that showed success. Microorganisms are primarily used in the creation of bioremediation techniques. The exclusion of xenobiotics from water, soil, or sediments results in the production of CO_2 and H_2O, which are environmentally safe byproducts. Research findings show that various types of microorganisms can deteriorate compounds.

6.1 Laccase and Xenobiotics

The enzymes found in laccases are crucial to many industries because they aid in the biodegradation of xenobiotics and the bleaching of paper, among other processes (Ademakinwa, 2021). Their low substrate particularity and capacity to oxidize a variety of substrates give laccase the essential ingredients it needs to carry out xenobiotic biodegradation (Giardina et al., 2010). Even with this, there are still some disadvantages to utilizing laccase because studies have shown that they possess elevated charges, low yields, reduced recyclability, and devastation of enzyme function (Bilal et al., 2019). Scientists examined the use of laccase enzymes in connection with several contaminants. Using fungi that produce ligninolytic extracellular enzymes (Basidiomycetes), more research was done on laccase activity for xenobiotic breakdown. Scientists have shown that white-rot fungi demonstrated greater effectiveness in the breakdown of lignin compared to other laccases found in fungi (Al Farraj et al., 2019). The capacity of fungi to produce laccase activity was continuously monitored and examined in connection to the breakdown of lignin. Therefore, it can be noticed that this depicts a decisive part in the breakdown of substances that are considered xenobiotics, illustrating this polycyclic aromatic hydrocarbon. Polycyclic aromatic hydrocarbons (PAHs) are among these environmental pollutants, regardless of whether we are talking about natural or human-caused activity. The authors of the research examined several strains from various Sarawakian locations that showed excellent laccase activity in culture. Laccase efficiency was maximum in *Coriolopsis caperata* BM-172 (880 U/L) (Al Farraj et al., 2019).

Heavy metals, which can originate from a variety of sectors, are ancillary pollutants that have consequences for the ecosystem and human health (textile, plastics, microelectronics, etc.). In light of this, the authors of the article Zdarta et al. (2022) discuss the effects of lead and nickel, two heavy metals, on the laccase enzyme activity of *Trametes pubescens*. In solutions containing lead along with nickel, the laccase action did not change. They also gave a conclusion that laccase activity decreased with increasing Pb concentration, whereas laccase activity increased with increasing Ni concentration (10–25 ppm).

Certain fungal laccases are frequently employed for breaking down pesticides since they do not create any other potentially dangerous compounds. Research on this topic is still being done, and one of the projects attempts to describe the laccase of *Tricholoma giganteum* AGDR1. It is common knowledge that using pesticides excessively may pollute the environment and affect the ecosystem as a whole. Many people are concerned about attempting to eliminate pesticide residues en route to prevent pollution. Researchers use a variety of techniques, including combining adsorption and degradation using biodegradable microporous laccase immobilized in starch (LA–MPS). Addressed to get satisfactory results, atrazine and the herbicide prometryn were removed from the water using LA–MPS in conjunction with mediators. The elimination recorded was 61% higher than natural degradation, indicating that the suggested strategy may be implemented, according to the data (Chen et al., 2021).

7 Degradation of Pharmaceutical Products

Pharmaceutical items are made of chemicals, either natural or manufactured, that are used to cure and prevent illnesses or to restore the biological functions of humans and other animals. They are used in significant quantities annually—thousands of tons—and comprise more than 4000 distinct compounds (Guardado et al., 2019). Pharmaceutical active compounds (PhACs) are classified as evolving contaminants since these contribute significantly to environmental contamination (soil and water); also, certain PhACs have been found in food and drinking water, which poses a direct risk to public health. Hospital aquatic habitats as well as home and commercial wastewaters include these organic microcontaminants (Guardado et al., 2019).

The most typical organisms that can break down the refractory components of wood (lignin, cellulose, and hemicellulose) are the group known as white-rot fungi (WRF). This is made feasible by ambiguous extracellular enzymatic aggregates, such as laccases enzymes, that are first and foremost released in a peripheral arrangement. In light of this, a large number of researchers have investigated the WRF and demonstrated their capability on the way to cessation of an extensive array of organic molecules, comprising newly discovered contaminants resembling herbicides, industrial chemicals, colorants, and medications.

Suda et al. conducted research on antibiotic degradation in 2012. In their investigations, laccase produced by *Trametes versicolor* was used to break down tetracycline (TC), chlortetracycline (CTC), oxytetracycline (OTC), and doxycycline (DC) along with the addition of 1-hydroxybenzotriazole (HBT), a redox mediator. The findings showed that tetracycline and chlortetracycline could be eliminated in 15 minutes, whereas doxycycline and oxytetracycline required 1 hour.

The study conducted by Guardado et al. (2019) also examined the breakdown of antibiotics by laccases. The report compares the deprivation efficiency of these three laccases on antibiotics (amoxicillin, ciprofloxacin, and sulfamethoxazole): *Myceliophthora thermophila*, *T. versicolor*, and *Pycnoporus sanguineus* CS43. All of these above-mentioned antibiotics failed to break down devoid of the usage of these redox intermediaries, except for amoxicillin, which did so in every instance. There are three redox mediators which include syringaldehyde, p-coumaric acid, and ABTS, they all were used in further studies to examine their effects on the laccase-mediated degradation process. After 3 hours of treatment, syringaldehyde obtained the greatest degradation percentages on amoxicillin, 100% in sulfamethoxazole, and 40% in ciprofloxacin, indicating that it was the best redox mediator. The investigation also found that, in comparison to the new laccase made by *P. sanguineus* CS43, the viable laccase enzyme generated by *M. thermophila* and *T. versicolor* has shown superior catalyst activity.

Aracagök et al. (2018) investigated the biological degradation of diclofenac utilizing raw laccase of *Trametes trogii*. The research revealed that *T. trogii* crude laccase degraded diclofenac at a rate of 97% in 48 hours. The researchers also mentioned that temperature and pH have an impact on the biodegradation process. Therefore, at pH levels of 4.5 and 5, the greatest diclofenac percentage elimination (95% and 97%) was achieved. When the temperature was adjusted between 15 °C and 55 °C at pH 5, the greatest diclofenac elimination (99%) was discovered at 30 °C.

Zdarta et al. (2019) scrutinized different uses of laccases enzyme throughout the breakdown associated with two popular anti-inflammatory drugs, diclofenac and naproxen, in 2019. In this investigation, poly(l-lactic acid)-co-poly(ε-caprolactone) (PLCL) electrospun nanofibers served in the same way as the sustenance for laccase immobilization. These findings demonstrated that when performed in batch mode, both anti-inflammatory substances were removed with an over 90% clearance rate by encapsulated laccase at 25 °C and pH values of 3 for diclofenac and 5 for naproxen. The studies also demonstrated an extraordinary frequency of reuse for the restrained laccase enzyme; following five different cure phases, clearance proportions which are around 40% for diclofenac along with 60% for naproxen remained achieved.

Alharbi et al. (2019) investigated the breakdown of trimethoprim, diclofenac, carbamazepine, and sulfamethoxazole when treated employing laccases enzyme derived from *Trametes versicolor*. Two approaches were utilized to carry out the experiments: either the medications were cast off alone, or a combination stayed prepared utilizing different deliberations (1.25 and 5 mg/L on behalf of each component). When testing each of the four different substances alone as opposed to in mixes, the removal effectiveness of the compounds was greater. Another finding was that, in contrast just before the findings achieved on behalf of complex beginning absorptions, in both the entities and blend techniques, an inferior preliminary deliberation will result in a poorer effectiveness intended for the breakdown of trimethoprim, carbamazepine, and sulfamethoxazole when mixed.

Using laccase, Taheran et al. (2017) looked into the enzymatic breakdown of a few medicinal molecules. On behalf of the tests, the laccases enzymes stood placed immobile upon a polyacrylonitrile–biochar composites membrane which is nanofibrous. Following this, the laccase enzyme structure was employed to degrade diclofenac, carbamazepine, and tetracycline in batch mode. These three pharmacological classes are the most significant ones: antidepressants, antibiotics, and anti-inflammatory medications. Immobilized laccase was used to produce the following degradation rates after the experiments: 72.7% for DCF, 63.3% for CTC, and 48.6% for CBZ.

The research's finding demonstrates how several variables, including substrate type, reaction parameters, and enzyme characteristics, affect these laccases enzymes' ability on the way to trigger the elimination of medicines. Furthermore, the place of origin of the enzymes has an impact on the process performance; this means that varied degradation rates result from biodegrading the same medicine utilizing different sources of enzymes (Taheran et al., 2017).

8 Conclusion

Enzymes play a significant role in biowaste management. Enzyme laccase plays a crucial part during the breakdown of lignin and complex pollutants. During this instance, the benefits are directly related to the correlation between biological treatment administered to physical and organic. This can assist the industry in applying the treatments and biodegradation procedures more effectively. Enzymes laccases remain highly significant in many diverse kinds of businesses, but uncertainty we gaze onward toward the deprivation of xenobiotic compounds, we are able to conclude that there are certain drawbacks which must be revealed, for instance, in elevation rates, reduced reutilization, lack of enzyme action, and little yields. Furthermore, xenobiotics like pharmaceuticals, synthetic dyes, aromatic amines, phenolic compounds, and additional substrates are acted upon by fungal laccases, which transform them into less hazardous molecules.

Chitinases are used in bioremediation, shellfish waste treatment, and insect waste treatment. Chitin is resistant to decay as compared to cellulose and lignin. Chitinases provide degradation of chitin, especially fungus cell wall. Chitinases are involved in bioremediation. As pollution is the international issue, chitinases provide solution to control pollution. Cellulose is the most abundant carbohydrate in nature. Municipal solid waste contains cellulose, hemicellulose, and lignin. MSW contain high ratio of cellulose in addition to variety of microbes. This needs to be treated with different enzymes. Cellulases are the effective tool for the management of municipal solid waste and lignocellulosic biomass. As agricultural waste, food processing waste, and textile industry waste contain high content of cellulose, all these wastes require cellulases for treatment. As we are processing fossil fuels for our daily requirements soon we will face fossil fuels depletion. World is going for alternate option for fossil fuels that is bioethanol. Bioethanol productions require pretreatment of lignocellulosic biomass. So, cellulases are utilized in bioethanol production and is of prior importance in future perspective.

These enzymes can possibly be used for an extensive array of compounds, from agriculture to commerce, because of their low substrate specificity. They are able to expand more in reaction toward alterations in the environmental effect of human activity throughout time. Optimizing the biosynthesis process and the activity of these kinds of enzymes from an insignificant laboratory level to commercial uses is one of the problems that must be taken into account. Furthermore, the prospects in this instance appear to align with the 2030 European Strategy's goal of achieving a supportable atmosphere within the framework of a globular economic system. However, ongoing research into methodologies, technologies, processes, parameters, chemicals, etc., is necessary to identify workable and effective solutions that can be used in any sector and guarantee environmental sustainability. Enzyme laccase has been extensively researched and used as green and ecologically acceptable options for pollution degradation because of the advancements in technology over the past 30 years.

References

Abadulla, E., Tzanov, T., Costa, S., Robra, K. H., Cavaco-Paulo, A., & Gübitz, G. M. (2000). Decolorization and detoxification of textile dyes with a laccase from *Trametes hirsuta*. *Applied and Environmental Microbiology, 66*(8), 3357–3362.

Ademakinwa, A. N. (2021). A heat-resistant intracellular laccase immobilized via cross-linked enzyme aggregate preparation: Characterization, application in bisphenol A removal and phytotoxicity evaluation. *Journal of Hazardous Materials, 419*, 126480.

Alharbi, S. K., Nghiem, L. D., Van De Merwe, J. P., Leusch, F. D., Asif, M. B., Hai, F. I., & Price, W. E. (2019). Degradation of diclofenac, trimethoprim, carbamazepine, and sulfamethoxazole by laccase from *Trametes versicolor*: Transformation products and toxicity of treated effluent. *Biocatalysis and Biotransformation, 37*(6), 399–408.

Al Farraj, D. A., Elshikh, M. S., Al Khulaifi, M. M., Hadibarata, T., Yuniarto, A. & Syafiuddin, A. (2019). Biotransformation and detoxification of antraquione dye green 3 using halophilic Hortaea sp. *International Biodeterioration & Biodegradation, 140*, 72–77.

Aracagök, Y. D., Göker, H., & Cihangir, N. (2018). Biodegradation of diclofenac with fungal strains. *Archives of Environmental Protection, 44*(1), 55–62.

Arregui, L., Ayala, M., Gómez-Gil, X., Gutiérrez-Soto, G., Hernández-Luna, C. E., De Los, H., Santos, M., Levin, L., Rojo-Domínguez, A., Romero-Martínez, D., Saparrat, M. C., & Trujillo-Roldán, M. A. (2019). Laccases: Structure, function, and potential application in water bioremediation. *Microbial Cell Factories, 18*, 1–33.

Bassanini, I., Ferrandi, E. E., Riva, S., & Monti, D. (2020). Biocatalysis with laccases: An updated overview. *Catalysts, 2021*(11), 26.

Bilal, M., Iqbal, H. M., & Barceló, D. (2019). Mitigation of bisphenol A using an array of laccase-based robust bio-catalytic cues–a review. *Science of the Total Environment, 689*, 160–177.

Campos, R., Kandelbauer, A., Robra, K. H., Cavaco-Paulo, A., & Gübitz, G. M. (2001). Indigo degradation with purified laccases from *Trametes hirsuta* and *Sclerotium rolfsii*. *Journal of Biotechnology, 89*(2–3), 131–139.

Chang, K. L., Liu, C. H., Phitsuwan, P., Ratanakhanokchai, K., Lin, Y. C., Dong, C. D., Lin, M. H., & Yang, G. C. (2021). Enhancement of biological pretreatment on rice straw by an ionic liquid or surfactant. *Catalysts, 11*(11), 1274.

Chen, X., Zhou, Q., Liu, F., Peng, Q., & Bian, Y. (2021). Performance and kinetic of pesticide residues removal by microporous starch immobilized laccase in a combined adsorption and biotransformation process. *Environmental Technology & Innovation, 21*, 101235.

Christopher, L. P., Yao, B., & Ji, Y. (2014). Lignin biodegradation with laccase-mediator systems. *Frontiers in Energy Research, 2*, 12.

Dana, M., Bakhshi Khaniki, G., Mokhtarieh, A. A., & Davarpanah, S. J. (2017). Biotechnological and industrial applications of laccase: A review. *Journal Of Applied Biotechnology Reports, 4*(4), 675–679.

Das, R., Liang, Z., Li, G., & An, T. (2020). A non-blue laccase of Bacillus sp. GZB displays manganese-oxidase activity: A study of laccase characterization, Mn (II) oxidation and prediction of Mn (II) oxidation mechanism. *Chemosphere, 252*, 126619.

Fillat, Ú., Ibarra, D., Eugenio, M. E., Moreno, A. D., Tomás-Pejó, E., & Martín-Sampedro, R. (2017). Laccases as a potential tool for the efficient conversion of lignocellulosic biomass: A review. *Fermentation, 3*(2), 17.

Fithri, L., Puspaningsih, N. N. T., Asmarani, O., Dewi, G. D. F., & Arizandy, R. Y. (2020). Characterization of fungal laccase isolated from oil palm empty fruit bunches (OPEFB) and its degradation from the agriculture waste. *Biocatalysis and Agricultural Biotechnology, 27*, 101676.

Giardina, P., Faraco, V., Pezzella, C., Piscitelli, A., Vanhulle, S., & Sannia, G. (2010). Laccases: A never-ending story. *Cellular and Molecular Life Sciences, 67*, 369–385.

Guardado, A. L. P., Belleville, M. P., Alanis, M. D. J. R., Saldivar, R. P. & Sanchez-Marcano, J. (2019). Effect of redox mediators in pharmaceuticals degradation by laccase: A comparative study. *Process Biochemistry, 78*, 123–131.

Gunasekaran, M., Kumar, G., Karthikeyan, O. P., & Varjani, S. (2021). Lignocellulosic biomass as an optimistic feedstock for the production of biofuels as valuable energy source: Techno-economic analysis, environmental impact analysis, breakthrough and perspectives. *Environmental Technology & Innovation, 24*, 102080.

Janusz, G., Pawlik, A., Świderska-Burek, U., Polak, J., Sulej, J., Jarosz-Wilkołazka, A., & Paszczyński, A. (2020). Laccase properties, physiological functions, and evolution. *International Journal of Molecular Sciences, 21*(3), 966.

Kaltschmitt, M., & Janczik, S. (2015). Biomass to power is on the rise globally. *Renewable Energy Focus, 16*(5–6), 174–176.

Khelifi, E., Ayed, L., Bouallagui, H., Touhami, Y., & Hamdi, M. (2009). Effect of nitrogen and carbon sources on Indigo and Congo red decolourization by *Aspergillus alliaceus* strain 121C. *Journal of Hazardous Materials, 163*(2–3), 1056–1062.

Martín, M., Taifouris, M., & Galán, G. (2023). Lignocellulosic biorefineries: A multiscale approach for resource exploitation. *Bioresource Technology, 385*, 129397.

Mishra, S., Lin, Z., Pang, S., Zhang, W., Bhatt, P., & Chen, S. (2021). Recent advanced technologies for the characterization of xenobiotic-degrading microorganisms and microbial communities. *Frontiers in Bioengineering and Biotechnology, 9*, 632059.

Palaniswamy, S., Ashoor, S., Eskasalam, S. R., & Jang, Y. S. (2023). Harnessing lignocellulosic biomass for butanol production through clostridia for sustainable waste management: Recent advances and perspectives. *Frontiers in Bioengineering and Biotechnology, 11*, 1272429.

Pang, S. (2016). Fuel flexible gas production: Biomass, coal and bio-solid wastes. In *Fuel flexible energy generation* (pp. 241–269). Woodhead Publishing.

Qian, E. W. (2014). Pretreatment and saccharification of lignocellulosic biomass. In Tojo, S. and Hirasawa, T., Eds., *Research approaches to sustainable biomass systems* Elsevier, Amsterdam, (pp. 181–204).

Qureshi, N., Singh, V., Liu, S., Ezeji, T. C., Saha, B. C., & Cotta, M. A. (2014). Process integration for simultaneous saccharification, fermentation, and recovery (SSFR): Production of butanol from corn stover using *Clostridium beijerinckii* P260. *Bioresource Technology, 154*, 222–228.

Qu, Y., Shen, E., Ma, Q., Zhang, Z., Liu, Z., Shen, W., Wang, J., Li, D., Li, H. & Zhou, J. (2015). Biodegradation of indole by a newly isolated Cupriavidus sp. *Journal of Environmental Sciences, 34*, 126–132.

Singh, G., & Arya, S.K. (2019). Utility of laccase in pulp and paper industry: A progressive step towards the green technology. *International Journal of Biological Macromolecules, 134*, 1070–1084.

Sondhi, S., Kaur, R., & Madan, J. (2021). Purification and characterization of a novel white highly thermo stable laccase from a novel Bacillus sp. MSK-01 having potential to be used as anticancer agent. *International Journal of Biological Macromolecules, 170*, 232–238.

Suda, T., Hata, T., Kawai, S., Okamura, H., & Nishida, T. (2012). Treatment of tetracycline antibiotics by laccase in the presence of 1-hydroxybenzotriazole. *Bioresource Technology, 103*(1), 498–501.

Taheran, M., Naghdi, M., Brar, S. K., Knystautas, E. J., Verma, M., & Surampalli, R. Y. (2017). Covalent immobilization of laccase onto nanofibrous membrane for degradation of pharmaceutical residues in water. *ACS Sustainable Chemistry & Engineering, 5*(11), 10430–10438.

Tripathi, N., Hills, C. D., Singh, R. S., & Atkinson, C. J. (2019). Biomass waste utilisation in low-carbon products: Harnessing a major potential resource. *npj Climate and Atmospheric Science, 2*(1), 35.

Wancura, J. H., Brondani, M., Vezaro, F. D., Martins-Vieira, J. C., Moreira, B. P., dos Santos, M. S., Abaide, E. R., de Castilhos, F., & Mayer, F. D. (2023). Motivations to produce biofuels from rice bran: An overview involving a recent panorama. *Industrial Crops and Products, 203*, 117170.

Wang, F., Ouyang, D., Zhou, Z., Page, S. J., Liu, D., & Zhao, X. (2021a). Lignocellulosic biomass as sustainable feedstock and materials for power generation and energy storage. *Journal of Energy Chemistry, 57*, 247–280.

Wang, H., Deng, W., Shen, M., Yan, G., Zhao, W., & Yang, Y. (2021b). A laccase Gl-LAC-4 purified from white-rot fungus *Ganoderma lucidum* had a strong ability to degrade and detoxify the alkylphenol pollutants 4-n-octylphenol and 2-phenylphenol. *Journal of Hazardous Materials, 408*, 124775.

Wong, D. W. (2009). Structure and action mechanism of ligninolytic enzymes. *Applied Biochemistry and Biotechnology, 157*, 174–209.

Xu, F., & Li, Y. (2017). Biomass digestion. In: M. A. Abraham (Ed.), *Encyclopedia of sustainable technologies* (pp. 197–204). Elsevier: Amsterdam, The Netherlands.

Yadav, M., & Vivekanand, V. (2020). Biological treatment of lignocellulosic biomass by *Curvularia lunata* for biogas production. *Bioresource Technology, 306*, 123151.

Younes, S. B., & Sayadi, S. (2013). Detoxification of Indigo carmine using a combined treatment via a novel trimeric thermostable laccase and microbial consortium. *Journal of Molecular Catalysis B: Enzymatic, 87*, 62–68.

Zdarta, J., Jankowska, K., Wyszowska, M., Kijeńska-Gawrońska, E., Zgoła-Grześkowiak, A., Pinelo, M., Meyer, A. S., Moszyński, D., & Jesionowski, T. (2019). Robust biodegradation of naproxen and diclofenac by laccase immobilized using electrospun nanofibers with enhanced stability and reusability. *Materials Science and Engineering: C, 103*, 109789.

Zdarta, J., Jesionowski, T., Pinelo, M., Meyer, A. S., Iqbal, H. M., Bilal, M., Nguyen, L. N., & Nghiem, L. D. (2022). Free and immobilized biocatalysts for removing micropollutants from water and wastewater: Recent progress and challenges. *Bioresource Technology, 344*, 126201.

Zheng, Y., Shi, J., Tu, M., & Cheng, Y. S. (2017). Principles and development of lignocellulosic biomass pretreatment for biofuels. *Advances in Bioenergy, 2*, 1–68.

Production of Hemicellulose from Biomass and Agricultural/Food Waste Contains Lignocellulosic Materials

Shraddha Awasthi, Ambneesh Mishra, and Dan Bahadur Pal

Abstract The potential for long-term synthesis of very valuable chemicals and materials has sparked a lot of interest in the valorisation process of lignocellulosic material and agricultural/food waste. Hemicellulose is a complex polymer found in biomass that has great potential for extraction and application in a range of industries. Because of its renewable nature and carbon neutrality, lignocellulosic biomass has garnered a lot of attention in recent years for energy production. There are several methods for converting lignocellulose to gaseous, liquid, and solid fuels, including thermochemical, chemical, and biological processes. This research chapter provides a comprehensive description of the techniques and strategies utilised to manufacture hemicellulose from multiple lignocellulosic sources, including biomass, agriculture, and food waste. Acid hydrolysis, alkaline extraction, steam explosion, microwave-assisted methods, enzyme procedures, etc., are all covered. Furthermore, the classification of hemicellulose-derived chemicals substances and current challenges along with strategies for lignocellulosic biomass valorisation are discussed. The consequences of extracting hemicellulose from biomass are also discussed, with a focus on sustainable and efficient practices. Overall, this study chapter provides insight into the intriguing possibilities of hemicellulose extraction from biomass including food and agricultural waste.

Keywords Hemicellulose · Lignocellulosic biomass valorisation · Agricultural waste · Food waste · Extraction techniques · Classification · Hemicellulose applications

---

S. Awasthi (✉)
Department of Environmental Studies, MLNCE, University of Delhi, Delhi, India

A. Mishra
Faculty of Sciences and Technology, University of Algarve, Faro, Portugal

D. B. Pal
Department of Chemical Engineering, Harcourt Butler Technical University, Kanpur, Uttar Pradesh, India

D. B. Pal, S. L. Bansal (eds.), *Fungal Waste Biomass Management for Energy, Environment and Value-Added Products*, Fungal Biology,
https://doi.org/10.1007/978-3-031-82599-6_12

1 Introduction

With the expansion of businesses and population, fossil fuel resources have been overused for a very long time, which has resulted in ongoing environmental degradation and a growing energy deficit. It now acts as a barrier to the sustained advancement of human society. Thus, the development of new non-polluting energy with affordable costs, safety, and sustainable use is imperative. With a stockpile of over 1800 billion tons, lignocellulosic biomass is the planet's most plentiful renewable resource. It is regarded as a potential replacement for fossil fuels and is equal to 64 billion tons of petroleum. For lignocellulosic biomass to be used as a source of energy, it must be converted. Various techniques have been devised, including pretreatment and biorefinery (Mujtaba et al., 2023). Banu et al. (2021) state that a robust bioeconomy is made possible by the incorporation of lignocellulosic biomass into the circular economy, which also supports a number of other sustainable development goals (SDGs): (a) By producing biofuels like bioethanol, biodiesel, and biogas from lignocellulosic biomass, we can lessen our reliance on fossil fuels and move towards renewable energy sources (SDG 7: Affordable and Clean Energy); (b) according to SDG 9: Industry, Innovation, and Infrastructure, the growth of lignocellulosic biorefineries fosters resource efficiency, innovation in biomass processing technology, and the adoption of sustainable industrial practices; (c) using bio-based substitutes made from lignocellulosic biomass in place of fossil fuel-based goods can help reduce greenhouse gas emissions and lessen the effects of climate change (SDG 13: Climate Action). Rao et al. (2023) investigated that hemicellulose and lignocellulosic biomass have shown a plethora of possible uses as well as difficulties in developing bio-based economies. The vital function that hemicellulose plays in the value-adding of food and agricultural waste has been thoroughly covered in this chapter. This process is becoming more and more recognised as a sustainable substitute for fossil fuels and petroleum-based products.

1.1 Overview of Lignocellulosic Materials

Cellulose, a polymer that is of glucose molecules, serves as the primary structural component of plant cell walls. Because of its crystalline structure, it cannot be hydrolysed, resulting in it challenging to transform into compounds that can be utilised. On the other hand, hemicellulose is an amorphous polymer consisting of various monomers of sugar, including galactose, xylose, mannose, and glucose (Kumari & Singh, 2018). Lignocellulosic compounds are complex biopolymers primarily composed of cellulose, hemicellulose, and lignin. These materials, which comprise the structural components of plants, constitute the planet's most abundant renewable biological resource (Sun & Cheng, 2002). The composition and structure of lignocellulosic biomass might vary depending on the source, but in general, cellulose makes up 40–50%, hemicellulose 20–30%, and lignin 20–30% of the biomass (Kumar et al., 2009). Hemicellulose is more hydrolysable than cellulose, but its heterogeneity also gives it a more complex structure (Nguyen & Trinh, 2022). According to Zhu et al. (2009), complex aromatic polymeric lignin adds stiffness and durability towards oxidative stress and microbial attack. Lignin commodities can be valued and transformed into valuable items like resources, chemicals, and biofuels by breaking down these complex structures into their constituent elements. This method not only adds value to what is often discarded but also reduces greenhouse gas emissions and the need for fossil fuels (Menon & Rao, 2012).

1.2 Importance of Hemicellulose

Because of its versatility and supply, hemicellulose is an essential element in the process of evaluating lignocellulosic biomass. Because hemicellulose is easier to hydrolyse than cellulose, it is a more accessible source of fermentable sugars. These sugars can be fermented into biofuels, as well as bioplastics and other biochemicals, using a range of techniques (Saha, 2003). To extract the hemicellulose from lignin-based biomass, three steps are involved: pretreatment, hydrolysis, and purification. Pretreatment is required to reduce the complex structure of lignocellulosic materials and improve the accessibility of hemicellulose. Numerous pretreatment methods, such as hydrolysis with acid, alkaline extraction, and enzymatic treatment, have been developed to improve the efficacy of hemicellulose extraction (Mosier et al., 2005). The technique of using saturated or diluted acids to dissolve the hemicellulose into its monomers sugars is known as acid hydrolysis. Despite its high efficiency, this process may lead to the breakdown of carbohydrates and the production of inhibitory byproducts (Chandra et al., 2007). Alkaline extraction dissolves hemicellulose in alkaline solutions, facilitating the following precipitation and purification. This method may not be as effective for all types of biomasses, while being less harsh than acid hydrolysis (Xu et al., 2013). In an enzymatic treatment, hemicellulose is specially hydrolysed into its monomer's sugars by hemicellulose enzymes. Even though this method is extremely selective and generates fewer byproducts, it may still be more expensive due to the higher cost of enzymes (Zhang et al., 2007). In the alkaline pulping process, more complete hemicellulose can be obtained, but they are degraded into small molecules and discharged into the black liquor. The alkali method is the most common separation method for hemicellulose; the method not only swells the cellulose but also breaks the connecting bonds (ester bonds, ether bonds) between hemicellulose, lignin, and hydroxy cinnamic acid, and hence dissolves hemicellulose from the cell wall (Laca et al., 2019). Several products can be made by further processing the recovered hemicellulose. Hemicellulose extraction offers significant environmental advantages. By utilising food and agricultural waste, this technique helps to reduce emissions of greenhouse gases and landfill garbage. By lowering dependency on fossil fuels, the use of renewable biomass resources contributes to the development of a circular economy (Klemm et al., 2005). Hemicellulose extraction is economically feasible due to the possibility of producing high-value products from low-cost biomass, which can yield substantial returns on investment (Taherzadeh and Karimi, 2000). Hemicellulose-derived sugars can be fermented to yield butanol, ethanol, and other biofuels utilised in the biofuel industry. According to Ragauskas et al. (2006), these sugars can be used in the bioplastics industry to produce biodegradable polymers such as polylactic acid (PLA) and polyhydroxyalkanoates (PHAs). Furthermore, according to Jiang et al. (2014), hemicellulose can be used to create high-value chemicals, pharmaceuticals, and functional food additives.

1.3 Valorisation of Biomass and Agricultural/Food Waste

The ability of biomass valorisation to create biofuels, a renewable energy source, is one of its main benefits. By lowering greenhouse gas emissions, the production of bioethanol, biogas, and biodiesel from biomass and waste materials not only offers a sustainable substitute for fossil fuels but also aids in the fight against climate change. For example, the production of bioethanol from agricultural residues such as maize stover, rice husk, and sugarcane bagasse has been well researched and shows great potential for widespread application (Sarkar et al., 2012). Moreover, managing organic waste while producing renewable energy is made possible by the anaerobic digestion of food waste to create biogas (Li et al., 2013). Burning biomass and food/agricultural waste is an important strategy in the fight for sustainable development and the circular economy. 'Valorisation', the act of converting waste materials into higher-value products, reduces the negative environmental effects and increases resource efficiency. Examples of lignocellulosic substances that are prevalent in biomass and food and agricultural waste are cellulose, hemicellulose, and lignin. These materials can be transformed into a range of useful products, such as chemicals, biofuels, bioplastics, and functional materials, using a variety of biochemical and thermochemical processes (Cherubini, 2010).

One of the key advantages of biomass valorisation is its capacity to produce biofuels, which are renewable energy source. The generation of ethanol, biogas, and biodiesel from feedstock and waste products not only provides a sustainable replacement for fossil fuels but also contributes to the campaign against climate change by reducing greenhouse gas emissions. For instance, extensive research has demonstrated the substantial opportunity for widespread application in the synthesis of ethanol from biomass residues, such as sugarcane bagasse, rice husk, and maize stover, from agriculture (Sarkar et al., 2012). Furthermore, anaerobic metabolism of food waste to produce biogas makes it feasible to manage organic waste while generating renewable energy (Li et al., 2013). The process of biomass valorisation not only produces energy but also allows for the production of valuable chemicals and goods. Hemicellitate, for example, can be separated and applied to the synthesis of furfural, a base chemical used in the production of resins, chemicals, and biomass fuels, or xylitol, a sugar alcohol used in the food and medicine sectors (Saha, 2003). Moreover, a different component of lignocellulosic biomass, lignin, can be used to make phenolic compounds, which are intermediates in the synthesis of adhesives, polymers, along with additional industrial goods. Valorising food and farm waste has further advantages for garbage management and environmental protection. Reusing trash to make valuable products reduces the need for landfill space and mitigates environmental dangers associated with the disposal of waste, such as leachate formation and methane emissions.

2 Sources of Hemicellulose (Lignocellulosic Materials)

Hemicellulose is an essential part of lignocellulosic biomass and is used to make materials, chemicals, and biofuels. It is a heteropolymer made up of several sugar monomers, such as galactose, glucose, mannose, xylose, and arabinose, making it a desirable feedstock for various industrial uses. The sources of hemicellulose, which are mostly various forms of biomass, are intimately related to the processes of extraction and application. The several biomass sources of hemicellulose are covered in this part, with special attention paid to their availability, content, and prospective applications in industry.

2.1 Biomass Sources

Hemicellium-producing biomass sources can be broadly categorised into four groups: industrial byproducts, forestry residues, dedicated energy crops, and agricultural residues. Each source has particular benefits and difficulties for the extraction and value-adding of hemicellulose.

2.1.1 Agricultural Residues

Hemicellulosic biomass derived from agricultural leftovers is abundant, and thus should be the focus of hemicellulose extraction efforts. Examples of these remnants are corn stover, rice straw, wheat straw, cotton stalks, and sugarcane bagasse. For instance, producing between 100 and 150 million tons annually, corn stover is one of the primary sources of lignocellulosic biomass in the United States (Kim & Dale, 2004). Hemicellulose, primarily xylan, makes up about 18–25% of maize stover. Xylose can be hydrolysed to produce xylitol and bioethanol, among other valuable compounds (Saha, 2003). Similarly, straws consisting of 20–30% dry weight composed of rice and wheat are valuable sources of hemicellulose. Typically found in nations that grow wheat and rice, these residues offer an underutilised resource for the production of biofuels and biochemicals. Sugarcane bagasse, a byproduct of the sugar industry, is another significant source of hemicellulose. The hemicellulose content of 25–35% in sugarcane bagasse has drawn a lot of attention because of its potential to produce bioethanol and be used as a source for furfural and other compounds (Carvalheiro et al., 2008).

2.1.2 Forestry Residues

Forestry wastes are particularly attractive for hemicellulose extraction because they are easily accessible and fairly priced. These materials are often generated as byproducts of the paper and wood industries, providing an affordable and sustainable source of lignin-based biomass. Furthermore, hemicellulose can be extracted from forestry leftovers to reduce waste in the forest and make effective utilisation of natural resources. For example, forestry residues such as sawdust, chips of wood bark, and logging leftovers are rich sources of hemicellulose. Most of these materials are composed of hardwood and softwood species, and each having a different amount of hemicellulose. Hardwood species like poplar, birch, and eucalyptus have higher amounts of xylan-based hemicellulose, which can account for 15–30% of the dry weight (Peng et al., 2012). On the other hand, softwoods are rich in hemicellulose derived from glucomannan, which accounts for 20–25% of biomass on average (Carvalheiro et al., 2008).

2.1.3 Dedicated Energy Crops

Some energy crops, such as hybrid poplar, switchgrass, and miscanthus, are specifically bred to produce bioenergy and biochemicals. These crops stand out because of their high biomass yield, low input requirements, and adaptability to various growth conditions. Miscanthus is a native Asian C4 perennial grass and a high-hemicellulose energy crop. It can yield up to 30 tons of dry biomass per hectare and has a hemicellulose content of 25–28% (Heaton et al., 2008). Woody plants that grow quickly, such as hybrid poplar, are an important source of hemicellulose. A common crop used to produce biofuels and biochemicals is hybrid poplar, which has a 20–25% hemicellulose content (Jiang et al., 2014). The application of particular energy crops to hemicellulose extraction offers advantages in controlled cultivation and increased biomass yield, hence enhancing the effectiveness and resilience of bio-based production techniques. Switchgrass is a perennial grass native to North America that can produce up to 15–20 tons of dry biomass per hectare annually. It is a promising feedstock for the synthesis of biofuel since it contains about 25–30% hemicellulose.

2.1.4 Industrial Byproducts

The majority of hemicellulose is found in industrial byproducts, particularly those made in the paper and food industries. Sludge from pulp and paper mills, peels from fruits and vegetables, and grain waste from brewers are a few examples of these byproducts. For example, 20–30% of spent grains, or the biomass left behind after brewing, are made of hemicellulose (Mussatto et al., 2006). Hydrolysing these grains can produce the sugars that ferment and can be utilised to produce bioethanol, or processing them can result in the production of significant chemicals such as xylitol. Pulp and paper mill sludge, rich in hemicellulose and lignin, is a major industrial byproduct. By converting waste into useful resources, the use of industrial byproducts for hemicellulose extraction is consistent with the concepts of a circular economy. The hemicellulose content of this sludge might vary depending on the source material and the pulping process, although it typically falls between 15% and 25% (Gutiérrez et al., 2009). In addition to providing the chemical industry with a steady source of feedstock, hemicellulose extraction—which is used to value wastewater from pulp and paper mills—also contributes to waste reduction and environmental protection. Fruit and vegetable peels, though often discarded as waste, are an excellent source of hemicellulose. Citrus peels, for instance, have a hemicellulose content of 15–20%. This hemicellulose can be extracted and used to create value-added goods like pectin and bioethanol.

2.2 *Agricultural Waste Sources*

Agricultural waste is a plentiful supply of lignocellulosic biomass that is produced as a result of various farming practices. Materials that are commonly thrown away, such as straw, husks, stalks, and shells, are included in this biomass. Significant amounts of hemicellulose, an important polysaccharide that may be separated and converted into biofuels, biochemicals, and various other high-value products, are found in these wastes. Utilising waste from agriculture for hemicellulose extraction reduces the adverse environmental impacts of waste disposal while also promoting the circular economy.

2.2.1 Rice Straw

Rice straw is one of the most common agricultural wastes, especially in regions like Southeast Asia. There is plenty of space for hemicellulose extraction given that 730 million tons of rice straw are generated annually worldwide. The two main sugars in rice straw, in accordance with Binod et al. (2012), are the xylose and arabinose, whereas the hemicellulose content usually ranges from 20% to 25%. Because rice straw has a high hemicellulose content, it is a suitable feedstock for the synthesis of bioethanol as well as other chemicals with added value. Hemicelloidal materials such as rice straw are usually extracted using alkaline extraction techniques, diluted acid hydrolysis, and steam explosion. These techniques work well to break up the lignin-carbohydrate matrix, which increases the yield of hemicellulose (Agbor et al., 2011). The hemicellulose that has been extracted can be utilised as a raw material for bioplastics or processed further into fermentable sugars for the synthesis of bioethanol.

2.2.2 Wheat Straw

Mostly prevalent in North America and Europe, wheat straw is a byproduct of growing wheat and a significant source of lignocellulosic biomass. According to Zhang et al. (2018), 400 million tons of wheat straw are produced worldwide each year. While wheat straw contains 25–30% hemicellulose, the majority of its sugar content is made up of xylose, according to Zhang et al. (2018). Typically, methods like alkaline extraction and enzymatic hydrolysis are used to extract hemicellitate from wheat straw. The ester bonds created by these processes bind hemicellulose to lignin, making the extraction process more efficient (Sun & Cheng, 2002). Wheat straw hemicellulose can be converted into bioethanol or utilised to create biocomposites and other bio-based goods.

2.2.3 Corn Stover

The leftover leaves, stalks, and cobs from harvesting maize are combined to create corn stover, one of the most prevalent agricultural wastes in the United States. 200 million tons are thought to be produced there annually (Kim & Dale, 2004). According to Mosier et al. (2005), corn stover contains 20–28% hemicellulose, the vast majority of which is made up of xylose. The availability and content of corn stover make it a great option for the extraction of hemicellulose. Hemicellulosic material from corn stover is commonly extracted using pretreatment methods such steam explosion and alkaline extraction. The recovered hemicellulose can be utilised to make bioethanol, a sustainable alternative to fossil fuels, according to Chen et al. (2011).

2.2.4 Sugarcane Bagasse

In regions that are tropical or subtropical, sugarcane bagasse—the fibrous residue that remains following sugarcane juice is extracted—is a typical agricultural waste. According to Pandey et al. (2000), the global production of sugarcane bagasse is projected to be 500 million tons annually. With a hemicellulose content of 25–30%, sugarcane bagasse is a viable feedstock for the production of biofuels and biochemicals. Techniques like acid hydrolysis, steam explosion, and alkaline extraction are used to remove hemicellulose from sugarcane bagasse. By aiding in the dissolution of the lignin-hemicellulose matrix, these pretreatment methods, according to Saha (2003), enhance the synthesis of hemicellulose. According to Pandey et al. (2000), the extracted hemicellulose can be used to create bioethanol, xylitol, and bioplastics.

2.3 Food Waste Sources

Peels, seeds, and pulp from wasted fruits and vegetables make up food waste, which is a significant source of lignocellulosic biomass. Biofuels, bioplastics, and several other goods of high value can be made from these materials by valorising them. Hemicellitate materials are often abundant. By employing waste from food for hemicellulose extraction—which also provides an environmentally friendly source of raw materials—food waste disposal is handled globally (Magama et al., 2022).

2.3.1 Fruit Peels

Due to their high hemicellulose content, fruit peels such as those from oranges, bananas, and apples provide suitable biomass sources for hemicellulose recovery. According to Mandalari et al. (2006), hemicellulose, which mostly consists of xyloglucan and arabinan, comprises 15–25% of orange peels. The method for extraction frequently uses pretreatment methods including steam explosion, enzymatic hydrolysis, or alkaline extraction to release the cellulose molecules (Farhat et al., 2017). Hydrolysing fruit peel hemicellulose yields fermentable sugars that can then be converted to bioethanol or various biochemicals. Furthermore, bioplastics, additives to foods, and other bio-based products can be made from fruit peel hemicellulose, which contributes to sustainability and waste reduction (Farhat et al., 2017).

2.3.2 Vegetable Peels

Vegetable peels such as potatoes, carrots, and onions are good sources of hemicellulose. These peels are often discarded during food processing, leading to substantial waste. For example, potato peels contain 10–20% hemicellulose, that is mostly composed of xyloglucan and arabinogalactan (Kumar et al., 2022). Hemicellulose is frequently extracted from vegetable peels using pretreatment methods such as acid hydrolysis, alkaline extraction, or enzymatic hydrolysis. According to Schieber et al. (2001), the obtained hemicellulose can be used to make biocomposites and biofuels, or biochemical substances. By obtaining hemicellulose from vegetable peels, value-adding them reduces the environmental effect of discarding of food waste and provides a sustainable supply of basic ingredients.

2.3.3 Spent Grains

Spent grains, one of the hemicellulose byproducts of brewing, contains a lot of arabinoxylan. According to Mussatto et al. (2006), brewer's discarded grains comprise 20–30% hemicellulose and are a significant source of lignocellulosic biomass. The pretreatment techniques used to extract hemicellulose through spent grains include the alkaline extraction process, acid hydrolysis, and enzymatic hydrolysis. Several processes, including the manufacture of xylitol, bioethanol, and bio-based goods, can make use of the obtained hemicellulose. Harvesting hemicellulose from leftover grains is a helpful strategy for cutting waste in the brewing industry and encouraging a more environmentally friendly production method (Mussatto et al., 2006).

3 Extraction Techniques

3.1 Acid Hydrolysis

The second most abundant component of lignocellulosic biomass is hemicellulose. To achieve the practical applications of biomass, it is imperative to develop extraction techniques for hemicellulose that are both economical and efficient. Hemicellulose can be hydrolysed with a high conversion rate using diluted acid hydrolysis technology, which is then fermented to make fuel ethanol. Moreover, hemicellulose can be immediately hydrolysed to create compounds like furfural and useful meals like oligosaccharides. A conventional method for hydrolysing hemicellulose is acid hydrolysis technology. On the other hand, issues like proton affinity, active site, and chemical reactivity need to be resolved soon. The method is preferred because it effectively depolymerises hemicellulose, which makes it an essential step in the conversion of biomass into biofuels and other products with added value. Acid hydrolysis is the process of subjecting lignocellulosic biomass to an acidic media for a predetermined amount of time and temperature. Usually, there are two stages to the process: a gentle pre-hydrolysis step and a more vigorous hydrolysis step. The biomass is treated with diluted acid (often sulphuric acid) at comparatively low temperatures (around 100–130 °C) for a brief amount of time during the pre-hydrolysis step. This stage mainly breaks down the hemicellulose portion into certain monomeric sugars and oligosaccharides (Sun & Cheng, 2002). The mild weather reduces the amount of sugars that break down into undesirable byproducts like furfural and hydroxymethylfurfural (HMF), which can impede fermentation operations further down the line (Jiang et al., 2014). The second stage of hydrolysis is characterised by harsher circumstances, including increased acid concentrations and temperatures (usually between 150 and 200 °C). This phase guarantees that hemicellulose fully depolymerises into its component sugars, which are mostly mannose, xylose, and arabinose (Carvalheiro et al., 2008). After that, the sugar solution can be neutralised, concentrated, and used to make additional chemicals or biofuels like bioethanol. Numerous variables affect acid hydrolysis efficiency, such as biomass type, acid concentration, reaction time, and temperature. To maximise hemicellulose yield while reducing sugar breakdown and the production of inhibitory byproducts, these parameters must be optimised (Kumar et al., 2009).

3.2 *Alkaline Extraction*

A common technique for extracting hemicellulose, causing cellulose to swell, rupturing the links between hemicellulose and lignin, and dissolving hemicellulose from the cell wall, is the alkali method, in addition to acid hydrolysis technology. The yield of hemicellulose has been discovered to be influenced by temperature, extraction time, and alkali type and concentration. For instance, Jin et al. (2009) used alkali treatments to remove hemicellulose from the mildly ball-milled cell walls of lignified barley straw and maize stems. It achieved a 94.6% dissolve rate for the original hemicellulose from barley straw and a 96.4% dissolution rate from maize stems. The authors used various alkali solutions to determine the various components. The separation of hemicellulose was enhanced by the high alkali concentration. Three organic solvents (90% neutral dioxane, 80% HCl-containing dioxane, and dimethyl sulfoxide) and an aqueous alkaline solution (8% KOH) were used in 2009 to extract various hemicelluloses from the slightly ball-milled cell walls of lignified barley straw and maize stems (Jin et al., 2009). Next, a comparative analysis was performed on the acquired hemicelluloses. The arabino-(4-O-methyl-D-glucurono) xylans that made up the hemicelluloses that were extracted using 8% KOH showed signs of partially depolymerised polysaccharides that were dissolved by cleaving the polymer glycosidic linkages and saponifying the polymer ester groups. Another method used to extract hemicellulose from wheat straw was cold alkaline extraction, which involved 90 minutes at 40 °C, then ethanol precipitation to separate the hemicellulose (García et al., 2013). The hemicellulose in the raw material was removed in 56.1% and 39.4% of cases, respectively, by the 100 g/L cold alkaline solution and the ethanol precipitation that followed. Hemicelluloses were also extracted from the raw materials by alkaline extraction and direct delignification (Egues et al., 2013). Colourless films were made from the delignified hemicelluloses, and the formability of the films was enhanced by the high Ara/Xyl ratio. Using KOH extraction, Zhao et al. (2017) recently separated hemicelluloses from *Neolamarckia cadamba*'s apical, middle, and basal segments. From the middle and basal segment stems, the researchers isolated (4-O-methyl) glucuronoxylan of the major hemicellulose and trace amounts of high-molecular-weight mannan and xylan. More recently, *Salvia miltiorrhiza* roots were used to separate hemicellulose-based polysaccharides using an 8% NaOH solution (Zhao et al., 2020). The main chain of 4-β-D-Xylp showed excellent inhibitory actions, indicating that physiologically active polysaccharides based on hemicellulose have potential uses in medicine.

3.3 *Enzyme Techniques*

The utilisation of enzyme approaches to extract hemicellulose from lignocellulosic biomass has garnered considerable interest owing to their ability to produce high-grade hemicellulosic products and their environmental friendliness. These approaches provide a more sustainable option to chemical extraction processes by utilising the power of enzymes to hydrolyse hemicellulose alone, without causing severe destruction of other components within the biomass.

Utilising particular enzymes such as xylanases, mannanases, and arabinofuranosidases that target the different glycosidic linkages within the hemicellulose structure is the main method used in the enzymatic extraction of hemicellulose. These enzymes may easily extract the individual sugars from the biomass after dissolving the complex polysaccharides in hemicellulose into their constituent forms (Gírio et al., 2010). To improve enzyme accessibility, lignocellulosic biomass is usually pretreated before the process starts. The lignocellulosic matrix can be broken down by mechanical, chemical, or thermal pretreatment techniques, which increases the accessibility of hemicellulose to enzymatic action (Biely et al., 2016). Once the biomass is adequately pretreated, it is subjected to enzymatic hydrolysis, where the specific enzymes catalyse the breakdown of hemicellulose. To maximise the effectiveness of enzymatic hydrolysis, reaction parameters, including temperature, pH, and enzyme concentration, are carefully tuned. Since most hemicellulolytic enzymes function best at near-neutral pH levels (pH 5–7) and moderate temperatures (30–50 °C), enzymatic reactions are typically carried out at these conditions (Saha, 2003). The kind of biomass and the enzymes utilised will determine how long the reaction takes, but it usually lasts anywhere from a few hours to a few days. Filtration or centrifugation are used to separate the hemicellulosic sugars that are produced during enzymatic hydrolysis from the leftover biomass. After that, these sugars can go through additional processing or purification for a variety of uses, like making biofuels, bioplastics, or functional food additives (Ebringerová & Heinze, 2000).

3.4 Steam Explosion Extraction

High-temperature, high-pressure steam was utilised to extract biomass, creating an 'explosion' effect by instantaneous pressure release to separate the components of biomass. Hemicellulosic polysaccharides from *Lespedeza crytobotrya* stalks were isolated using steam explosion (Wang et al., 2010). Following the steam explosion, the hemicellulosic fractions were visibly deteriorated. Hemicellulose was extracted from slash pine sawdust using a combination of alkaline deresination and steam explosion catalysed by sulphuric acid (Stoffel et al., 2017). According to the data, 90% of the hemicelluloses were extracted at 200 °C for 5 minutes of deresination, in conjunction with a high glucan recovery from the steam explosion. The effect of steam explosion conditions, such as the severity factor and pH of the presoaking medium, on the selective extraction of hemicelluloses from spruce sawdust was examined by Chadni et al. (2019a, b). The impregnation medium was responsible for the great extraction selectivity of high-molecular-weight hemicelluloses, and the maximum yield was observed at a severity factor of 3.65. Mihiretu et al. (2019) created a steam explosion pretreatment to remove xylan-rich biopolymers from aspen wood and sugarcane waste at 176–204 °C for 3–17 minutes. At 204 °C for

10 minutes, it produced the highest xylan yields of 51% and 24% for aspen wood and sugarcane waste, respectively. The authors suggested that a workable biorefinery method for coproducing xylan biopolymers and bioethanol was the steam explosion pretreatment with alkali impregnation of lignocelluloses. Furthermore, it was proposed that a viable biorefinery method for producing hemicellulose with a high yield and efficiency would be the steam explosion pretreatment.

3.5 Ultrasonic-Assisted Extraction

Ultrasonic-assisted extraction, which has been widely employed in the extraction of hemicellulose, has the benefits of high yield, short time, and low temperature due to the cavitation effect of mechanical effect, thermal effect, and chemical effect. By using ultrasonic assistance during the alkali treatment process, the extraction impact of hemicellulose can be enhanced. The first 27.1–28.1% hemicelluloses from wheat straw were extracted as early as 2003 using the ultrasonic-assisted alkali treatment (Sun & Tomkinson, 2003). It was discovered that all of the hemicellulose had xylose as the main sugar, with a high molecular weight and a discernible concentration of 4-O-methyl-D-glucuronic acid (6.0–7.0%). Subsequently, the hemicellulosic fractions of the partially delignified fast-growing poplar wood were separated using the ultrasound-assisted alkali extraction method (Yuan et al., 2010). By utilising an alkali treatment with ultrasonic assistance, 75.5% of the original hemicelluloses were liberated. Following that, it was reported that the hemicelluloses from *Eucalyptus grandis* could be extracted by ultrasonic-assisted alkali treatment with a 5% KOH solution at 50 °C for 3 hours (Xu et al., 2018). An increase in ultrasonic time from 5 to 35 minutes was shown to enhance the production of hemicelluloses from 2.6% to 19.6%. The maximum yield of hemicelluloses (95.2%) was attained after 30 minutes of ultrasonic treatment. The hemicelluloses with somewhat lower molecular weights (74510–66770 g/mol) and higher xylose contents (83.95–84.96%) were produced by the alkali treatment with ultrasonic assistance. The extraction of hemicelluloses from grape pomace at 20 °C was also done with the use of ultrasound assistance (Minjares-Fuentes et al., 2016). The greatest extraction yield of hemicelluloses was attained by the authors, who discovered a strong correlation between the experimental and predicted values. It was proposed that an effective method for extracting hemicellulosic polysaccharides from grape pomace at an industrial scale would be ultrasound assisted extraction.

3.6 Microware-Assisted Extraction

The advantages of microwave-assisted extraction—namely, its speed, high efficiency, affordability, and large yield—have made it a popular method for hemicellulose extraction. For the coproduction of xylan-based biopolymers and bioethanol from aspen wood sawdust and sugarcane waste, microwave-induced pressurised hot water was created (Mihiretu et al., 2017). It produced a monomeric xylose extraction of less than 7% for both feedstocks and a maximum xylan extraction of 66% for aspen wood sawdust and 50% for sugarcane garbage. The selective extraction of high-molecular-weight hemicelluloses from spruce sawdust was also accomplished by microwave-assisted extraction (Chadni et al., 2019a, b). In both cases, the yield of hemicellulose rose as the microwave power was raised. The sawdust impregnation media, which separated high-molecular-mass arabinoglucuronoxylans (Mw ≈ 66 kDa) in the basic medium and high-molecular-mass acetylated galactoglucomannans (Mw = 41 kDa) in water, was responsible for the high extraction selectivity. Hemicelluloses obtained from biomass based on tobacco were also extracted using a microwave-assisted hydrothermal process (Yuan et al., 2019). For the leaf and stem, the maximum yields were 120.33 mg/g and 118.57 mg/g, respectively. At 200 °C, it yielded 105.15 mg/g of hemicellulose without holding. The Mw values of hemicelluloses that precipitated declined as the temperature and holding period increased, from 143.5 to 13.25 kDa.

4 Properties and Composition

4.1 Chemical Composition

The bulk of plant cell walls are composed of complex polysaccharides called hemicelluloses, along with cellulose and lignin. They frequently feature a backbone composed of xylose, mannose, galactose, glucose, and arabinose with diverse replacements such as uronic acids and acetyl groups, though their precise chemical composition varies widely according to the parent plant. According to the review of Rao et al. (2023), the bulk of hemicelluloses are composed of xyloglucans, xylans, glucomannans, and galactoglucomannans. B-(1-4)-linked xylopyranosyl subunits are the primary structural element of xylans, which are abundant in grasses and hardwoods. Conversely, softwoods are more likely to contain glucomannans, which are a combination of glucose and mannose units. Apart from their quantity and presence, these sugars' acetylation and branching patterns determine the hemicellulose's general chemical composition and characteristics.

4.2 Structural Properties

The structural properties of hemicelluloses are critical to their application in various sectors as well as in plant cell walls. Hemicelluloses have an even more flexible structure than cellulose, which has a crystalline structure. This amorphousness results from the random and inconsistent bifurcation of sugar units in the hemicellulose chains. They have less crystallinity than cellulose because to their physical irregularity, which affects how they react with both water and other polysaccharides. Rao et al. (2023) investigate the effects of hemicelluloses' structural properties on their mechanical properties and relationships with the cell wall matrix. For example, the division and substitution patterns affect the extent of bonds of hydrogen and the structure of network development with cellulose and lignin, which all in turn affect the durability and elasticity of the cell.

4.3 Functional Properties

Hemicelluloses have a variety of functional properties that make them valuable for both natural and commercial settings. Among their functional attributes are their ability to dissolve in water and their role as a dietary fiber source. The level of significance of bifurcation and acetylation of hemicelluloses significantly affects their solubility. Soluble hemicelluloses can gel and have thickening qualities that are useful for food and medicine applications. Moreover, the role hemicelluloses play in maintaining the structural strength of plant cell walls and their prospective application in biotechnological processes, such as biofuel generation, is significant. Their relationships with other cell wall constituents like cellulose and lignin also affect their functional properties, which include their ability to act as natural stabilisers and emulsifiers in a range of industrial processes (Rao et al., 2023).

5 Classification of Hemicellulose

According to Mathura et al. (2024), the four major classes of hemicelluloses can be listed:

5.1 Xylans

Homoxylans and heteroxylans are two types of xylans, with β-(1 → 4)-linked xylose residues typically making up their backbone. The backbone of homoxylans is composed of d-xylopyranose residues connected by β(1 → 4) glycosidic bonds. Mainly, homoxylans serve structural purposes. Heteroxylans with a backbone of d-xylopyranose and short carbohydrate branches made up of various sugars are called glucuronoxylans, glucuronoarabinoxylans, and complicated heteroxylans.

5.2 *Mannans*

Mannans are categorised into two groups according to the presence or absence of hexoses in their chain: galactomannans and glucomannans. The backbone of galactomannans solely contains β-(1 → 4) connected d-mannopyranose residues. The primary backbone of glucomannans contains both β-(1 → 4) linked d-mannopyranose and β-(1 → 4) linked d-glucopyranose residues. For both kinds, the branches contain β-(1 → 6) connected d-galactopyranose residues.

5.3 *Xyloglucans*

β(1 → 4)-linked d-glucose makes up the backbone of xyloglucans, which is similar to cellulose except that α-d-xylose replaces three of the four glucose residues at position 6. G is an unbranched Glc residue; X is α-d-Xyl-(1 → 6)-Glc; L is β-Gal; S is α-l-Araf; and F is α-l-Fuc, these are the various sorts of side chains. XXXG and XXGG are the two most prevalent forms of xyloglucans found in plant cell walls (X = xylopyranose; G = glucopyranose).

5.4 *Mixed Linkage β-Glucans*

A single β-(1 → 3) d-glucopyranose residue separates strings of consecutive β-(1 → 4) d-glucopyranose residues in mixed linkage β-glucans. β-(1 → 4) and β-(1 → 3) glucopyranoses have relative abundances of roughly 70% and 30%, respectively. Mostly, the segments of cellotriosyl and cellotetrasyl are arranged in an arbitrary order to form these glucans.

6 Current Strategies and Challenges in Lignocellulosic Biomass Valorisation

6.1 *From an Oil/Fossil-Based Society to a Biobased Society*

The world's population is expanding, which has led to an intensive use of land and natural resources for the production of food, commodities, and energy. While this has raised living standards in industrialised nations to levels never seen before, it has also had negative effects on the environment and public health. Because of this disruption to the planet's natural equilibrium, we are seeing things like diseases, famine, drought, biodiversity loss, and climate change. The use of petroleum and other non-renewable, finite fossil resources by modern society for energy, chemicals, and materials poses a hazard to the environment. Oil demand, particularly for energy for transportation, is expected to rise in 2024 despite efforts to reduce consumption and use renewable resources, according to a Deloitte analysis (Oil & Gas Outlook, 2024). According to the World Meteorological Organisation (WMO), there is a 66% chance that between 2023 and 2027, the yearly average near-surface global temperature will be more than 1.5 °C above pre-industrial levels due to global warming. Biomass is a sustainable alternative with benefits for the ecology and economy. It includes biodiversity, lignocellulosic biomass, and residual biomass. The amount of biomass wastes from agriculture, forestry, the agroindustry, and the food business is rising along with the global population and the need for commodities, energy, and food. There is a growing recognition that biomass wastes have potential as a resource for the future generation of biofuels, biochemicals, and biomaterials. Bioresources would help save financial resources by reducing reliance on oil while also providing the advantage of proper waste disposal (Awasthi et al., 2020).

6.2 *Food Supply Chain Waste (FSCW) Valorisation*

Human survival requires access to food, water, sanitary conditions, and shelter; yet, low-income societies, particularly those in developing nations, often struggle to provide these necessities. On the other hand, in middle-class and wealthy nations, the prevalence of nutrition-related illnesses like diabetes and obesity is rising. Research shows that food processing, which uses a lot of refined, energy-dense chemicals with no nutritional value, is to blame for these metabolic illnesses. Food supply chain waste (FSCW) is becoming a problem for both food security and the growth of a sustainable bioeconomy because the food processing industries are energy and water intensive and still inefficient (Van Der Goot et al., 2016). As a result, there is an urgent need to change behaviour towards the FSCW towards its valorisation and utilisation. This will make the carbon footprint across the FSC an eco-friendly, more efficient, and sustainable process. Approximately one-third of all food produced worldwide is wasted in the food value chain. Food leftovers should be redesigned, reused, recovered, and recycled using cutting-edge, environmentally friendly technology as the main emphasis of efficient and advanced waste management disposal (Mathura et al., 2024).

7 Conclusion

The chapter has discussed a wide range of hemicellulose sources, including food waste materials such as fruit and vegetable peels and spent grains, as well as agricultural leftovers such as wheat, corn, rice husk, and sugarcane bagasse. These sources provide a sustainable and economically viable feedstock for hemicellulose extraction and usage, offering opportunities to reduce waste and improve sustainability. Although hemicellulose extraction from lignocellulosic biomass has been extensively researched, greater attention must be devoted to how extraction and modification parameters influence the properties of the isolated hemicelluloses and their modified derivatives. To develop simple, efficient, and eco-friendly processes, it is important to combine the advantages of various extraction techniques to effectively separate lignocellulosic biomass from hemicelluloses. This can be achieved by limiting degradation, preserving structural integrity, enhancing purity, and increasing efficiency. Furthermore, it is anticipated that universal pretreatment techniques would advance in relation to the intricate structure and source-dependent composition of hemicellulose.

The chapter finishes by emphasising the importance of lignocellulosic biomass and hemicellulose as resources in the effort to build a sustainable and bio-based economy. Hemicellulose, a renewable resource with high biodegradability, bioactivity, and biocompatibility, ranks second only to cellulose in lignocellulosic biomass. The process of isolating superior hemicellulose from lignocellulosic biomass and then modifying the separated hemicellulose encourages the expansion of its uses. We can realise the full potential of these resources by improving extraction technology, refining valorisation tactics, and tackling issues with waste management and resource efficiency. It is important that forthcoming studies persist in investigating inventive resolutions and improving current techniques in order to augment the sustainability and financial feasibility of hemicellulose utilisation.

References

Agbor, V. B., Cicek, N., Sparling, R., Berlin, A., & Levin, D. B. (2011). Biomass pretreatment: Fundamentals toward application. *Biotechnology Advances, 29*(6), 675–685. https://doi.org/10.1016/j.biotechadv.2011.05.005

Awasthi, M. K., Sarsaiya, S., Patel, A., Juneja, A., Singh, R. P., Yan, B., Awasthi, S. K., Jain, A., Liu, T., Duan, Y., Pandey, A., Zhang, Z., & Taherzadeh, M. J. (2020). Refining biomass residues for sustainable energy and bio-products: An assessment of technology, its importance, and strategic applications in circular bio-economy. *Renewable and Sustainable Energy Reviews, 127*, 109876. https://doi.org/10.1016/j.rser.2020.109876

Banu, J. R., Preethi, N., Kavitha, S., Tyagi, V. K., Gunasekaran, M., Karthikeyan, O. P., & Kumar, G. (2021). Lignocellulosic biomass based biorefinery: A successful platform towards circular bioeconomy. *Fuel, 302*, 121086. https://doi.org/10.1016/j.fuel.2021.121086

Biely, P., Singh, S., & Puchart, V. (2016). Towards enzymatic breakdown of complex plant xylan structures: state of the art. *Biotechnology Advances, 34*(7), 1260–1274. https://doi.org/10.1016/j.biotechadv.2016.09.001

Binod, P., Kuttiraja, M., Archana, M., Janu, K. U., Sukumaran, R. K., & Pandey, A. (2012). High-temperature pretreatment and hydrolysis of rice straw for the production of ethanol. *Biomass and Bioenergy, 38*, 68–78. https://doi.org/10.1016/j.biombioe.2011.05.005

Carvalheiro, F., Duarte, L. C., & Gírio, F. M. (2008). Hemicellulose biorefineries: A review on biomass pretreatments. *Journal of Scientific & Industrial Research, 67*(11), 849–864. https://nopr.niscair.res.in/handle/123456789/2858

Chadni, M., Bals, O., Ziegler-Devin, I., Brosse, N., & Grimi, N. (2019a). Microwave-assisted extraction of high-molecular-weight hemicelluloses from spruce wood. *C R Chimie, 22*(8), 574–584.

Chadni, M., Grimi, N., Bals, O., Ziegler-Devin, I., & Brosse, N. (2019b). Steam explosion process for the selective extraction of hemicelluloses polymers from spruce sawdust. *Industrial Crops and Products, 140*, 111757. https://doi.org/10.1016/j.indcrop.2019.111757

Chandra, R., Bura, R., Mabee, W., Berlin, A., Pan, X., & Saddler, J. (2007). Substrate pretreatment: The key to effective enzymatic hydrolysis of lignocellulosics? *Advances in Biochemical Engineering/Biotechnology, 108*, 67–93. https://doi.org/10.1007/10_2007_064

Chen, M. L., Yang, D. J., & Liu, S. C. (2011). Effects of drying temperature on the flavonoid, phenolic acid and antioxidative capacities of the methanol extract of citrus fruit (Citrus sinensis (L.) Osbeck) peels. *International Journal of Food Science & Technology, 46*(6), 1179–1185. https://doi.org/10.1111/j.1365-2621.2011.02605.x

Cherubini, F. (2010). The biorefinery concept: Using biomass instead of oil for producing energy and chemicals. *Energy Conversion and Management, 51*(7), 1412–1421. https://doi.org/10.1016/j.enconman.2010.01.015

Ebringerová, A., & Heinze, T. (2000). Xylan and xylan derivatives–biopolymers with valuable properties, 1. Naturally occurring xylans structures, isolation procedures and properties. *Macromolecular Rapid Communications, 21*(9), 542–556. https://doi.org/10.1002/1521-3927(20000701)21:9<542::AID-MARC542>3.0.CO;2-7

Egues, I., Eceiza, A., & Labidi, J. (2013). Effect of different hemicelluloses characteristics on film forming properties. *Industrial Crops and Products, 47*, 331–338.

Farhat, W., Venditti, R., Quick, A., Taha, M., Mignard, N., Becquart, F., & Ayoub, A. (2017). Hemicellulose extraction and characterization for applications in paper coatings and adhesives. *Industrial Crops and Products, 107*, 370–377. https://doi.org/10.1016/j.indcrop.2017.05.055

García, J. C., Díaz, M. J., Garcia, M. T., & López, F. (2013). Search for optimum conditions of wheat straw hemicelluloses cold alkaline extraction process. *Biochemical Engineering Journal, 71*(1), 127–133.

Gírio, F. M., Fonseca, C., Carvalheiro, F., Duarte, L. C., Marques, S., & Bogel-Łukasik, R. (2010). Hemicelluloses for fuel ethanol: A review. *Bioresource Technology, 101*(13), 4775–4800. https://doi.org/10.1016/j.biortech.2010.01.088

Gutiérrez, A., Del Río, J. C., & Martínez, A. T. (2009). Microbial and enzymatic control of pitch in the pulp and paper industry. *Applied Microbiology and Biotechnology, 82*(6), 1005–1018. https://doi.org/10.1007/s00253-009-1905-z

Heaton, E. A., Dohleman, F. G., & Long, S. P. (2008). Meeting US biofuel goals with less land: The potential of Miscanthus. *Global Change Biology, 14*(9), 2000–2014. https://doi.org/10.1111/j.1365-2486.2008.01662.x

Jiang, H., Chen, Q., Ge, J., & Zhang, Y. (2014). Efficient extraction and characterization of polymeric hemicelluloses from hybrid poplar. *Carbohydrate Polymers, 101*, 1005–1012. https://doi.org/10.1016/j.carbpol.2013.10.030

Jin, A. X., Ren, J. L., Peng, F., Xu, F., Zhou, G. Y., Sun, R. C., & Kennedy, J. F. (2009). Comparative characterization of degraded and non-degradative hemicelluloses from barley straw and maize stems: Composition, structure, and thermal properties. *Carbohydrate Polymers, 78*, 609–619.

Kim, S., & Dale, B. E. (2004). Global potential bioethanol production from wasted crops and crop residues. *Biomass and Bioenergy, 26*(4), 361–375. https://doi.org/10.1016/j.biombioe.2003.08.002

Klemm, D., Heublein, B., Fink, H.-P., & Bohn, A. (2005). Cellulose: Fascinating biopolymer and sustainable raw material. *Angewandte Chemie International Edition, 44*(22), 3358–3393. https://doi.org/10.1002/anie.200460587

Kumar, P., Barrett, D. M., Delwiche, M. J., & Stroeve, P. (2009). Methods for pretreatment of lignocellulosic biomass for efficient hydrolysis and biofuel production. *Industrial & Engineering Chemistry Research, 48*(8), 3713–3729. https://doi.org/10.1021/ie801542g

Kumar, H., Bhardwaj, K., Dhanjal, D. S., Sharma, R., Nepovimova, E., Verma, R., Kumar, D., & Kuča, K. (2022). Fruit and vegetable peel waste: Applications in food and environmental industries. In *Fruits and vegetable wastes* (pp. 259–287). https://doi.org/10.1007/978-981-16-9527-8_11

Kumari, D., & Singh, R. (2018). Pretreatment of lignocellulosic wastes for biofuel production: A critical review. *Renewable and Sustainable Energy Reviews, 90*, 877–891. https://doi.org/10.1016/j.rser.2018.03.111

Laca, A., Laca, A., & Díaz, M. (2019). Hydrolysis: From cellulose and hemicellulose to simple sugars. In *Elsevier eBooks* (pp. 213–240). https://doi.org/10.1016/b978-0-12-815162-4.00008-2

Li, Y., Park, S. Y., & Zhu, J. (2013). Solid-state anaerobic digestion for methane production from organic waste. *Renewable and Sustainable Energy Reviews, 15*(1), 821–826. https://doi.org/10.1016/j.rser.2010.07.042

Magama, P., Chiyanzu, I., & Mulopo, J. (2022). A systematic review of sustainable fruit and vegetable waste recycling alternatives and possibilities for anaerobic biorefinery. *Bioresource Technology Reports, 18*, 101031. https://doi.org/10.1016/j.biteb.2022.101031

Mandalari, G., Faulds, C. B., Sancho, A. I., Saija, A., Bisignano, G., Lo Curto, R. B., Waldron, K. W., & Rich, G. T. (2006). Fractionation and characterization of arabinoxylans from brewers' spent grain. *Journal of Agricultural and Food Chemistry, 54*(16), 5812–5818. https://doi.org/10.1021/jf061086z

Mathura, S. R., Landázuri, A. C., Mathura, F., Sosa, A. G. A., & Orejuela-Escobar, L. M. (2024). Hemicelluloses from bioresidues and their applications in the food industry—Towards an advanced bioeconomy and a sustainable global value chain of chemicals and materials. *Sustainable Food Technology., 2*, 1183. https://doi.org/10.1039/d4fb00035h

Menon, V., & Rao, M. (2012). Trends in bioconversion of lignocellulose: Biofuels, platform chemicals & biorefinery concept. *Progress in Energy and Combustion Science, 38*(4), 522–550. https://doi.org/10.1016/j.pecs.2012.02.002

Mihiretu, G. T., Brodin, M., Chimphango, A. F., Øyaas, K., Hoff, B. H., & Görgens, J. F. (2017). Single-step microwave-assisted hot water extraction of hemicelluloses from selected lignocellulosic materials—A biorefinery approach. *Bioresource Technology, 241*, 669–680.

Mihiretu, G. T., Chimphango, A. F., & Görgens, J. F. (2019). Steam explosion pre-treatment of alkali-impregnated lignocelluloses for hemicelluloses extraction and improved digestibility. *Bioresource Technology, 285*, 122121. https://doi.org/10.1016/j.biortech.2019.122121

Minjares-Fuentes, R., Femenia, A., Garau, M. C., Candelas-Cadillo, M. G., Simal, S., & Rosselló, C. (2016). Ultrasound-assisted extraction of hemicelluloses from grape pomace using response surface methodology. *Carbohydrate Polymers, 138*, 180–191.

Mosier, N., Wyman, C., Dale, B., Elander, R., Lee, Y. Y., Holtzapple, M., & Ladisch, M. (2005). Features of promising technologies for pretreatment of lignocellulosic biomass. *Bioresource Technology, 96*(6), 673–686. https://doi.org/10.1016/j.biortech.2004.06.025

Mujtaba, M., Fraceto, L. F., Fazeli, M., Mukherjee, S., Savassa, S. M., De Medeiros, G. A., Pereira, A. D. E. S., Mancini, S. D., Lipponen, J., & Vilaplana, F. (2023). Lignocellulosic biomass from agricultural waste to the circular economy: A review with focus on biofuels, biocomposites and bioplastics. *Journal of Cleaner Production, 402*, 136815. https://doi.org/10.1016/j.jclepro.2023.136815

Mussatto, S., Dragone, G., & Roberto, I. (2006). Brewers' spent grain: Generation, characteristics and potential applications. *Journal of Cereal Science, 43*(1), 1–14. https://doi.org/10.1016/j.jcs.2005.06.001

Nguyen, A. Q., & Trinh, L. T. P. (2022). Thermochemical conversion of cellulose and hemicellulose. In N. P. Nghiem, T. H. Kim, & C. G. Yoo (Eds.), *Biomass utilization: Conversion strategies*. Springer. https://doi.org/10.1007/978-3-031-05835-6_6

Oil and gas industry outlook. (2024). https://www2.deloitte.com/us/en/insights/industry/oil-and--gas/oil-and-gas-industry-outlook.html. Accessed 15th August 2024.

Pandey, A., Soccol, C. R., Nigam, P., & Soccol, V. T. (2000). Biotechnological potential of agro-industrial residues. *Bioresource Technology, 74*(1), 69–80. https://doi.org/10.1016/S0960-8524(99)00142-X

Peng, F., Peng, P., Xu, F., & Sun, R. C. (2012). Fractional purification and bioconversion of hemicelluloses. *Biotechnology Advances, 30*(4), 879–903. https://doi.org/10.1016/j.biotechadv.2012.01.018

Ragauskas, A. J., Williams, C. K., Davison, B. H., Britovsek, G., Cairney, J., Eckert, C. A., et al. (2006). The path forward for biofuels and biomaterials. *Science, 311*(5760), 484–489. https://doi.org/10.1126/science.1114736

Rao, J., Lv, Z., Chen, G., & Peng, F. (2023). Hemicellulose: Structure, chemical modification, and application. *Progress in Polymer Science, 140*, 101675. https://doi.org/10.1016/j.progpolymsci.2023.101675

Saha, B. C. (2003). Hemicellulose bioconversion. *Journal of Industrial Microbiology and Biotechnology, 30*(5), 279–291. https://doi.org/10.1007/s10295-003-0049-x

Sarkar, N., Ghosh, S. K., Bannerjee, S., & Aikat, K. (2012). Bioethanol production from agricultural wastes: An overview. *Renewable Energy, 37*(1), 19–27. https://doi.org/10.1016/j.renene.2011.06.045

Schieber, A., Stintzing, F. C., & Carle, R. (2001). By products of plant food processing as a source of functional compounds—Recent developments. *Trends in Food Science & Technology, 12*(11), 401–413. https://doi.org/10.1016/S0924-2244(02)00012-2

Stoffel, R. B., Neves, P. V., Felissia, F. E., Ramos, L. P., Gassa, L. M., & Area, M. C. (2017). Hemicellulose extraction from slash pine sawdust by steam explosion with sulfuric acid. *Biomass and Bioenergy, 107*, 93–101.

Sun, R. C., & Tomkinson, J. (2003). Characterization of hemicelluloses isolated with tetraacetylethylenediamine activated peroxide from ultrasound irradiated and alkali pre-treated wheat straw. *European Polymer Journal, 39*, 751–759.

Sun, Y., & Cheng, J. (2002). Hydrolysis of lignocellulosic materials for ethanol production: A review. *Bioresource Technology, 83*(1), 1–11. https://doi.org/10.1016/s0960-8524(01)00212-7

Van Der Goot, A. J., Pelgrom, P. J., Berghout, J. A., Geerts, M. E., Jankowiak, L., Hardt, N. A., Keijer, J., Schutyser, M. A., Nikiforidis, C. V., & Boom, R. M. (2016). Concepts for further sustainable production of foods. *Journal of Food Engineering, 168*, 42–51. https://doi.org/10.1016/j.jfoodeng.2015.07.010

Wang, K., Xu, F., Sun, R. C., & Jones, G. L. (2010). Influence of incubation time on the physicochemical properties of the isolated hemicelluloses from steam-exploded lespedeza stalks. *Industrial & Engineering Chemistry Research, 49*(18), 8797–8804.

Xu, F., Yu, J., Tesso, T., Dowell, F., & Wang, D. (2013). Qualitative and quantitative analysis of lignocellulosic biomass using infrared techniques: A mini-review. *Applied Energy, 104*, 801–809. https://doi.org/10.1016/j.apenergy.2012.12.019

Xu, J. Y., Yuan, T. Q., Xiao, L., & Sun, R. C. (2018). Effect of ultrasonic time on the structural and physicochemical properties of hemicelluloses from Eucalyptus grandis. *Carbohydrate Polymers, 195*, 114–119.

Yuan, T. Q., Xu, F., He, J., & Sun, R. C. (2010). Structural and physicochemical characterization of hemicelluloses from ultrasound-assisted extractions of partially delignified fast-growing poplar wood through organic solvent and alkaline solutions. *Biotechnology Advances, 28*(5), 583–593.

Yuan, Y., Zou, P., Zhou, J. H., & Zhang, C. (2019). Microwave-assisted hydrothermal extraction of non-structural carbohydrates and hemicelluloses from tobacco biomass. *Carbohydrate Polymers, 212*, 115043. https://doi.org/10.1016/j.carbpol.2019.115043

Zhang, Y. P., Ding, S., Mielenz, J. R., Cui, J., Elander, R. T., Laser, M., Himmel, M. E., McMillan, J. R., & Lynd, L. R. (2007). Fractionating recalcitrant lignocellulose at modest reaction conditions. *Biotechnology and Bioengineering, 97*(2), 214–223. https://doi.org/10.1002/bit.21386

Zhang, C., Ho, S. H., Chen, W. H., Xie, Y., Liu, Z., & Chang, J. S. (2018). Torrefaction performance and energy usage of biomass wastes and their correlations with torrefaction severity index. *Applied Energy, 220*, 598–604.

Zhao, K., Li, B., He, D. M., Zhao, C., Shi, Z., Dong, B., Pan, D., Patil, R. R., Yan, Z., & Guo, Z. (2020). Chemical characteristic and bioactivity of hemicellulose-based polysaccharides isolated from salvia miltiorrhiza. *International Journal of Biological Macromolecules, 165*, 2475–2483.

Zhao, X. H., Tong, T. T., Li, H. L., Lu, H., Ren, J., Zhang, A., Deng, X., Chen, X., & Wu, A. M. (2017). Characterization of hemicelluloses from *Neolamarckia cadamba* (Rubiaceae) during xylogenesis. *Carbohydrate Polymers, 156*, 333–339.

Zhu, J., Pan, X., Wang, G., & Gleisner, R. (2009). Sulfite pretreatment (SPORL) for robust enzymatic saccharification of spruce and red pine. *Bioresource Technology, 100*(8), 2411–2418. https://doi.org/10.1016/j.biortech.2008.10.057

Extraction and Purification of Essential Oils from Food Waste-Derived Waste Biomass

Praveen Kumar, Ravikumar Harikrishnan, Velayutham Jeevakanth, Ashish Kapoor, Dan Bahadur Pal, and Muthukumar Raghunathan

Abstract The chapter delves into innovative methods for the extraction and purification of essential oils from food waste biomass, highlighting a sustainable approach to valorizing agricultural and food industry byproducts. Food waste, a significant environmental and economic issue, offers untapped potential as a renewable resource for essential oils, which are valuable in pharmaceuticals, health care, cosmetics, and food industries. A comprehensive overview of various extraction techniques, including steam distillation, solvent extraction, microwave-assisted extraction, and supercritical fluid extraction, emphasizing their efficiency, scalability, and environmental impact, is provided. The purification processes, crucial for achieving high-quality essential oils, are explored with a focus on distillation, chromatography, and membrane separation technologies. Advanced techniques such as green solvents and biocatalysis are discussed for their roles in enhancing the purity and yield of essential oils while minimizing ecological footprints. The optimization of extraction parameters and the incorporation of these processes into current waste management systems, promoting a circular economy, are also discussed. Case studies illustrate successful applications and economic feasibility, reinforcing the practicality of these methods in industrial settings. By providing a comprehensive overview of current advancements and future prospects, this chapter aims to inspire further research and industrial adoption, contributing to sustainable development goals. The chapter concludes with a discussion on the regulatory, safety, and market aspects that influence the commercialization of essential oils derived from food waste, advocating for policy frameworks that support sustainable practices.

P. Kumar · R. Harikrishnan · V. Jeevakanth · M. Raghunathan (✉)
Department of Biotechnology, School of Bioengineering, SRM Institute of Science and Technology,
Chennai, Tamil Nadu, India
e-mail: muthukur@srmist.edu.in

A. Kapoor · D. B. Pal
Department of Chemical Engineering, Harcourt Butler Technical University, Kanpur, Uttar Pradesh, India

D. B. Pal, S. L. Bansal (eds.), *Fungal Waste Biomass Management for Energy, Environment and Value-Added Products*, Fungal Biology,
https://doi.org/10.1007/978-3-031-82599-6_13

Keywords Essential oils · Food waste · Extraction techniques · Purification processes

1 Introduction

Essential oils (EOs) are liquid extracts generated from diverse components of medicinal or herbal plants. They find significant usage in a range of fields. Concerns about the harmful effects of synthetic alternatives have led to a surge in the use of these chemicals as food preservatives and antimicrobials (Pandey et al., 2017). In order to attack bacteria, fungi, insects, nematodes, weeds, and mites, these natural oils have been employed as environmentally benign pesticides (Batish et al., 2008). Essential oils have demonstrated as one of the promising antimicrobial preservatives in the food sector, notably in the cheese-making process, where they efficiently suppress conventional spoilage and pathogenic microorganisms (Khorshidian et al., 2018). Purification of essential oils is necessary for industrial usages. The processing of essential oils for taste applications requires a range of separation methods, stressing the relevance of distillation principles and alternative processing procedures (Berdis et al., 2003). Study has been done on the use of nanoencapsulation technologies to protect food bioactive elements—such as essential oils—from harmful conditions and ensure the safe delivery of these chemicals to targeted organs and cells (Assadpour & Mahdi Jafari, 2019). Research has focused on extracting essential oils from forest waste, namely from plants in the Pinaceae and Cupressaceae families. The emphasis has been on studying the bioactive components of these oils and their potential uses (Park et al., 2023).

Essential oils are employed in the rubber business in addition to the food industry. It has been discovered that scanning transmission X-ray microscopy (STXM) is a beneficial approach for exploring multicomponent elastomeric systems, highlighting the numerous uses of essential oils in many sectors. Furthermore, research has been undertaken on the influence of edible oils on cardiovascular illnesses, with an emphasis on the nutritional advantages and their economic implications in the food preparation business (Soundararajan et al., 2023). Extraction and purification of essential oils from food waste biomass present current challenges and advancements in the field of biorefineries and waste valorization. Essential oils, rich in bioactive compounds, are valuable byproducts that can be obtained from different sources such as agriculture food waste and citrus peel waste. These oils include essential fatty acids like linoleic acid and oleic acid, making them nutritionally significant for various applications in the food, cosmetic, and biofuel industries (Bhatti et al., 2022; Ochsenreither et al., 2016). Recent developments in biorefineries have focused on utilizing waste lignocellulosic biomass to generate biofuels and platform chemicals, highlighting the importance of sustainable waste management practices.

The valorization of agriculture food waste biomass for bioactive compounds has led to advancements in essential oil extraction techniques, including the use of green solvents and supercritical fluid extraction methods. These techniques aim to improve the efficiency of extraction processes while reducing environmental impact (Paini et al., 2022; Torres-Valenzuela et al., 2020). The production of microbial single cell oils from oleaginous microorganisms like bacteria, algae, and fungi has emerged as a potential and significant source of essential fatty acids and valuable chemicals for various applications (Galán et al., 2019; Ochsenreither et al., 2016). The production and purification of microbial lipids require efficient downstream processing methods, including cell disruption and lipid extraction techniques. These methods play an important role in determining the yield and composition of microbial lipids, influencing the overall extraction efficiencies and product quality (Sun et al., 2023; Zainuddin et al., 2021). Overall, the extraction and purification of essential oils from food waste biomass involve a combination of innovative technologies and sustainable practices to address the challenges of waste management, resource utilization, and economic efficiency. Additional investigation and study are required to optimize the extraction methods, improve the quality of the products, and fully exploit the advantages of essential oils obtained from biomass waste generated by food.

2 Collection and Preparation of Food Waste Biomass

The treatment of food waste biomass is vital for fostering sustainability and circular economy concepts. Negri et al. (2020) underline the need for collecting food waste biomass from various sources to assure high-quality feedstock for anaerobic digestion, especially in China and Southeast Asia. This strategy is crucial for bioenergy generation and corresponds with the objective of establishing a circular economy in food waste management. Similarly, Govindasamy et al. (2020) emphasize the possibility of transforming food waste into biodegradable and recyclable materials, solving environmental pollution and nutrient loss concerns. They specialize on the valorization of food waste for biopolymer manufacture. In the perspective of food waste biomass usage, Spalvins et al. (2020) study the use of waste cooking oil as a good source for single-cell protein synthesis by the yeast *Yarrowia lipolytica*. Their work illustrates the viability of employing industrial waste as a resource for biomass production, obtaining significant biomass concentrations with waste cooking oil as a carbon source. This research adds to the sustainable utilization of food waste biomass for value goods. Furthermore, Xiong et al. (2021) studied the conversion of humins derived from food waste biorefinery into Lewis acid catalysts assisted by biochar. This research demonstrates the potential for recycling byproducts from food waste valorization, namely starch-rich rice trash, by exploiting humins as an

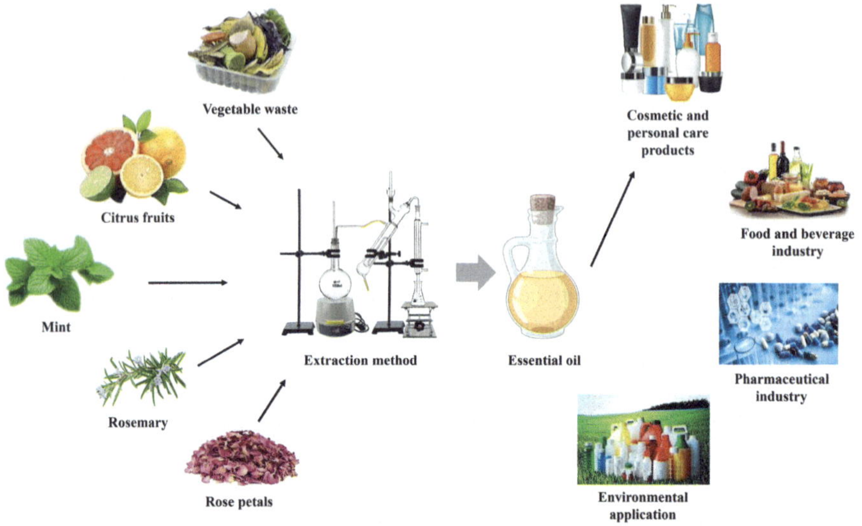

Fig. 1 Extraction of essential oil from various food waste biomass and their potential applications

inexpensive carbon source. This approach promotes circular economy practices and sustainable development. Figure 1 illustrates the overall procedure for extracting essential oil from different food waste biomass and its potential application in various industrial sectors. Their research underlines the necessity to reduce food waste amounts and associated economic losses in hospital food services to increase environmental sustainability.

3 Extraction Techniques

Food waste is a substantial reservoir of biomass that can be converted into commercially valuable essential oils. The composition of food waste often contains a significant proportion of lipids, which can be recovered and transformed into valuable products. Food waste, which includes a variety of organic materials such as vegetable peels, seeds, and oils, can be effectively utilized to extract essential oils through various methods, including supercritical CO_2 extraction, ultrasound-assisted extraction (UAE), and microwave-assisted extraction (MAE). Supercritical CO_2 extraction is an advanced method that utilizes carbon dioxide at supercritical conditions to extract essential oils. This technique is known for its ability to selectively extract compounds without the use of harmful solvents, making it an environmentally friendly option. Various food waste materials, such as citrus peels and other biomass, have been effectively subjected to supercritical CO_2 extraction, leading to the production of significant quantities of essential oils that retain their beneficial properties. The approach enables the extraction of a variety of chemicals, including those that may exhibit sensitivity to heat, therefore preserving the intrinsic quality of the essential oils (Houng et al., 2023).

Ultrasound-assisted extraction (UAE) is another innovative technique that has been shown to improve the efficiency of essential oil extraction. By applying ultrasonic waves to the biomass, UAE enhances the mass transfer of volatile compounds, leading to higher yields in shorter extraction times. Miljanović et al. (2020) research has shown that the integration of ultrasonic technology with conventional hydrodistillation (HD) may enhance the production of essential oils by about 33%. This approach is especially beneficial for the extraction of oils from food waste, as it may be carried out at lower temperatures, therefore reducing the likelihood of thermal deterioration of delicate chemicals. This method is particularly advantageous for extracting oils from food waste, as it can be performed at lower temperatures, thereby minimizing the risk of thermal degradation of sensitive compounds (Patrascu & Radoiu, 2016).

Microwave-assisted extraction (MAE) is an innovative approach that use microwave radiation to raise the temperature of the plant material and solvent, resulting in the fast extraction of essential oils. This approach has been acknowledged for its efficacy and capacity to provide extracts of superior quality within a reduced time period. Akşit and Gençcelep (2021) study has shown that MAE can efficiently extract vital oils from biomass derived from food waste, such as kaffir lime peel, resulting in oils that possess notable antioxidant and antibacterial characteristics. Maximum extraction efficiency (MAE) offers the benefits of decreased extraction time and reduced energy consumption, making it a feasible choice for extensive implementations in the food industry (Lin et al., 2023).

A combination of these advanced extraction methods makes it possible to achieve improved recovery of essential oils from biomass derived from food waste. This makes a valuable contribution to the implementation of sustainable waste management and resource utilization as shown in Table 1. The integration of supercritical CO_2 extraction with ultrasonic or microwave methods has the potential to achieve even greater extraction efficiencies and improved quality oils (Uwineza & Waśkiewicz, 2020).

4 Purification Processes

Food waste biomass can contain a variety of unwanted compounds, such as pesticides, waxes, and organic solvents (if used during extraction). Purification ensures these are removed, leaving a cleaner and safer product. Purification helps to concentrate the desired aromatic and therapeutic components of the essential oil, leading to a more potent and effective product. The purification of essential oils from food waste biomass using steam distillation, chromatography, membrane separation, and adsorption methods presents a promising avenue for both waste valorization and the production of valuable compounds. The utilization of food waste can effectively resolve environmental issues linked to waste disposal and also enhances the circular economy by converting waste into valuable products.

Table. 1 Various techniques for extracting essential oils from biomass derived from food waste

S. No	Food waste type	Botanical name	Extraction method	Major constituent essential oil	Yield (wt.%)	Beneficial effects	Reference
1	Orange peel	*Citrus sinensis*	Microwave-assisted hydro-distillation	D-Limonene (97.38%)	1. 8	The essential oil derived from orange peel is often used as a perfume in natural skincare, bath, and body care products.	(Bustamante et al., 2016)
2	Lime peel	*Citrus aurantiifolia*	Hydro-distillation	Limonene, β-pinene (16.5%), and p-cymene (13.9%)	1	Antimicrobial activity of lime peel essential oil was shown to be the most potent against *Staphylococcus aureus*, *Candida albicans*, *Escherichia coli*, and *Aspergillus flavus*.	(Abd-Elwahab et al., 2016)
3.	Lemonpeel	*Citrus limon*	Microwave-assisted hydro-distillation	B-pinene (14.517 wt.%), Limonene (65.082 wt.%), and γ-terpinene (9.743 wt.%)	2	Lemon essential oil (LEO) has antimicrobial activity against *Staphylococcus* and *Escherichia coli* bacteria.	(Martínez-Abad et al., 2020)
4	Pomelo peel	*Citrus maxima*	Hydro-distillation	β-sitosterol (17.99%), D-limonene (33.61%), α-pinene (4.32%) and α-sitosterol (12.19%)	0.58	Pomelo exhibited notable larvicidal activity against *Culex tritaeniorhynchus* and *Aedes aegypti* mosquito species. *Armigeres subalbatus* shown greater resistance.	(Visakh et al., 2022)
5	Sweet lime peel	*Citrus limetta*	Hydro-distillation	D-limonene (85%) and α-myrcene (4.85%)	0.63	It is beneficial as a potent antibacterial and cytotoxic agent.	(Narayanankutty et al., 2022)

6	Grapefruit peel	*Citrus Paradisi.* L	Solvent-free microwave extraction (SFME)	B-pinene (0.8%), limonene (91.5%), and linalool (1.1%)	0.44	The essential oil extracted from grapefruit peel exhibited a broad range of antimicrobial effects against several bacteria, including *Staphylococcus aureus*, *Salmonella typhimurium*, *Serratia marcescens*, *Staphylococcus epidermidis*, *Enterococcus faecalis*, *Escherichia coli*, and *Proteus vulgaris*. The inhibition zones observed for these bacteria ranged from 11 to 53 mm.	(Uysal et al., 2011)
7	Pomelo peel	*Citrus maxima*	Hydro-distillation	Limonene (71.33%) and β-myrcene (5.31%)	0.73	It has antioxidant property.	(Badalamenti et al., 2022)
8	Pineapple waste	*Ananas comosus* (L.) Merril	Supercritical fluid extraction	2-heptanol (15.02%), lactic acid ethyl ester (19.35%), propanoic acid ethyl ester (40.25%), 3-hexanone (2.60%), and propanol (8.18%)	0.17	Pineapple essential oil has the potential to be used as medicinal substances, preservatives, substitutes for synthetic antioxidants, and in the development of functional foods and industrial food applications.	(M. Hikal et al., 2021)
9	Onion	*Allium cepa*	Hydro-distillation	S-propyl methane thiosulfonate (13.74%)	0.16	The results of the antioxidant activity analysis indicate that the essential oil provides protection against oxidative stress, a crucial factor for maintaining good health.	(Taiwo et al., 2024)
10	Carrot seeds	*Daucus carota*	Hydro-distillation	Sabinene (10.16%), carotol (34.84%), and α-pinene (8.02%)	0.73	The results suggest that carrot seed essential oil has potent fungistatic activity against *Candida yeasts* and Gram-positive bacteria.	(Smigielski et al., 2014)
11	Rosemary leaves	*Lavandulaangustifolia*	Steam distillation	α-pinene (29.73%), 1,8-cineole (27.14%), and verbenone (12.81%)	2.8	Rosemary essential oil has antibacterial action against the pathogens *E. coli* and *S. aureus*.	(Teshale et al., 2022)

(continued)

Table. 1 (continued)

S. No	Food waste type	Botanical name	Extraction method	Major constituent essential oil	Yield (wt.%)	Beneficial effects	Reference
12	Rosemary leaves	*Rosmarinus officinalis*	Water-steam distillation	Camphor (12.40%), α-pinene (12.94%), Eucalyptol (1,8-cineole) (40.10%), and β -pinene (8.94%)	–	Rosemary essential oil has potent antiviral, antifungal, and antioxidant activities, as well as significant antibacterial characteristics, particularly efficient against *S. aureus*.	(Christopoulou et al., 2021)
13	Sea fennel (flower)	*Crithmum maritimum* L., Apiaceae	Hydro-distillation	Sabinene (44.94%) and limonene (43.58%)	1.35	A significant amount of essential oil is derived from sea fennel, which may be used as appropriate raw materials for the production of culinary, pharmaceutical, or cosmetic substances and products.	(Politeo et al., 2023)
	Sea fennel (leaf)			sabinene (51.47%) and limonene (36.28%)	0.63		
	Sea fennel (stem)			Sabinene (42.55%) and limonene (36.45%)	0.62		
14	Mint leaves	*Mentha piperita*	Steam distillation	1-decanol, 2-hexyl-(3.8471%), and L-menthol (3.54%)	1.36	Levo-menthol is used for the control, regulation, prevention, and improvement of many conditions, such as occasional minor irritation, discomfort, mouth sores, throat soreness, and coughing.	(Ibrahim et al., 2021)
15	Basil leaves	*Ocimum canum* Sims	Hydro-distillation	(E)-methyl cinnamate (51.28%), camphor (10.35%), 1,8-cineole (5.58%), and trans-β-caryophyllene (4.11%)	0.83	It is used as both food flavorings and antioxidants agents.	(Morsy & Hammad, 2021)

4.1 Distillation

The extraction and purification of essential oils from food waste biomass have been an important topic lately. Food waste biomass is rich with useful components including proteins, amino acids, sugars, fibers, organic acids, minerals, lipids, and notably essential oils like d-limonene (Mahato et al., 2019). Different processes, including stream distillation, fractional distillation, and vacuum distillation, are employed to isolate these essential oils from the biomass. Fractional and vacuum distillation have proved effective in extracting essential oils from several forms of biomass, including pine, helping the circular bioeconomy by converting waste into useful goods (Mora et al., 2022). Steam distillation is a well-established and extensively used technique for obtaining essential oils. This process involves the application of steam to plant material, resulting in the vaporization of volatile chemicals. The steam containing the essential oils is cooled and transformed back into a liquid state, enabling the oil to be separated from the water. This method is particularly effective for extracting oils from citrus peels and other aromatic plants. Research has shown that steam distillation can yield high-quality essential oils from citrus waste, which contain valuable phytochemicals with antimicrobial properties. The efficiency of steam distillation can be influenced by factors such as temperature and duration of distillation, which can be optimized to enhance oil yield and quality (Grover et al., 2023). Moreover, vacuum fractional distillation is adaptable enough to extract bioactive chemicals from medicinal plants and herbs, proving its extensive utility for separating volatile molecules like essential oils (Fotsing et al., 2021). Adding some additives may optimize the distillation process, lowering water content and acidity in the fractions produced during short-path vacuum distillation. This innovation makes the approach significantly more successful for separating essential oils and other important chemicals from biomass, boosting the sustainable use of resources (Li et al., 2020). Overall, employing fractional and vacuum distillation to extract and purify essential oils from food waste biomass provides a sustainable and effective technique to get high-value chemicals.

4.2 Chromatography

The extraction and purification of essential oils from food waste biomass using column chromatography and gas chromatography have been explored in various studies. One study highlighted the efficiency of overpressured layer chromatography (OPLC) in isolating antimicrobial components, such as essential oils, from different matrices (Choma & Jesionek, 2015). Another study discussed the extraction of bioactive components and essential oils on a large scale from plants using supercritical fluid extraction (SFE) technology (Khoddami et al., 2013). Additionally, the biosynthetic process strategies for sustainable essential oil production have been investigated, with cembratrienol (CBTol) being identified through gas chromatography. Green chemistry has also been employed in the extraction and purification of essential oils, highlighting the use of gas chromatography for analysis (Cannavacciuolo et al., 2023). Additionally, enzymatic hydrolysis of lignocellulosic polysaccharides has been studied using various methods, including alumina column chromatography and liquid-liquid extraction (enzymatic hydrolysis of lignocellulosic polysaccharides). Overall, these studies demonstrate the diverse approaches and techniques employed in the extraction and purification of essential oils from food waste biomass, emphasizing the importance of chromatography methods in this process.

4.3 Membrane Separation

Membrane technology enables the targeted isolation of components according to their size and chemical characteristics, which can be very advantageous in enhancing the quality of essential oils obtained from processed food waste. This method has the capability to efficiently eliminate contaminants and enhance the concentration of the targeted volatile chemicals, therefore enhancing the overall quality of the acquired essential oils. The use of membrane technology into conventional distillation techniques has the potential to improve extraction strategies, thus reducing energy usage and enhancing the quality of essential oil (Sniegoňová et al., 2023). Cassano et al. (2018) showed that tight ultrafiltration and nanofiltration membranes with enormous pore diameters, ranging from 100 to 10,000 nm, are employed to recover essential oils. This reveals that the pore size of the membrane is a critical aspect for the extraction of essential oil from the aqueous distillates. The developments in membrane technology, as highlighted by Qasem et al. (2021), have led to greater usage of membranes for filtering and extraction activities. Membrane technology, according to research on materials and membrane technologies for water and energy, is crucial for water treatment and the removal of diverse components.

4.4 Adsorption Methods

Recently, there has been growing interest in using activated charcoal and silica gel in adsorption and absorption techniques to extract essential oils from food waste biomass. Activated carbons made from biomass and waste residues have showed tremendous potential in increasing adsorption efficiency (Dziejarski et al., 2023). Research has shown the efficacy of activated carbon obtained from biomass in eliminating heavy metals, underscoring its potential use (Wang et al., 2023). Furthermore, it is noteworthy that activated carbon is an economical and effective material for eliminating synthetic dyes (Rápó & Tonk, 2021). Recent developments in the field of adsorptive removal have included the successful removal of siloxanes from biogas via the use of phosphoric acid absorption and activated carbon adsorption techniques (Vali et al., 2023).

5 Quality Analysis: Chemical, Physical, and Bioactivity Assessments

The quality analysis of essential oils extracted from food waste biomass has gained significant attention recently. Viuda-Martos et al. (2010) studied the impact of adding rosemary essential oil and citrus fiber washing water to bologna sausage, finding flavonoids like hesperidin and narirutin. Similarly, Alparslan et al. (2016) looked at the quality and shelf life of shrimp preserved with orange leaf essential oil in gelatin, highlighting how gelatin coatings enriched with essential oils can be effective. In terms of bioactivity, Lesjak et al. (2017) evaluated the bioactivity and chemical composition of *Juniperus excelsa* leaves and cones, suggesting its potential as a health-promoting food additive. Essential oils possess several pharmacological activities, including antiviral, antioxidant, anti-inflammatory, anticancer, antidiabetic, antibacterial, anti-hyperpigmentation, cardioprotective, hepatoprotective, and neuroprotectant effects. Figure 2 shows various pharmacological properties of essential oil. These features support their potential for biomedical uses and contribute to their overall health benefits (Jugreet et al., 2020). Fonseca et al. (2020) demonstrated the high encapsulation efficiency and antioxidant activity of thyme

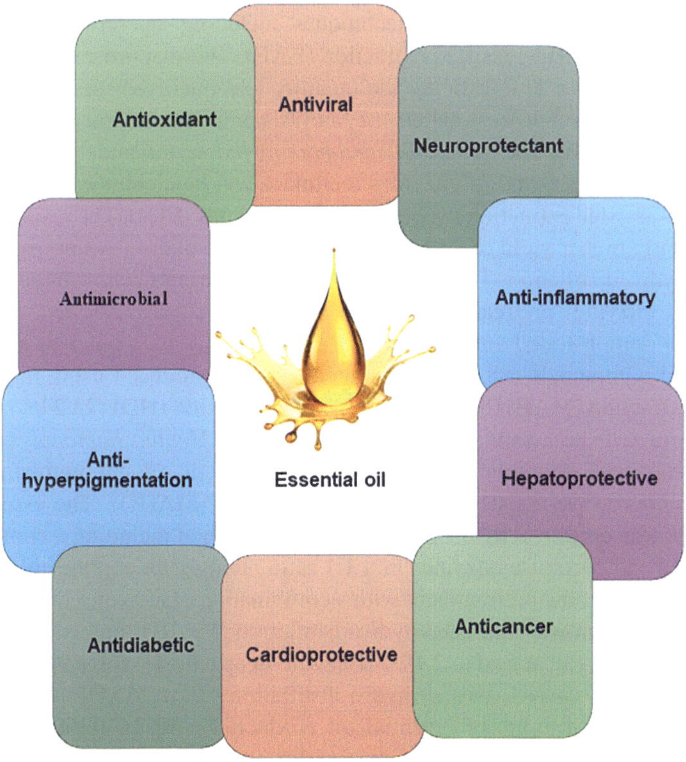

Fig. 2 Schematic representation of various pharmacological properties of essential oil

essential oil encapsulated in electrospun potato starch nanofibers. Extraction techniques for essential oils have also been explored. Memarzadeh et al. (2020) compared microwave-assisted steam hydro-diffusion and traditional methods for extracting Bakhtiari savory essential oil, finding the former more efficient in terms of both quantity and quality. Carmona Cabello et al. (2018) highlighted the feasibility of using oil from solid food waste to produce biodiesel, emphasizing the importance of valorizing food waste. Additionally, Prates-Valério et al. (2019) comprehensively characterized *Acrocomia aculeata* oil for potential food and industrial applications, noting its consumer acceptance and nutraceutical value. Arun et al. (2022) focused on incorporating cellulose nanofibers from coconut industry waste into biodegradable packaging films to enhance their properties.

6 Case Studies: Industrial Implementations and Product Development

A recent study emphasizes the practical uses and advancement of extracting essential oils from biomass obtained from food waste biomass in industrial settings and for product development (Stashenko et al., 2023). The research focuses on using food waste via advanced extraction techniques, including supercritical CO_2 extraction (SCO_2) and enzyme-assisted extraction (EAE). These approaches have shown considerable promise in enhancing the quantity and quality of essential oils. Liu et al. (2021) devised a process called enzymolysis pretreatment microwave-assisted extraction (EP-MAE). The leaves of *Cinnamomum burmannii* undergo enzymolysis pretreatment using a blend of enzymes (cellulase and hemicellulase), followed by microwave-assisted extraction to obtain the essential oils (EOs). The results indicate a significantly higher yield of essential oil compared to microwave-assisted extraction and hydro-distillation. The essential oil extracted from *Cinnamomum burmannii* using enzyme pretreatment followed by microwave-assisted extraction (EP-MAE) demonstrated superior quality, as shown by its greater proportion of oxygenated compounds (88.31%) compared to those produced using microwave-assisted extraction (MAE) (80.41%) and hydro-distillation (HD) (73.24%). Li et al. (2022) extracted the essential oil from the leaves of *Mentha haplocalyx* Briq by using an enhanced technique called enzyme-deep eutectic solvent pretreatment followed by microwave-assisted hydro-distillation (EDP-MAHD). The essential oil production was enhanced by using a pretreatment method including a combination of cellulase and pectinase enzymes in a 1:1 ratio, along with choline chloride/glucose. After undergoing pretreatment with a combination of enzymes and then being subjected to a microwave-assisted hydro-distillation (MAHD) procedure, the yield of essential oil was increased to 2.19%. Enzyme-deep eutectic solvent pretreatment followed by microwave-assisted hydro-distillation (EDP-MAHD) outperforms other approaches in terms of essential oil production and its ability to enhance inhibitory action on acetylcholinesterase and α-amylase. Research conducted by Wei et al. (2022) found that the essential oil derived from the *Platycladus orientalis*

(L.) Franco plant was obtained by a combination of enzymatic pretreatment and solvent-free microwave extraction technique (EP-SME). The essential oil yield reached 1.13% under the optimum conditions, representing a 24.18% increase compared to the yield achieved by the conventional steam distillation (SD) process which yielded 0.91%. Additionally, the essential oil included a greater amount of polyphenols and flavonoids. Furthermore, the essential oil obtained by the EP-SFME approach exhibited superior overall antioxidant activity and more potent acetylcholinesterase inhibitory activity compared to the essential oil obtained through the hydro-distillation method.

7 Industrial Applications of Essential Oils (EOs) from Food Waste Biomass

Industries are increasingly focusing on sustainable practices and alternative resources, leading to increased interest in extracting and using essential oils from food waste. Researchers are actively studying the possibilities of extracting essential oils from food waste for use in different industrial applications. This approach provides an environmentally acceptable answer to waste management and resource recovery.

7.1 Cosmetic and Health Care Products

Essential oils (EOs) are extensively used in the cosmetic and health care sectors because of their fragrant qualities and possible advantages for health. Extracting essential oils from biomass derived from food waste, such as citrus peels, has yielded a sustainable supply of natural perfumes and active compounds for use in cosmetics and haircare products. Limonene is the primary chemical component found in different oils derived from citrus species (Karne et al., 2023; Weng et al., 2024). It is characterized by its lemon-like fragrance.

Limonene, due to its pleasant smell, is used in the perfumery industry to enhance fragrances. It is also employed in the food industry to add taste to different food products. Additionally, limonene finds applications in the cosmetic industry. Moreover, it is considered to have minimal toxicity towards cells (Paw et al., 2020). Dinh et al. (2020) research demonstrates the use of microwave-assisted distillation for extracting essential oil from rosemary. The primary constituents found in rosemary essential oil are α-pinene (22%), verbenole (16%), 1,8-cineole (16%), and geraniol (5%). The extracts of lavender essential oil include a significant amount of phenolic compounds and gentisic acid. These chemicals are often used in cosmetics for their skin-lightening properties (Turrini et al., 2021). These chemicals have the potential to be used in the cosmetics industry due to their beneficial effects on health. Utilizing essential oils generated from food waste is in line with the increasing customer preference for environmentally friendly and organic goods.

7.2 Pharmaceutical and Therapeutic Applications

The pharmaceutical sector has shown interest in using essential oils derived from food waste biomass due to their potential medicinal characteristics. Herbaceous food waste, such as basil or oregano stems, contains bioactive chemicals that have anti-inflammatory, antibacterial, and antioxidant activities. The bioactive compounds extracted from *Mentha piperita* (peppermint), *Piper nigrum* (black pepper), *Cinnamomum zeylanicum* (true cinnamon), *Ocimum basilicum* (basil), *Origanum vulgare* (oregano), *Rosmarinus officinalis* (rosemary), and *Thymus vulgaris* (thyme) essential oils show their potential applications in pharmaceutical industry (Alizadeh Behbahani et al., 2020; Camele et al., 2021; Eid et al., 2023; Moghaddam & Mehdizadeh, 2020; Oussaid et al., 2020; Zhao et al., 2021). Medicinal plants serve as a valuable reservoir of natural antioxidants that have the potential to be used in the food and pharmaceutical industries. Lemon balm, mint, and sweet basil species have a greater concentration of valuable compounds that may be used to produce goods with additional value. These compounds have strong antioxidant properties and can be utilized as organic preservatives (Ilić et al., 2022). In addition to herbs and medicinal plants, vegetable food waste biomass may also be used for pharmaceutical purposes. Vegetable and fruit wastes, including leaves, stems, seeds, peels, pulps, and roots, as well as wastes from the agricultural and food industries, include a variety of useful components such as oils, proteins, fatty acids, fiber, isoprenoids, lipids, antioxidants, saponins, and phytoestrogens. These bioactive compounds have many applications, such as being utilized as food additives and pharmaceutical excipients (Jiménez-Moreno et al., 2020). These essential oils are suitable for topical treatments, supplements, and alternative medicine products.

7.3 Food and Beverage Industry

Essential oils (EOs) are utilized in the food and beverage industry as natural additives for taste enhancement and preservation purposes. Extracting essential oils from food waste, such as herb trimming and fruit peels, provides a cost-efficient and environmentally friendly method to improve the taste and longevity of food items. Various studies have shown that citrus peels contain vital oils that have diverse applications in the fields of food preservation, food safety, and nutraceutical sectors. The majority of food firms use these food wastes as coating components to safeguard food goods from diverse environmental elements. Over the last 10 years, researchers have explored the potential of polysaccharides as a suitable substance for creating edible films. Anis et al. (2021) elaborated in their study that the essential oils (EOs) from various food waste were combined with polysaccharide matrices to improve the functional properties of edible films. Citrus waste biomass helps to facilitate the production of high-quality food and drinks that are rich in nutrients. These essential oils have versatile applications in beverages, food additives, and food items due to their ability to enhance taste and produce a pleasant fragrance. Additionally, these compounds may be used as biofertilizers to enhance the freshness of food and drinks (Padilla-de la Rosa et al., 2021). Moreover, the antibacterial characteristics of these oils may be used to produce organic food preservatives, hence minimizing the reliance on artificial chemicals.

7.4 Environmental Application

Food waste-derived essential oils are now being investigated for their potential use in environmental applications, namely in the fields of bioremediation and pest management. Due to their inherent insecticidal qualities, they are well suited for inclusion in environmentally benign insecticides, therefore decreasing the dependence on synthetic chemicals that have detrimental effects on the environment. Govindarajan et al. (2016) conducted a study on the use of essential oil derived from *Pinus kesiya*, an Asian pine species, as a highly effective insecticide against three mosquito species (*Aedes aegypti* (which transmits dengue), *Anopheles stephensi* (which transmits malaria), and *Culex quinquefasciatus* (which transmits lymphatic filariasis)). Their findings indicated that at a dosage of 125 mg/L, essential oils (EOs) exhibited a death rate ranging from 96% to 100% for all species tested, suggesting their potential efficacy in reducing harmful mosquito larvae. Essential oils (EOs) are a blend of volatile oils that develop as a byproduct in medicinal plants. They possess properties that may combat pests, insects, and pathogenic fungus. Several variables, including plant species or cultivar, geographical origin, ambient circumstances, farming techniques, and extraction process, influence the chemical composition of the plants (Assadpour et al., 2024).

8 Conclusion

The extraction and purification of essential oils from food waste biomass present a compelling opportunity to address several pressing challenges, including waste management, sustainability, and resource utilization. Essential oils, rich in bioactive compounds, have significant applications across industries such as food, cosmetics, health care, pharmaceuticals, and biofuels. The valorization of food waste not only contributes to environmental sustainability by reducing waste but also adds economic value through the production of high-demand essential oils. This chapter has highlighted the comprehensive methodologies involved in the extraction and purification processes, including various pretreatment, extraction, and purification techniques. Each method, whether traditional or innovative, offers unique advantages and challenges. Techniques such as steam distillation, supercritical fluid extraction, and membrane separation have shown promise in efficiently isolating high-quality essential oils from diverse food waste sources. Quality analysis through chemical, physical, and bioactivity assessments ensures that the extracted oils meet industry standards and retain their functional properties. These assessments are crucial for confirming the suitability of essential oils for various applications, ensuring their effectiveness as preservatives, antimicrobials, and therapeutic agents.

9 Future Prospects

The future of essential oil extraction from food waste biomass is poised for significant advancements driven by technological innovation and increased environmental consciousness. Optimization of extraction techniques, such as ultrasound-assisted extraction (UAE) and microwave-assisted extraction (MAE), will focus on improving yield, efficiency, and purity while minimizing environmental impact. Emphasis on sustainable practices, including the use of green chemistry and environmentally friendly solvents, will further reduce the ecological footprint of extraction processes. Integration with biorefineries is a promising area, aiming to develop models that utilize food waste to produce a range of value-added products, including essential oils, biofuels, and biopolymers, enhancing both economic viability and sustainability. Advanced purification methods, like innovative chromatography and membrane technologies, are expected to achieve higher purity and better separation of essential oil components. Expanding the application of essential oils into emerging fields such as nanotechnology, pharmaceuticals, and advanced materials will open new markets and opportunities. Addressing regulatory challenges and enhancing consumer acceptance through rigorous safety testing and transparent communication will be crucial. Finally, fostering interdisciplinary research involving chemists, biotechnologists, environmental scientists, and industry stakeholders will drive innovation and tackle the complex challenges in this field. By leveraging cutting-edge research and innovative practices, the extraction and purification of essential oils from food waste biomass can significantly contribute to a circular economy, turning waste into valuable resources while supporting environmental and economic sustainability.

References

Abd-Elwahab, S. M., El-Tanbouly, N. D., Moussa, M. Y., Abdel-Monem, A. R., & Fayek, N. M. (2016). Antimicrobial and antiradical potential of four agro-waste citrus peels cultivars. *Journal of Essential Oil Bearing Plants, 19*, 1932–1942. https://doi.org/10.1080/0972060X.2016.1232609

Akşit, Z., & Gençcelep, H. (2021). Investigation of tomato, quince and grapefruit waste; compositions and functional properties. *Hittite Journal of Science & Engineering, 8*, 10.17350/hjse19030000208.

Alizadeh Behbahani, B., Falah, F., Lavi Arab, F., Vasiee, M., & Tabatabaee Yazdi, F. (2020). Chemical composition and antioxidant, antimicrobial, and antiproliferative activities of Cinnamomum zeylanicum bark essential oil. *Evidence-based Complementary and Alternative Medicine, 2020*, 5190603. https://doi.org/10.1155/2020/5190603

Alparslan, Y., Yapıcı, H., Metin, C., Baygar, T., Günlü, A., & Baygar, T. (2016). Quality assessment of shrimps preserved with orange leaf essential oil incorporated gelatin. *LWT - Food Science and Technology, 72*, 457. https://doi.org/10.1016/j.lwt.2016.04.066

Anis, A., Pal, K., & Al-Zahrani, S. M. (2021). Essential oil-containing polysaccharide-based edible films and coatings for food security applications. *Polymers, 13*, 575. https://doi.org/10.3390/polym13040575

Arun, R., Shruthy, R., Preetha, R., & Sreejit, V. (2022). Biodegradable nano composite reinforced with cellulose nano fiber from coconut industry waste for replacing synthetic plastic food packaging. *Chemosphere, 291*, 132786. https://doi.org/10.1016/j.chemosphere.2021.132786

Assadpour, E., Can Karaça, A., Fasamanesh, M., Mahdavi, S. A., Shariat-Alavi, M., Feng, J., Kharazmi, M. S., Rehman, A., & Jafari, S. M. (2024). Application of essential oils as natural biopesticides; recent advances. *Critical Reviews in Food Science and Nutrition, 64*, 6477–6497. https://doi.org/10.1080/10408398.2023.2170317

Assadpour, E., & Mahdi Jafari, S. (2019). A systematic review on nanoencapsulation of food bioactive ingredients and nutraceuticals by various nanocarriers. *Critical Reviews in Food Science and Nutrition, 59*, 3129–3151. https://doi.org/10.1080/10408398.2018.1484687

Badalamenti, N., Bruno, M., Schicchi, R., Geraci, A., Leporini, M., Tundis, R., & Loizzo, M. R. (2022). Reuse of food waste: The chemical composition and health properties of pomelo (*Citrus maxima*) cultivar essential oils. *Molecules, 27*, 3273. https://doi.org/10.3390/molecules27103273

Batish, D. R., Singh, H. P., Kohli, R. K., & Kaur, S. (2008). Eucalyptus essential oil as a natural pesticide. *Forest Ecology and Management, 256*, 2166–2174. https://doi.org/10.1016/j.foreco.2008.08.008

Berdis, E., Buckley, J., & Kraft, J. (2003). *Citrus flavor technologies: Citrus oils — Processing, separation, application*. https://doi.org/10.1115/CEC2003-4906.

Bhatti, S. A., Hussain, M. H., Mohsin, M. Z., Mohsin, A., Zaman, W. Q., Guo, M., Iqbal, M. W., Siddiqui, S. A., Ibrahim, S. A., Ur-Rehman, S., & Korma, S. A. (2022). Evaluation of the antimicrobial effects of Cap*sicum*, *Nigella sativa*, *Musa paradisiaca* L., and *Citrus limetta*: A review. *Frontiers in Sustainable Food Systems, 6*. https://doi.org/10.3389/fsufs.2022.1043823

Bustamante, J., van Stempvoort, S., García-Gallarreta, M., Houghton, J. A., Briers, H. K., Budarin, V. L., Matharu, A. S., & Clark, J. H. (2016). Microwave assisted hydro-distillation of essential oils from wet citrus peel waste. *Journal of Cleaner Production, 137*, 598–605. https://doi.org/10.1016/j.jclepro.2016.07.108

Camele, I., Gruľová, D., & Elshafie, H. S. (2021). Chemical composition and antimicrobial properties of *Mentha × piperita* cv. 'Kristinka' essential oil. *Plants, 10*, 1567. https://doi.org/10.3390/plants10081567

Cannavacciuolo, C., Pagliari, S., Frigerio, J., Giustra, C. M., Labra, M., & Campone, L. (2023). Natural deep eutectic solvents (NADESs) combined with sustainable extraction techniques: A review of the green chemistry approach in food analysis. *Food, 12*, 56. https://doi.org/10.3390/foods12010056

Carmona Cabello, M., Leiva, D., Castro-Cantarero, J., Pinzi, S., & Dorado, M. (2018). Valorization of food waste from restaurants by transesterification of the lipid fraction. *Fuel, 215*, 492–498. https://doi.org/10.1016/j.fuel.2017.11.096

Cassano, A., Conidi, C., Ruby-Figueroa, R., & Castro-Muñoz, R. (2018). Nanofiltration and tight ultrafiltration membranes for the recovery of polyphenols from agro-food by-products. *International Journal of Molecular Sciences, 19*, 351. https://doi.org/10.3390/ijms19020351

Choma, I. M., & Jesionek, W. (2015). TLC-direct bioautography as a high throughput method for detection of antimicrobials in plants. *Chromatography, 2*, 225–238. https://doi.org/10.3390/chromatography2020225

Christopoulou, S. D., Androutsopoulou, C., Hahalis, P., Kotsalou, C., Vantarakis, A., & Lamari, F. N. (2021). Rosemary extract and essential oil as drink ingredients: An evaluation of their chemical composition, genotoxicity, antimicrobial, antiviral, and antioxidant properties. *Food, 10*, 3143. https://doi.org/10.3390/foods10123143

Dinh, P. N., Thuy, L. M. T., & Quoc, T. P. (2020). The extraction of rosemary essential oil by microwave-assisted hydrodistillation method. *IOP Conference Series: Materials Science and Engineering, 991*, 012051. https://doi.org/10.1088/1757-899X/991/1/012051

Dziejarski, B., Serafin, J., Andersson, K., & Krzyżyńska, R. (2023). CO2 capture materials: A review of current trends and future challenges. *Materials Today Sustainability, 24*, 100483. https://doi.org/10.1016/j.mtsust.2023.100483

Eid, A. M., Jaradat, N., Shraim, N., Hawash, M., Issa, L., Shakhsher, M., Nawahda, N., Hanbali, A., Barahmeh, N., Taha, B., & Mousa, A. (2023). Assessment of anticancer, antimicrobial, antidiabetic, anti-obesity and antioxidant activity of Ocimum Basilicum seeds essential oil from Palestine. *BMC Complementary Medicine and Therapies, 23*, 221. https://doi.org/10.1186/s12906-023-04058-w

Fonseca, L. M., Radünz, M., Dos Santos Hackbart, H. C., da Silva, F. T., Camargo, T. M., Bruni, G. P., Monks, J. L., da Rosa Zavareze, E., & Dias, A. R. (2020). Electrospun potato starch nanofibers for thyme essential oil encapsulation: Antioxidant activity and thermal resistance. *Journal of the Science of Food and Agriculture, 100*, 4263–4271. https://doi.org/10.1002/jsfa.10468

Fotsing, Y. S. F., Kezetas, J. J. B., Batiha, G. E.-S., Ali, I., & Ndjakou, B. L. (2021). Extraction of bioactive compounds from medicinal plants and herbs. In *Natural medicinal plants*. IntechOpen. https://doi.org/10.5772/intechopen.98602

Galán, B., Santos-Merino, M., Nogales, J., de la Cruz, F., & García, J. L. (2019). Microbial oils as nutraceuticals and animal feeds. In H. Goldfine (Ed.), *Health consequences of microbial interactions with hydrocarbons, oils, and lipids* (pp. 1–45). Springer. https://doi.org/10.1007/978-3-319-72473-7_34-1

Govindarajan, M., Rajeswary, M., & Benelli, G. (2016). Chemical composition, toxicity and non-target effects of *Pinus kesiya* essential oil: An eco-friendly and novel larvicide against malaria, dengue and lymphatic filariasis mosquito vectors. *Ecotoxicology and Environmental Safety, 129*, 85–90. https://doi.org/10.1016/j.ecoenv.2016.03.007

Govindasamy, S., Muthukumaran, C., Kumar, M., Sivakumar, V. M., & Thirumarimurugan, M. (2020). *Food waste valorization for biopolymer production* (pp. 233–249). https://doi.org/10.1016/B978-0-444-64321-6.00012-4

Grover, S., Aggarwal, P., Kaur, S., Yadav, R., & Kumar, A. (2023). Physicochemical assessment, characterization, antioxidant and antimicrobial potential study of essential oil extracted from the peel of different galgal (Citrus Pseudolimon) cultivars. *Waste and Biomass Valorization, 15*, 3157. https://doi.org/10.1007/s12649-023-02321-2

Houng, P., Ly, K., & Lay, S. (2023). Valorization of kaffir lime peel through extraction of essential oil and process optimization for phenolic compounds. *Journal of Chemical Technology & Biotechnology, 98*, 2745. https://doi.org/10.1002/jctb.7354

Ibrahim, M., Ankwai, G. E., Gungshik, J. R., & Taave, P. (2021). Comparative extraction of essential oils of *Mentha piperita* (mint) by steam distillation and enfleurage. *Nigerian Journal of Chemical Research, 26*, 56–62. https://doi.org/10.4314/njcr.v26i2.2

Ilić, Z. S., Milenković, L., Tmušić, N., Stanojević, L., Stanojević, J., & Cvetković, D. (2022). Essential oils content, composition and antioxidant activity of lemon balm, mint and sweet basil from Serbia. *LWT, 153*, 112210. https://doi.org/10.1016/j.lwt.2021.112210

Jiménez-Moreno, N., Esparza, I., Bimbela, F., Gandía, L. M., & Ancín-Azpilicueta, C. (2020). Valorization of selected fruit and vegetable wastes as bioactive compounds: Opportunities and challenges. *Critical Reviews in Environmental Science and Technology, 50*, 2061–2108. https://doi.org/10.1080/10643389.2019.1694819

Jugreet, B. S., Suroowan, S., Rengasamy, R. R. K., & Mahomoodally, M. F. (2020). Chemistry, bioactivities, mode of action and industrial applications of essential oils. *Trends in Food Science & Technology, 101*, 89–105. https://doi.org/10.1016/j.tifs.2020.04.025

Karne, H., Kelkar, V., Mundhe, A., Ikar, M., Betawar, S., & Chaudhari, N. (2023). Essential oil extraction from orange and lemon peel. *E3S Web of Conferences, 455*, 01005. https://doi.org/10.1051/e3sconf/202345501005

Khoddami, A., Wilkes, M. A., & Roberts, T. H. (2013). Techniques for analysis of plant phenolic compounds. *Molecules, 18*, 2328. https://doi.org/10.3390/molecules18022328

Khorshidian, N., Yousefi, M., Khanniri, E., & Mortazavian, A. M. (2018). Potential application of essential oils as antimicrobial preservatives in cheese. *Innovative Food Science & Emerging Technologies, 45*, 62–72. https://doi.org/10.1016/j.ifset.2017.09.020

Lesjak, M., Beara, I., Orčić, D., Anačkov, G., Knežević, P., Mrkonjić, Z., & Mimica-Dukić, N. (2017). Bioactivity and chemical profiling of the Juniperus excelsa, which support its usage as a food preservative and nutraceutical. *International Journal of Food Properties, 20*, 1652–1663. https://doi.org/10.1080/10942912.2017.1352598

Li, S., Zhu, X., Li, S., & Zhu, X. (2020). Improved bio-oil distilling effect by adding additives to enhance downstream bio-oil processing and separation. *Separation and Purification Technology, 247*, 116982. https://doi.org/10.1016/j.seppur.2020.116982

Li, Z., Wang, H., Pan, X., Guo, Y., Gao, W., Wang, J., Dong, B., Duan, M., Yin, H., Zhang, Q., & Chen, F. (2022). Enzyme-deep eutectic solvent pre-treatment for extraction of essential oil from *Mentha haplocalyx* Briq. leaves: Kinetic, chemical composition and inhibitory enzyme activity. *Industrial Crops and Products, 177*, 114429. https://doi.org/10.1016/j.indcrop.2021.114429

Lin, H., Jiang, Y., Yi, Z., Zhang, L., Zhang, S., Ding, K., Li, B., Wang, S., & Hu, X. (2023). Involvement of cooking oils in hydrothermal carbonization of food waste. *ACS Food Science & Technology, 3*, 1098. https://doi.org/10.1021/acsfoodscitech.3c00110

Liu, Z., Li, H., Cui, G., Wei, M., Zou, Z., & Ni, H. (2021). Efficient extraction of essential oil from *Cinnamomum burmannii* leaves using enzymolysis pretreatment and followed by microwave-assisted method. *LWT, 147*, 111497. https://doi.org/10.1016/j.lwt.2021.111497

Hikal, W., Ahl, H. A. H. S.-A., Tkachenko, K. G., Bratovcic, A., & Szczepanek, M. (2021). Sustainable and environmentally friendly essential oils extracted from pineapple waste. *Biointerface Research in Applied Chemistry, 12*, 6833–6844. https://doi.org/10.33263/BRIAC125.68336844

Mahato, N., Sinha, M., Sharma, K., Koteswararao, R., & Cho, M. H. (2019). Modern extraction and purification techniques for obtaining high purity food-grade bioactive compounds and value-added co-products from citrus wastes. *Food, 8*, 523. https://doi.org/10.3390/foods8110523

Martínez-Abad, A., Ramos, M., Hamzaoui, M., Kohnen, S., Jiménez, A., & Garrigós, M. C. (2020). Optimisation of sequential microwave-assisted extraction of essential oil and pigment from lemon peels waste. *Food, 9*, 1493. https://doi.org/10.3390/foods9101493

Memarzadeh, S. M., Gholami, A., Ghasemi Pirbalouti, A., & Masoum, S. (2020). Bakhtiari savory (*Satureja bachtiarica* Bunge.) essential oil and its chemical profile, antioxidant activities, and leaf micromorphology under green and conventional extraction techniques. *Industrial Crops and Products, 154*, 112719. https://doi.org/10.1016/j.indcrop.2020.112719

Miljanović, A., Bielen, A., Grbin, D., Marijanović, Z., Andlar, M., Rezić, T., Roca, S., Jerković, I., Vikić-Topić, D., & Dent, M. (2020). Effect of enzymatic, ultrasound, and reflux extraction pretreatments on the yield and chemical composition of essential oils. *Molecules, 25*. https://doi.org/10.3390/molecules25204818

Moghaddam, M., & Mehdizadeh, L. (2020). Chemical composition and antifungal activity of essential oil of *Thymus vulgaris* grown in Iran against some plant pathogenic fungi. *Journal of Essential Oil Bearing Plants, 23*, 1072–1083. https://doi.org/10.1080/0972060X.2020.1843547

Mora, M., Fàbregas, E., Céspedes, F., Bartrolí, J., & Puy, N. (2022). Production and separation of value-added compounds from pine wood using pyrolysis and biorefinery techniques. *Fuel Processing Technology, 238*, 107509. https://doi.org/10.1016/j.fuproc.2022.107509

Morsy, N. F. S., & Hammad, K. S. M. (2021). Extraction of essential oil from methyl cinnamate basil (*Ocimum canum* Sims) with high yield in a short time using enzyme pretreatment. *Journal of Food Science and Technology, 58*, 2599–2605. https://doi.org/10.1007/s13197-020-04766-y

Narayanankutty, A., Visakh, N. U., Sasidharan, A., Pathrose, B., Olatunji, O. J., Al-Ansari, A., Alfarhan, A., & Ramesh, V. (2022). Chemical composition, antioxidant, anti-bacterial, and anti-cancer activities of essential oils extracted from Citrus limetta risso peel waste remains after commercial use. *Molecules, 27*, 8329. https://doi.org/10.3390/molecules27238329

Negri, C., Ricci, M., Zilio, M., D'Imporzano, G., Qiao, W., Dong, R., & Adani, F. (2020). Anaerobic digestion of food waste for bio-energy production in China and Southeast Asia: A review. *Renewable and Sustainable Energy Reviews, 133*, 110138. https://doi.org/10.1016/j.rser.2020.110138

Ochsenreither, K., Glück, C., Stressler, T., Fischer, L., & Syldatk, C. (2016). Production strategies and applications of microbial single cell oils. *Frontiers in Microbiology, 7*. https://doi.org/10.3389/fmicb.2016.01539

Oussaid, A., Azzouzi, M., Mansour, A. I., Azouagh, M., Koudad, M., & Oussaid, A. (2020). Assessment of the Chemical/biological activities of extracts and essential oil of *Rosmarinus officinalis* L. from the Oriental region of Morocco. *Moroccan Journal of Chemistry, 8*(3), 732–744. https://doi.org/10.48317/IMIST.PRSM/morjchem-v8i3.20151

Padilla-de la Rosa, J. D., Manzano-Alfaro, M. D., Gómez-Huerta, J. R., Arriola-Guevara, E., Guatemala-Morales, G., Cardador-Martínez, A., & Estarrón-Espinosa, M. (2021). Innovation in a continuous system of distillation by steam to obtain essential oil from Persian lime juice (Citrus latifolia Tanaka). *Molecules, 26*, 4172. https://doi.org/10.3390/molecules26144172

Paini, J., Benedetti, V., Ail, S. S., Castaldi, M. J., Baratieri, M., & Patuzzi, F. (2022). Valorization of wastes from the food production industry: A review towards an integrated agri-food processing biorefinery. *Waste Biomass Valor, 13*, 31–50. https://doi.org/10.1007/s12649-021-01467-1

Pandey, A. K., Kumar, P., Singh, P., Tripathi, N. N., & Bajpai, V. K. (2017). Essential oils: Sources of antimicrobials and food preservatives. *Frontiers in Microbiology, 7*. https://doi.org/10.3389/fmicb.2016.02161

Park, C., Woo, H., & Park, M.-J. (2023). Development of *Pinaceae* and *Cupressaceae* essential Oils from forest waste in South Korea. *Plants, 12*, 3409. https://doi.org/10.3390/plants12193409

Patrascu, M., & Radoiu, M. (2016). Rose essential oil extraction from fresh petals using synergetic microwave & ultrasound energy: Chemical composition and antioxidant activity assessment. *Journal of Chemistry and Chemical Engineering, 10*, 10.17265/1934-7375/2016.03.004.

Paw, M., Begum, T., Gogoi, R., Pandey, S. K., & Lal, M. (2020). Chemical composition of Citrus limon L. burmf peel essential oil from north East India. *Journal of Essential Oil Bearing Plants, 23*, 337–344. https://doi.org/10.1080/0972060X.2020.1757514

Politeo, O., Popović, M., Veršić Bratinčević, M., Kovačević, K., Urlić, B., & Generalić Mekinić, I. (2023). Chemical profiling of sea fennel (*Crithmum maritimum* L., Apiaceae) essential oils and their isolation residual waste-waters. *Plants, 12*, 214. https://doi.org/10.3390/plants12010214

Prates-Valério, P., Celayeta, J. M. F., & Cren, E. C. (2019). Quality parameters of mechanically extracted edible Macauba oils (Acrocomia aculeata) for potential food and alternative industrial feedstock application. *European Journal of Lipid Science and Technology, 121*, 1800329. https://doi.org/10.1002/ejlt.201800329

Qasem, N. A. A., Mohammed, R. H., & Lawal, D. U. (2021). Removal of heavy metal ions from wastewater: A comprehensive and critical review. *npj Clean Water, 4*, 1–15. https://doi.org/10.1038/s41545-021-00127-0

Rápó, E., & Tonk, S. (2021). Factors affecting synthetic dye adsorption; desorption studies: A review of results from the last five years (2017–2021). *Molecules, 26*, 5419. https://doi.org/10.3390/molecules26175419

Smigielski, K. B., Majewska, M., Kunicka-Styczyñska, A., & Gruska, R. (2014). The effect of ultrasound-assisted maceration on the bioactivity, chemical composition and yield of essential oil from waste carrot seeds (*Daucus carota*). *Journal of Essential Oil Bearing Plants, 17*, 1075–1086. https://doi.org/10.1080/0972060X.2014.931253

Sniegoňová, P., Szotkowski, M., Holub, J., Sikorová, P., & Márová, I. (2023). The effect of oil-rich food waste substrates, used as an alternative carbon source, on the cultivation of microalgae—A pilot study. *Microorganisms, 11*. https://doi.org/10.3390/microorganisms11071621

Soundararajan, P., Parthasarathy, S., Sakthivelu, M., Karuppiah, K. M., Velusamy, P., Gopinath, S. C. B., & Raman, P. (2023). Effects of consuming repeatedly heated edible oils on cardiovascular diseases: A narrative review. *Current Medicinal Chemistry, 31*, 6630. https://doi.org/10.2174/0109298673250752230921090452

Spalvins, K., Geiba, Z., Kusnere, Z., & Blumberga, D. (2020). Waste cooking oil as substrate for single cell protein production by yeast *Yarrowia lipolytica*. *Environmental and Climate Technologies, 24*, 457–469. https://doi.org/10.2478/rtuect-2020-0116

Stashenko, E., Martínez, J. R., Stashenko, E., & Martínez, J. R. (2023). Essential oils and the circular bioeconomy. In *Essential oils - Recent advances, new perspectives and applications*. IntechOpen. https://doi.org/10.5772/intechopen.112958

Sun, H., Gao, Z., Zhang, L., Wang, X., Gao, M., & Wang, Q. (2023). A comprehensive review on microbial lipid production from wastes: Research updates and tendencies. *Environmental Science and Pollution Research, 30*, 79654–79675. https://doi.org/10.1007/s11356-023-28123-6

Taiwo, E. A., Sanda, O., Odesola, O. O., Aransiola, E. F., & Ehinmitola, F. (2024). Influence of enzymatic pretreatment and optimization on hydro distillation recovery of essential oils from Allium cepa. *Chemistry Africa, 7*, 1223–1233. https://doi.org/10.1007/s42250-023-00828-8

Teshale, F., Narendiran, K., Beyan, S. M., & Srinivasan, N. R. (2022). Extraction of essential oil from rosemary leaves: Optimization by response surface methodology and mathematical modeling. *Applied Food Research, 2*, 100133. https://doi.org/10.1016/j.afres.2022.100133

Torres-Valenzuela, L. S., Ballesteros-Gómez, A., & Rubio, S. (2020). Green solvents for the extraction of high added-value compounds from agri-food waste. *Food Engineering Reviews, 12*, 83–100. https://doi.org/10.1007/s12393-019-09206-y

Turrini, F., Beruto, M., Mela, L., Curir, P., Triglia, G., Boggia, R., Zunin, P., & Monroy, F. (2021). Ultrasound-assisted extraction of lavender (Lavandula angustifolia Miller, Cultivar Rosa) solid by-products remaining after the distillation of the essential oil. *Applied Sciences, 11*, 5495. https://doi.org/10.3390/app11125495

Uwineza, P. A., & Waśkiewicz, A. (2020). Recent advances in supercritical fluid extraction of natural bioactive compounds from natural plant materials. *Molecules, 25*. https://doi.org/10.3390/molecules25173847

Uysal, B., Sozmen, F., Aktas, O., Oksal, B. S., & Kose, E. O. (2011). Essential oil composition and antibacterial activity of the grapefruit (Citrus Paradisi. L) peel essential oils obtained by solvent-free microwave extraction: Comparison with hydrodistillation. *International Journal of Food Science & Technology, 46*, 1455–1461. https://doi.org/10.1111/j.1365-2621.2011.02640.x

Vali, S. A., Moral-Vico, J., Font, X., & Sánchez, A. (2023). Adsorptive removal of siloxanes from biogas: Recent advances in catalyst reusability and water content effect. *Biomass Conversion and Biorefinery, 14*, 23259. https://doi.org/10.1007/s13399-023-04478-1

Visakh, N. U., Pathrose, B., Narayanankutty, A., Alfarhan, A., & Ramesh, V. (2022). Utilization of pomelo (Citrus maxima) peel waste into bioactive essential oils: Chemical composition and insecticidal properties. *Insects, 13*, 480. https://doi.org/10.3390/insects13050480

Viuda-Martos, M., Ruiz-Navajas, Y., Fernández-López, J., & Pérez-Álvarez, J. (2010). Effect of adding citrus fibre washing water and rosemary essential oil on the quality characteristics of a bologna sausage. *Lebensmittel-Wissenschaft und-Technologie, 43*, 958–963. https://doi.org/10.1016/j.lwt.2010.02.003

Wang, B., Lan, J., Bo, C., Gong, B., & Ou, J. (2023). Adsorption of heavy metal onto biomass-derived activated carbon: Review. *RSC Advances, 13*, 4275–4302. https://doi.org/10.1039/D2RA07911A

Wei, L., Pu, D., Mi, S., Yang, H., Wei, L., Lu, Q., Liu, M., & Zu, Y. (2022). Essential oil extraction and evaluation from the fresh *Platycladus orientalis* (L.) Franco seed peel waste by an environment-friendly method. *Sustainable Chemistry and Pharmacy, 29*, 100771. https://doi.org/10.1016/j.scp.2022.100771

Weng, Y.-X., Wang, H.-C., Chu, Y.-L., Wu, Y.-Z., Liao, J.-A., & Su, Z.-Y. (2024). Essential oil from Citrus depressa peel exhibits antimicrobial, antioxidant and cancer chemopreventive effects. *Journal of the Science of Food and Agriculture, 104*, 3982–3991. https://doi.org/10.1002/jsfa.13280

Xiong, X., Yu, I. K. M., Dutta, S., Mašek, O., & Tsang, D. C. W. (2021). Valorization of humins from food waste biorefinery for synthesis of biochar-supported Lewis acid catalysts. *Sci Total Environ, 775*, 145851. https://doi.org/10.1016/j.scitotenv.2021.145851

Zainuddin, M. F., Fai, C. K., Ariff, A. B., Rios-Solis, L., & Halim, M. (2021). Current pretreatment/cell disruption and extraction methods used to improve intracellular lipid recovery from oleaginous yeasts. *Microorganisms, 9*, 251. https://doi.org/10.3390/microorganisms9020251

Zhao, Y., Yang, Y.-H., Ye, M., Wang, K.-B., Fan, L.-M., & Su, F.-W. (2021). Chemical composition and antifungal activity of essential oil from *Origanum vulgare* against *Botrytis cinerea*. *Food Chemistry, 365*, 130506. https://doi.org/10.1016/j.foodchem.2021.130506

Index

D. B. Pal, S. L. Bansal (eds.), *Fungal Waste Biomass Management for Energy, Environment and Value-Added Products*, Fungal Biology,
https://doi.org/10.1007/978-3-031-82599-6

The manufacturer's authorised representative in the EU is Springer Nature Customer Service Centre GmbH, Europaplatz 3, 69115 Heidelberg, Germany. If you have any concerns regarding our products, please contact ProductSafety@springernature.com

Printed and bound by CPI Group (UK) Ltd, Croydon, CR0 4YY
30/09/2026
02225596-0008